Practical Histochemistry

Practical Histochemistry

Second Edition

J. Chayen and L. Bitensky
Unit of Cellular Pharmacology and Toxicology, Robens Institute,
University of Surrey, UK

JOHN WILEY & SONS

Chichester · New York · Brisbane · Toronto · Singapore

Other Wiley Editorial Offices

John Wiley & Sons, Inc., 605 Third Avenue,
New York, NY 10158-0012, USA

Jacaranda Wiley Ltd, G.P.O. Box 859, Brisbane,
Queensland 4001, Australia

John Wiley & Sons (Canada) Ltd, 22 Worcester Road,
Rexdale, Ontario M9W 1L1, Canada

John Wiley & Sons (SEA) Pte Ltd, 37 Jalan Pemimpin 05-04,
Block B, Union Industrial Building, Singapore 2057

Library of Congress Cataloging-in-Publication Data:

Chayen, J. (Joseph), 1924–
    Practical histochemistry/J. Chayen, L. Bitensky.—2nd ed.
      p.   cm.
    Includes bibliographical references and index.
    ISBN 0 471 92931 X
      1. Histochemistry—Technique.   I. Bitensky, L. (Lucille), 1931–.
  II. Title.
  QH613.C47   1991
  574.8′212—dc20   90-19606
                                              CIP

British Library Cataloguing in Publication Data:

Chayen, J. Joseph
    Practical histochemistry.—2nd ed
    1. Histochemistry
    I. Title   II. Bitensky, Lucille
    574.8212

    ISBN 0 471 92931 X

Typeset by Dobbie Typesetting Limited, Tavistock, Devon
Printed in Great Britain by Courier International, Tiptree

Contents

Acknowledgements ix

Preface to First Edition xi

Introduction xiii

1 **Preparation of Sections** 1
   Introduction................................ ,,,,,,,,,,,,,,,, 1
   Freeze-sectioning procedure 1
   General advice on sectioning 7

2 **Methods of Incubating Sections** 9
   Need for tissue stabilizers 11
   Use of tissue stabilizers 13
   Cytochemistry of isolated cells 14
   Ions for stabilizing sections 16
   To prepare a stock solution of polyvinyl alcohol 16

3 **Methods of Quantification** 17
   Elution techniques 18
   Precise quantification: scanning and integrating microdensitometry. 21
   Polarized light microscopy 29

4 **Common Histological Stains** 35
   Rapid diagnostic sections and storage of tissue 35
   Special methods of preparing tissue 36
   Fixatives 36
   Staining procedures 37

5   Analysis of Chemical Components of Cells and Tissues 45
General analysis 45
Practical methods 54
Stage 2:   Reactions for protein 54
    Dinitrofluorobenzene method 54
    Tetra-azotized dianisidine method 56
    Baker's method for tyrosine 58
    Quantitative staining for total protein by amido black 59
    Picrosirius red method for collagen 61
    Fast Green method for basic protein 63
    Deamination method 64
    Assessment of 'pK' values 64
    The histochemical Sakaguchi reaction for arginine 66
    Methods for thiol and disulphide groups 68
    DDD method for −SH groups 69
    Prussian blue reactions for −SH and −S−S− groups 71
    Demonstration of total −SH and −S−S− 73
    To demonstrate −S−S− groups 74
    Method for glutathione 75
    Immunocytochemistry for detecting proteins 76
Stage 3:   Reactions for nucleic acids and polyphosphate 77
    The Feulgen reaction for deoxyribonucleic acid (DNA) 77
    DNA-synthesis index 85
    Kurnick's methyl green method for DNA 86
    The use of deoxyribonuclease 89
    The methyl green–pyronin method for DNA and RNA 91
    The use of ribonuclease 93
    Hydrochloric acid method for RNA 94
    Differentiation of acidic moieties 94
    Toluidine blue method for basophilia 94
    Metachromatic staining with toluidine blue 96
    Ebel's test for polyphosphate 98
    Acridine orange method for DNA and RNA 98
Stage 4:   Reactions for polysaccharides 99
    Periodic acid–Schiff (PAS) method 99
    Benzoylation 104
    Acetylation 104
    Alcian blue method 105
    Induced birefringence of proteoglycans 107
    Diamine methods for mucosubstances 107
    Hale's colloidal iron method 108
    Best's carmine method for glycogen 109
    Lugol's iodine 110
    The use of diastase and amylases 111

Stage 5:   Reactions for lipids 111
    Sudan black B method 112
    Modified Sudan black methods 114
    Oil red O method for colouring fats.................... 114
    Nile blue method for lipids........................... 115
    Benzpyrene method for lipids.......................... 116
    Other fluorescent methods for lipids................... 118
    Acid haematein method for phospholipids 118
    Methods for steroids 121
    Methods for carotenoids 121
Stage 6:   Minor components 122
    Iron 122
    Calcium 123
    Pigments 124
    Reducing substances................................. 124
    Autofluorescent substances 126
    Induced fluorescence of catecholamines and related
        substances...................................... 127
    Autoradiography..................................... 128

6 Enzyme Histochemistry 135
General introduction 135
Phosphatases 145
    Alkaline phosphatase 145
        The Gomori–Takamatsu procedure................... 147
        The naphthol phosphate method 149
    Acid phosphatase 151
        The Gomori lead phosphate method................. 152
        Acid phosphatase fragility test 154
        Azo-dye acid phosphatase method 154
        The post-coupling method........................ 156
5′-Nucleotidase 158
    Calcium method for 5′-nucleotidase activity.............. 160
    Lead method for 5′-nucleotidase activity................. 161
Glucose-6-phosphatase..................................... 162
Adenosine triphosphatases 164
    Calcium-activated (myosin-type) ATPase 164
    Magnesium-activated ATPases (including Na^+-K^+-
        ATPase) 166
    Calcium adenosine triphosphatase 169
Phosphamidase 171

Esterases 174
    Naphthol acetate method for esterase 180
    Indoxyl acetate method for esterase 181
    Cholinesterase 183
Lipases 185
Proteases 186
    Aminopeptidases and naphthylamidases................... 188
β-Glucuronidase 193
    Post-coupling method 195
    Modified post-coupling method 196
N-Acetyl-β-glucosaminidase 198
Phosphorylases 201
Carbonic anhydrase 205
Aryl sulphatase 207
Decarboxylases: ornithine decarboxylase 208
Oxidases 211
    Cytochrome oxidase 211
    DOPA-oxidase (phenolase) 215
    Monoamine oxidase 219
        Peroxidase method 222
        Tetrazolium method 223
Peroxidases 224
Dehydrogenases... 225
    Succinate dehydrogenase 234
    Phenazine methosulphate method 238
    α-Glycerophosphate dehydrogenase..................... 239
    Glyceraldehyde 3-phosphate dehydrogenase.................. 241
    Lactate dehydrogenase................................. 243
    Glutamate dehydrogenase 247
    β-Hydroxyacyl CoA dehydrogenase 249
    Uridine diphosphoglucose dehydrogenase 252
    Steroid dehydrogenases 254
    NADP-dependent dehydrogenases: glucose 6-phosphate
        and 6-phosphogluconate dehydrogenases 255
NADPH-diaphorase 263
NAD(P)H-oxidase... 264
Transhydrogenases 265

References 267

Appendix 1   Effect of Fixation on Enzymes 293

Appendix 2   Buffers 301

Index 307

Acknowledgements

We gratefuly acknowledge our indebtedness to many colleagues who have contributed to our understanding and practice of this subject. They are far too many to be mentioned individually. Many are referred to in the text; we hope that the others will recognize their contributions and that they will be able to approve of our use of them. In particular we must acknowledge our indebtedness to Mr A. A. Silcox, FIST, for his early development of cryostat microtomy; to the late Dr R. G. Butcher, to Dr F. P. Altman and to Dr N. Loveridge who did so much to establish 'scientific' histochemistry; and to Dr G. T. B. Frost for the chemical aspects of many of our more novel reagents. We are very grateful to Miss A. O'Farrell for coping so efficiently with an unruly manuscript.

Preface to First Edition

Over the past ten years several hundred academic and applied biological and medical scientists have visited us to ask how they could apply histochemistry to their particular studies. Often their questions were purely technical: they wanted to know how to stain for a particular enzyme or substance. They saw histochemistry as an extension of histology and they wanted a simple, well established recipe; they did not want to go into the minutiae of the science underlying the staining method any more than they did when applying special stains in histology. For such enquirers we saw our task as twofold: we had to give them a reliable technique that they could follow as simply as a routine histological method, but it had to be a method which, as far as we could ascertain, had a real scientific basis. Other workers came with problems in tissue metabolism and chemical dysfunction. For these, histochemistry was an extension of biochemistry. Its special advantages were that it allowed them to relate a specific activity to a particular tissue component and it was a form of biochemistry that could be done on relatively minute pieces of tissue such as could be removed safely by biopsy, or could be kept in either maintenance or proliferative culture. Some of these queries related to the mode of action of drugs, or of hormones, or of potentially toxic food additives; others were concerned with the biochemistry of disease. Such workers were primarily concerned with metabolic biochemistry; but they too did not want to delve into the minutiae of the many variants that can be found for almost every histochemical reaction (and which are discussed in detail in the specialist textbooks of histochemistry such as those by Pearse (1960) and by Barka and Anderson (1963)).

Hence, over the years, we have tested a range of histochemical reactions to provide simple techniques which, if followed precisely, will yield reactions where the reactive compounds or enzymes are present; which will give a measure of the degree of activity actually shown in the tissue at the time of its removal; which can also indicate the total activity of which the tissue could be capable; but all of which have a rational basis. These methods were collected in a laboratory manual and they have been used by the many visitors to our laboratory. They have now been put together, in rather more detail, first of all for the many visitors who have complained that they could not copy out all the manual, and then for wider use by whoever wishes to apply histochemistry in his or her studies, whether as an extension of histology or of biochemistry. **No attempt has been made to include techniques of electron histochemistry. This is a special field of its own and is completely outside these terms of reference.** The main criterion we have used in selecting these methods is their reliability (for a more complete review of all methods which have been described, the reader must be referred to the comprehensive books on histochemistry); provided the instructions are followed, and provided the tissue contains the substance or enzyme to be tested, you should obtain a visible reaction, the intensity of which should indicate the effective concentration or activity of the substance tested. We do not guarantee that these methods will give the most 'beautiful' stained preparations (although we believe that they often will); we have aimed at providing methods that will give the most beautiful preparations which are also rational and scientifically meaningful, should such meaning become required. For this is one of the hazards of histochemistry; initially a method may be selected deliberately and solely because it yields an excellent histological stain, but there seem to be few histochemists who can then resist the temptation to interpret tissue metabolism in the lurid light of such staining.

Introduction

Histochemistry, in its broadest definition, covers two quite distinct forms of investigation. The first, which is still the most used, may be considered primarily as an extension of histology, the chemical and enzymatic characteristics of cells, their matrix, or of tissues being used to cause them to become coloured distinctively after they have been subjected to a particular histochemical procedure. Thus, for example, different cell types may be distinguished by the fact that one will stain intensely for one type of esterase whereas another colours only in response to a different enzymatic reaction. An extension of this type of histochemistry is its use for localizing particular chemical entities. The most important aspect of this form of histochemistry is the current, powerful use of immunohistochemistry, in which particular substances, often peptides, are localized by their response to a specific antibody. The site of binding of the antibody is then visualized by means either of a fluorescent or of an enzymatic marker, the former being detected by fluorescence microscopy and the latter by the histochemical reaction that can be evoked by it. (Autoradiography can be used if the final complex contains a suitable radiolabel.)

The other aspect is typified by quantitative cytochemistry (Chayen, 1978a, 1984), which is an extension of conventional biochemistry down to the level of the individual cell. Because this is a fairly rigorous form of histochemical investigation, its principles and practice form the basis of the methods described in this book; the same methods can be used for more conventional histochemical investigations.

The past ten years have seen remarkable growth and development of quantitative cytochemistry. The success of this subject is seen in its many applications, reviewed in the book edited by Pattison *et al.* (1979). But possibly its most striking success has been in the development of the microbioassays of polypeptide hormones (Chayen, 1978a, 1980; Chayen and Bitensky, 1983). The basic problem addressed by these assays is one that is fundamental to current research in cellular biochemistry and histochemistry, namely, that although immunoassay can detect, with a fair degree of specificity, the presence of a particular type of polypeptide, it cannot determine whether or not that peptide is biologically functional. To choose only one example, immunoassays show that the circulating levels of 'parathyroid hormone' in the blood of normal individuals can be measured in hundreds of picograms per millilitre. Physiological studies indicated that the probable circulating level was around 10 pg/ml, and this is borne out by the cytochemical bioassay. The high values recorded by the various immunoassays are probably due to cleavage products of the peptide or to aberrant forms of it, neither of which possess the biological function of parathyroid hormone (as discussed in Chayen, 1980; Chayen and Bitensky, 1983). This example is pertinent to immunohistochemical demonstrations of the presence, for instance, of a particular enzyme; they may show that the proteinaceous material of the enzyme is present, but they rarely, if ever, demonstrate whether the enzymatic, or other biological, function is also present. Moreover, it must be remembered that many enzymes can possess very different degrees of activity. Consequently, the effect of a drug, or a disease process, or of a hormone, may be mainly to change the activity of a particular enzyme or active moiety. This cannot be determined by immunochemical procedures; it is the province of quantitative cytochemistry.

As in the previous edition, no attempt has been made to encompass the techniques of electron histochemistry, which is a specialized study in its own right and is completely outside the terms of reference of this book. The same applies to techniques involving histochemical reactions done with semipermeable membranes (Meijer, 1980).

1

Preparation of Sections

INTRODUCTION

To prepare sections of a tissue, it is necessary to harden the specimen and to halt enzymatic activity that can involve autolysis. In conventional histology, this is achieved by embedding the sample in paraffin wax, or some similar matrix. Since this involves the use of hydrophobic solvents, the specimen is chemically fixed, to allow the removal of water and to protect it against the effects of such solvents. Because the whole point of chemical fixation is to alter the physicochemical nature of the protoplasm, its use must obviously be eschewed for studies on the chemical nature of the tissue. Moreover, it may involve considerable inhibition of enzymatic activities (Appendix 1). Such inhibition can occur even after freeze drying (Appendix 1). The dangers of chemical fixation have been fully documented, for example by Danielli (1953), Wolman (1955) and Barka and Anderson (1963); the effect on enzymatic activity has been detailed by Nachlas et al. (1956), Shnitka and Seligman (1960), and Holt (1959; see also Holt et al., 1960); the loss of lipids during treatment with hydrophobic solvents has been measured by Ostrowski et al. (1962a, b and c).

The alternative way of hardening tissue, so that it can be sectioned, is by cooling it to a suitably low temperature. This was the basis of the methods of freeze-drying (e.g. Bell, 1956; Danielli, 1953; Symposium of the Institute of Biology, 1952) and of freeze-substitution (Simpson, 1941; Ostrowski et al., 1962a, b, c). However, these methods also cause loss of enzymatic activity (Appendix 1). On the other hand, it is not necessary to embed the chilled tissue; it can be sectioned in the solid state as is done in the freeze-sectioning procedure recommended for all studies.

FREEZE-SECTIONING PROCEDURE

Background

The original objection to freezing tissue was that, when cells are frozen slowly, ice forms first in the less concentrated fluid in the extracellular spaces and then later even inside the cells (depending on the rate of cooling). During this process, the cells shrink, giving up their water to the growth of extracellular ice. Increased ionic concentration causes denaturation of the protoplasm and ruptures lipid–protein complexes, destroying the integrity of membranes (Lovelock, 1957). With more rapid, or more prolonged, freezing the residual intracellular water will also freeze, mechanically shattering the cells. Then, when the frozen block of tissue is cut, even in the cold, the heat generated by the cutting (Thornburg and Mengers, 1957) will cause the ice to melt, giving the worst effects of freezing and thawing and refreezing.

However, it has been known for a very long time (e.g. Asahina, 1956; Luyet, 1951, 1960) that protoplasm can be supercooled; the water loses its mobility without forming crystals, and becomes set in an amorphous state (as discussed by Chayen, 1980). The temperatures that favour the formation of ice have been defined by Tammann (1898). His figures indicated that about $-18\,°C$ is a dangerous temperature and it is unfortunate that this temperature is much favoured by many cryostat histologists; it should not be used for quantitative studies.

The process of chilling and sectioning tissue has been investigated in some detail by the present authors. They have shown that the dangers can be avoided; measurements with thermocouples under controlled conditions showed that tissue can be supercooled (i.e. no ice is formed) and this supercooled state can be maintained even during sectioning provided that the process is carried out at sufficiently low temperature, and with the knife cooled further to $-70\,°C$ to dissipate the heat generated by the cutting (see Lynch et al., 1966; Silcox et al., 1965). The section is removed from the knife by apposing to it a glass slide taken from the ambient temperature of the laboratory and held by hand. The section jumps the gap between the knife (at $-70\,°C$) and the slide (at e.g. $+20\,°C$) and an imprint of the section is left as liquid water on the knife (it freezes rapidly of course). In this way the section is flash-dried by distillation of the supercooled water over a temperature gradient of nearly $100\,°C$ operating over a gap of a few millimetres. Such sections are then stable.

Ice damage is undetectable in such tissue and sections. The detecting device may be a thermocouple in the tissue (as studied by Lynch et al., 1966); it may be simple microscopic examination for ice crystals (or for the holes and damage produced by them, as in the study by Silcox et al., 1965); it may be dark-ground illumination for investigating the degree of denaturation–aggregation of the protoplasm; or it may involve a study of lipid–protein associations (as in Chayen, 1968a) or of the permeability of subcellular membranes (see Chayen, 1968b). Moreover, it has been shown that such flash-dried sections, subjected to ultrathin microtomy, show no evidence of ice artefact when examined by electron microscopy (Altman and Barrnett, 1975; Zoller and Weisz, 1980). However, probably the most powerful evidence that such sectioning and flash-drying retains the physicochemical integrity of the cells is their use in cytochemical

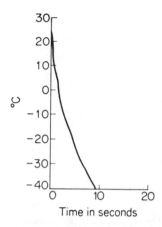

Figure 1. The cooling curve of a piece of tissue immersed in hexane at − 70 °C. Note that the temperature reaches − 40 °C in about 9 s and that there is no distortion (flattening) of the curve; therefore there is no indication of the formation of ice in this tissue.

section bioassays in which sections of the target tissue respond to the hormone with the same sensitivity as the original tissue (Chayen, 1980) even though the response may require many interrelated metabolic reactions to occur within the cells.

In the recommended freeze-sectioning procedure, the specimen is cooled by precipitate immersion in hexane at − 70 °C. The rate of chilling is shown in the thermocouple trace (Figure 1). It will be seen firstly that the rate of cooling is twice as fast as that recorded by Moline and Glenner (1964) when they used liquid nitrogen (which forms an insulating layer of gaseous nitrogen around the warm specimen), and secondly that there is no sign of ice formation (as is evident in Figure 2).

To chill tissue

Pretreatment

A wide range of tissues has been successfully chilled by this procedure. Usually the tissue is chilled directly it is removed from the animal or plant. However, delicate structures may benefit from immersion for 5–10 min in a 5% aqueous solution of polyvinyl alcohol (PVA) before being chilled. Very fatty tissue can be successfully chilled if it is first immersed for 5 min in a 7% solution of calcium chloride in 5% PVA in glycyl glycine buffer (0.05 M, pH 8.0) (Zoller and Weisz, 1980). Very hard tissue, such as bone, should be dipped in a 5% aqueous solution of PVA before being chilled; the layer of PVA helps to bond the specimen to the microtome chuck during sectioning.

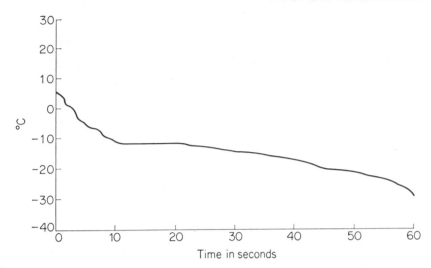

Figure 2. The cooling curve of a piece of tissue chilled at about − 20 °C. In contrast to the curve in Figure 1 there is extensive flattening of the curve, denoting ice formation.

Apparatus and materials

The main apparatus required is a chilling bath (a cheap polythene sandwich box is suitable if holes are cut in the lid). It should be heat-insulated by having expanded polystyrene or similar insulating matter put around it.

  Also required are:
A beaker (about 50–100 ml capacity), which can be covered
A low-temperature thermometer (recording down to − 100 °C)
CO_2 ice
Commercial spirit (or absolute alcohol)
Hexane (BDH Laboratory Reagent, boiling range 67–70 °C, low in aromatic
  hydrocarbons)
3×1 in corked specimen tubes
Dewar flasks (the 1-gallon (4.5 l) size is very useful for storage of tubes)

Method

The *chilling bath* is prepared by adding small chips of solid carbon dioxide (which cool the freezing mixture quicker than larger pieces) to alcohol in the chilling bath until a saturated solution is obtained. This state is obvious because (a) the alcohol–carbon dioxide mixture becomes viscous, (b) addition of more solid carbon dioxide does not cause bubbling, and (c) the thermometer records about − 70 °C.

The beaker, containing 30–50 ml hexane, is inserted into the bath, preferably through a close-fitting space cut in the lid. More carbon dioxide ice is added to the bath to maintain the temperature. The temperature of the hexane should be at least $-65\,^\circ C$ before it is used. (Some workers prefer to use isopentane in place of hexane. We have no experience of this, but there is no reason to expect it to behave markedly differently from hexane.)

The *tissue* is cut into suitable pieces (which can be as large as 5 mm³ in size) and these are dropped into the hexane at about $-70\,^\circ C$. Small pieces (e.g. needle biopsies or material from curettage) can be blown into the hexane from a hollow spatula fitted with a large rubber bulb. Care must be taken to ensure that the tissue is ejected straight into the hexane without touching the sides of the beaker. The specimen is left in the hexane for at least 30 s and preferably not longer than 2 min. It is then transferred, by the use of precooled forceps, to a dry tube at $-70\,^\circ C$. (*Note: from this stage onwards, the chilled tissue should be handled only with cold instruments.*) It is often advisable to shake the surface hexane off the specimen while transferring it. Care must be taken to ensure that the tissue does not warm up during this process; the specimen, in the corked dry tube, should be encased in solid carbon dioxide in the Dewar flask for storage.

To mount the tissue

This is the most hazardous operation in the whole procedure. It should be effected expeditiously, and care must be taken not to allow the tissue to be warmed at any stage.

1. Prepare another alcohol–solid carbon dioxide bath at about $-70\,^\circ C$.
2. Place the metal chuck in this bath with its top well clear of the alcohol; leave it to equilibrate with the bath.
3. Place a drop of water on top of the chuck. The water should begin to freeze rapidly. It is sometimes convenient to colour the water, for example with toluidine or methylene blue, to differentiate water from ice. For some purposes, especially when hard tissues are to be sectioned, it is best to use a 5% aqueous solution of PVA instead of ordinary water; it bonds to PVA-coated bone and holds it better during sectioning.
4. Remove the chilled tissue from the 3×1 in corked tube (in which it has been stored in the Dewar flask, packed with solid carbon dioxide) and place it in a cavity in a piece of carbon dioxide ice so that its orientation can be determined before it is mounted, while keeping it chilled.
5. Leave the water on the chuck to freeze until there is only a thin film of water, of comparable size to that of the specimen, left unfrozen. Then, expeditiously, transfer the specimen (with cold forceps) to this film of water. For cutting thicker sections (see later), it is advisable to mount the tissue eccentrically, so that it will become sited at the top side of the chuck relative to the knife. The residual water will freeze the tissue to the drop of ice on the chuck.

6. Remove the chuck (with the specimen) from the bath; wipe its sides free of adherent alcohol and stand it in the refrigerated cabinet of the cryostat.

To prepare the microtome for sectioning

The angle of the knife is critical. When using a Cambridge rocking microtome, the knife is positioned as follows. On the knife mounting of the microtome there are four screws: two threaded into lugs at the back and two into lugs at the front. The back pair are marked into ten divisions. Before the knife is placed in the mounting, the two back index screws should be unscrewed far enough to prevent them from protruding into the knife-mounting space.

The knife is then placed in the mounting and the two front screws are tightened up as far as they will go, forcing the knife flat against the back of the mounting. The two index screws are then screwed forward until they touch the back of the knife. The number on each index screw opposite the line marked on the top of the respective lugs at the back of the mounting is then read off and recorded.

The front screws are then unscrewed, releasing the knife. Both the index screws are then advanced 1 or 1.4 complete revolutions until the recorded number for each appears opposite the line on its respective lug again. The front screws are then tightened back onto the knife so that it is held firm, and it will now be at the right angle.

Once it has been set, the knife must be cooled to $-70\,°C$ by packing carbon dioxide ice around the handle. We suggest that the knife be left packed around with carbon dioxide ice for at least one hour before it is used. The ambient temperature in the cabinet must be maintained at $-30\,°C$ (Silcox et al., 1965).

To cut sections

The chuck is locked onto the front of the arm of the microtome. It is positioned in such a way that the smallest edge of the block faces downwards. Blocks that have an epithelium on one edge should be positioned with this epithelium on

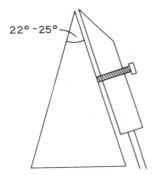

22° – 25°

Figure 3. Diagram of the sideview of the anti-roll plate in position against the knife, with the gap for the section adjusted by the screw.

one side and not at the top or bottom facing the knife. The screw that determines the thickness of the sections is set to the required size. In general, sections 8–10 μm thick are suitable for all histochemical work. However, tissue chilled and prepared in this way can be cut at any required thicknesses from 2 to 20 μm.

On the front of the microtome is the anti-roll plate assembly. As its name suggests, it is designed to prevent the sections from curling up as they are cut. The anti-roll plate is made of transparent Perspex and has two small nylon screws which allow the gap between the anti-roll plate and the knife to be adjusted (Figure 3). The angle of this plate to the knife is also critical but may have to be varied according to the tissue to be cut; consequently, the correct angle setting will be found only by experience. However, whatever angle of the anti-roll plate is used, the top of the plate should never be lower than the cutting edge or high enough to touch the block.

To pick up the sections

The anti-roll plate keeps the sections flat against the knife while they are being cut. To pick up the sections:

1. Swing the anti-roll plate away from the knife.
2. Take a glass slide from the ambient temperature of the laboratory and bring it up to, and parallel to, the section on the knife. There should be no need to *press* the slide on to the section; it should move onto the slide.
3. Store the sections, on the slides, in the cryostat until used.

GENERAL ADVICE ON SECTIONING

Knives

It is essential that the knife is kept sharp and free from ice particles. It should be sharpened to give an angle, at the tip, of 22–27°. Many instances of workers finding that a particular tissue is not cuttable have been shown to be due to the knives having angles considerably in excess of these values. Another cause of difficulty is the presence of fat. This can largely be overcome by cooling the block of tissue, and the knife, with solid carbon dioxide.

With rocking microtomes, it is necessary when cutting sections thicker than 12 μm to ensure that the block clears the back of the knife on the upward stroke before the advance mechanism starts to become effective. This may require the orientation of the specimen to be changed.

Automatic cutting

One of the well known problems in microtomy, even for paraffin sections, is the fact that serial sections may not be of equal thickness. This is serious enough for staining histochemistry because it may deceive the histochemist into believing that one section has more of a particular chemical entity than another, whereas in fact one section is merely thicker than the other. However, in quantitative cytochemistry, variation in section thickness would be disastrous because all microdensitometric measurement depends on a constant light path, i.e. a constant thickness of the section in which the coloured reaction product is situated. This problem has been overcome by fitting cryostats with an automatic drive which cuts sections automatically and at a constant speed, since it has been shown (Butcher, 1971a) that it is the speed of cutting that determines the thickness of the sections (subject to the thickness setting on the microtome). With such an automatic cutting device, the thickness of serial sections is constant to $\pm 5\%$; this is fundamental to cytochemical bioassays (Chayen, 1978a, 1980).

Adhesion of cryostat sections to the slide

Normally cryostat sections can be flash-dried onto clean glass slides, and they will adhere to the slides. However, sections of some materials (notably cartilage and bone) will not adhere firmly. It may be preferable to use slides which have been very lightly coated with either glycerin–albumin, as used commonly in histology, or a 5% solution of polyvinyl alcohol. The adhesive should be dry before the section is made to adhere to the slide. Alternatively a fine film of poly-L-lysine (Sigma), or of chrome–gelatin (5 g gelatin and 0.05 g chrome alum in 100 ml distilled water) can be used.

Two new adhesives (available from Molecular Design and Synthesis, University of Surrey, Guildford) may be of particular value. No. 8338 has been found to bind sections of most tissues; No. 8339, which is more cationic, is of especial use for some sections of some tissues.

If all else fails, it may be necessary to coat the sections with celloidin, as described for the arginine reaction (p. 67). However, this cannot be used when enzymatic activity is being investigated.

2

Methods of Incubating Sections

The simplest way to incubate sections is to immerse the section in a **Coplin jar**, very much as is done in conventional (but small-scale) histology. This has the following advantages:

1. The solutions, in the Coplin jars, can be left in a water bath, incubator or ice bath to equilibrate to a constant, predetermined temperature. (This can be checked with a thermometer.)
2. The sections are immersed in a constant volume of solution, i.e. there is no danger of drying out. Moreover, since there is such a large volume of solution relative to the volume of the sections, there is no danger that the reactants will be seriously depleted during the incubation, or that deleterious matter, produced by the sections, will become sufficiently concentrated to have inhibitory effects on the reaction.
3. It is simple to do.

The disadvantage of using Coplin jars is that they require a large volume of the incubation medium. Should this contain coenzymes (NAD, NADP, coenzyme A, etc.) the cost of these substances would make histochemistry prohibitively expensive.

The usual way of overcoming this, in the past, has been to prepare only small volumes of expensive incubating media (e.g. 1 ml) and to add this as a drop, to cover the section, which is laid horizontally in some sort of humidity chamber to stop evaporation of the drop (Figure 4).

The advantage of this procedure is that it avoids expense. The disadvantages are as follows:

1. Frequently part of the drop does evaporate, leaving some of the section free of fluid. It is difficult to be sure that a negative reaction in part of the slide is not due to this.
2. Evaporation can change the concentration of the reagents in the drop of incubation fluid.

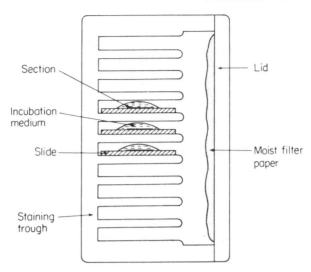

Figure 4. Diagram of a cross-section of a humidity chamber. An ordinary staining trough stands on edge so that the sections lie horizontally. The atmosphere is kept humid by a layer of moist filter paper.

3. The drop is not of uniform thickness over the section and so, depending on how it has spread, some parts may have a greater depth of solution above them than others.
4. The small volume of the drop could allow inhibitory substances, produced by the histochemical reaction, to reach a concentration that could seriously affect the result of the reaction.

The method recommended, whenever costly incubation fluid is used, is the **open-ring technique**. In this the section is picked up normally onto a slide and brought into the open laboratory (or into a room or cabinet kept at 37 °C). A Perspex ring just large enough to encircle the section, and of depth about 3 mm, is set around the section and held onto the slide by a thin film of Vaseline (Figure 5). The incubation medium (a known volume, e.g. 0.25 ml, can be pipetted into the space bounded by the ring if this precision is required) is added to the section and confined around the section by the open ring. The gaseous atmosphere can be controlled, especially if the incubation is done in specially designed incubation boxes (Figure 6). The rings and boxes can be made readily from Perspex. The whole incubation box can be put into a large incubator or a hot room so that the reactions can be done at 37 °C if necessary. It should be remembered, however, that the incubation medium, and preferably the atmosphere, should be equilibrated to the incubation temperature before they are used.

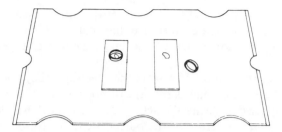

Figure 5. Diagram to illustrate the open ring incubation technique. On the right is a section on a glass slide with a ring of Perspex alongside just big enough to encompass the section. On the left, the ring has been held to the slide by Vaseline and can be filled with the incubation medium.

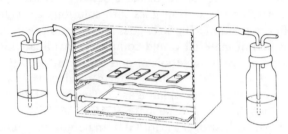

Figure 6. Diagram of the incubation box. The slides bearing sections surrounded by Perspex rings (as in Figure 5) are placed on Perspex trays, which can be stacked in layers. The bottom of the box is filled with moist filter paper. The front of the box is closed with a Perspex lid. The gaseous atmosphere is controlled by bubbling the appropriate gas through the wash bottle on the left; the moistened gas enters the box through the holes in the tube at the bottom of the box; the flow of gas is checked by the bubbling of the gas through the escape wash bottle on the right.

NEED FOR TISSUE STABILIZERS

The dilemma of histochemists has been as follows. If the tissues or sections are fixed, this causes such gross inactivation of active groups (and hence of enzymes) that it becomes difficult to determine whether the histochemistry has any meaning whatsoever. For example, suppose a disease or a toxic substance (such as food additive) or a hormone produces 50% reduction in the activity of a particular enzyme. It is not unusual for chemical fixation to cause 70% loss of activity of an enzyme. This 70% could be 70% of all the enzyme activity; in this case the normal tissue will show 30 units of activity (30% left after fixation) while the diseased or damaged tissue will show 15 units (30% of the 50% left after the disease process); this difference in the reduced activity might be detectable qualitatively. However, it is more likely that the 70% reduction in activity will represent 100% loss of activity of that 70% of the enzyme which is most labile. In this case the normal tissue will still show 30 units of activity but the abnormal tissue will show 30 units (of non-labile activity) + 0% of the reduced labile activity, i.e. $30 + [0 \times (70 - 50)] = 30$. Consequently it will seem as if the disease, or the toxic substance, has had no effect. (These figures are offered only to give some idea of the magnitude of changes that could be involved.)

However, if histochemists do *not* fix their sections, they know that much of the undenatured section may break up, or go into solution into the reaction medium, during the cytochemical reaction. Just how much will do so will depend on the pH, and on the nature of the buffer used. So, for example, a reaction done in acetate buffer at a pH of between 4 and 5 may retain most of the tissue because of the fixing effect of acetate ions at this pH. In contrast, when fresh cryostat sections are placed in a phosphate buffer at a pH of between 7 and 8, as for testing succinate dehydrogenase activity, as much as 70% of the nitrogenous matter of the section goes into solution; glycylglycine buffer is somewhat better in that only 50–60% goes into solution (Altman and Chayen, 1965), but this loss of material would completely invalidate histochemistry as a scientific study of tissue metabolism and function.

An example of one of the serious effects produced by solubilization of cytosolic soluble enzymes is shown in Figure 7. Two important cytosolic enzymes are glucose 6-phosphate dehydrogenase (G6PD) and its associated enzyme 6-phosphogluconate dehydrogenase (6PGD). In homogenate biochemistry, both are generally considered as soluble enzymes (Roodyn, 1965) in that they occur in the supernatant fractions—implying, at least, that they are not tightly bound within the cytoplasm. Both use NADP as coenzyme; their activity generates NADPH. Both enzymes are rapidly lost (within 1–2 min) when fresh sections are immersed in a normal histochemical reaction medium for demonstrating their activity (Figure 7a). Even in solution, they will react with the substrate (glucose 6-phosphate) and the coenzyme to generate NADPH *in solution*. When

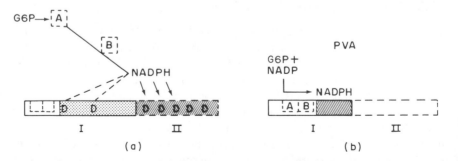

Figure 7. A serious effect of 'solubilization' of cytosolic 'soluble' enzymes (see text). Section I has strong glucose 6-phosphate and 6-phosphogluconate dehydrogenase activities with only moderate diaphorase activity. Section II does not have these dehydrogenases but it does have strong diaphorase activity. A, glucose 6-phosphate dehydrogenase; B, 6-phosphogluconate dehydrogenase; D, diaphorase; G6P, glucose 6-phosphate. Shading indicates degree of formazan produced.

the concentration of NADPH becomes suitably elevated, it will act as the substrate for the NADPH-diaphorase (or NADPH-oxidases) that may occur in the section. Consequently if a section (I in Figure 7) that contains these dehydrogenases, but only moderate diaphorase activity, is incubated together with a section of tissue (II in Figure 7) that does not contain these dehydrogenases (or has only low activity of them) but has strong diaphorase activity, it is this section that will show a strong deposition of the formazan that normally indicates the generation of NADPH. Consequently it would be thought that section II contained active G6PD and 6PGD. In contrast (Figure 7b), when these soluble dehydrogenases are retained within the section (as they are when a suitable colloid stabilizer is used), the NADPH generated by their activity will be passed immediately to the adjacent diaphorase system; under these conditions, only section I will be stained.

USE OF TISSUE STABILIZERS

At least as regards those reactions, for loosely bound enzymes, that must be done at near-neutral pH values, a fundamental requirement of quantitative cytochemistry was met by the finding that certain colloid stabilizers, when used at sufficient concentration, could stabilize fresh cryostat sections and maintain soluble enzymes within the sections during the histochemical reaction (Altman and Chayen, 1965, 1966; Chayen, 1978b). In practice, different colloid stabilizers may be more suitable for different types of investigation; different concentrations may be needed for isolated cells, reacted in suspension, than when used as a smear or after cytocentrifugation. A simple method for assessing how well a particular concentration of a colloid stabilizer preserves soluble enzymes is given below.

Method for assessing the efficacy of a colloid stabilizer

1. Cut eight serial sections, of uniform thickness (cut with the automatic cutting device), or prepare eight aliquots of the isolated cells.
2. React two with the normal reaction medium, lacking colloid stabilizer, for a standard time, e.g. 10 min. Suppose this gives 10 units of activity.
3. React another two, with the same medium, for twice that time. Suppose that this gives 20 units of activity.
4. React two with the reaction medium containing the expected required concentration of the colloid stabilizer for the first time (e.g. 10 min). Let us suppose that this gives 10 units of activity.
5. React another two in the medium containing the stabilizer for 10 min; then remove the reaction medium from the slides and replace it with fresh medium. Continue the reaction for another 10 min.

If the stabilizer has maintained the soluble enzyme within the sections, the results of the 10 min + 10 min split reaction (step 5) will be double those of the 10 min reaction. That is, the enzyme will have been transferred, together with the sections, to the second reaction. If, however, the stabilizer has not been effective, some or all of the enzyme will have been removed together with the reaction medium used in the first 10 min reaction; if all has been removed, the results of the 10 + 10 min reaction will be the same as for the 10 min reaction alone (i.e. 10 units of activity). Alternatively, the amount of activity may be intermediate between 10 and 20 units, implying that an insufficient concentration of the stabilizer has been used.

It may be found that the resultant activity after 10 min reaction in the medium containing the stabilizer (step 4) is less than that found when no stabilizer is present. The results from step 5 may be twice this diminished activity, indicating that the stabilizer may be effectively retaining the enzyme inside the sections. This type of result indicates that the concentration of all the reagents used for demonstrating the enzymatic activity are inadequate when dissolved in the colloid stabilizer, which can decrease the diffusion of reagents by a factor of 3 or 4. If this result is obtained, increase the concentration of all the reagents by up to four times, or until the activity obtained with step 4 is at least equal to that with step 2 (i.e. 10 units in this example). A full study, along these lines, is that of Henderson *et al.* (1978), assaying the efficacy of different grades of PVA.

Selection of the colloid stabilizer

Polyvinyl alcohol, of a correct grade and when used at the optimal concentration, has been found to be most generally effective in retaining soluble enzymes within sections. However, it has the disadvantage of being too good a stabilizer in that it may over-stabilize cellular and subcellular membranes. Consequently, for example, it may ensure that none of the histochemical reagents penetrate into intact isolated cells, or into whole cells in thick sections, or through lysosomal membranes in relatively thin sections. For this reason, it is sometimes preferable to use some other colloid stabilizer. For isolated cells, some workers have used Ficoll (Stuart and Simpson, 1970). However, it is usually more useful to use Polypep 5115 (Sigma) which is a partially degraded collagen (Butcher, 1971b). It has been used extensively for stabilizing sections in the cytochemical section bioassays where it is essential that the degree of stabilization should not render the cell membranes unresponsive to the influence of the polypeptide hormones (Chayen, 1978a, 1980). Recently it has been shown that the commercial preparation (Polypep 5115) contains sufficient calcium to interfere with some cytochemical reactions; a modified form of this preparation, with less contamination by calcium ions, is now available (Polypeptide 8350; Molecular Design and Synthesis, University of Surrey, Guildford).

CYTOCHEMISTRY OF ISOLATED CELLS

Cytochemical reactions can be done on isolated cells tested either as air-dried smears (or after cytocentrifugation) or as suspensions in a suitable fluid medium. The former may well require stabilizing with a suitable colloid stabilizer, since the process of smearing and drying may well make the cell membranes sufficiently permeable to allow loss of soluble enzymes and active moieties into a non-stabilizing reaction medium. On the other hand, it might have been thought that the cells studied in suspension would not require such stabilization because the cell membrane should ensure full retention of soluble contents. This has not been found in practice; different cells, in suspension, may require a particular concentration of a colloid stabilizer added to the suspension medium. This has to be determined for each cell type, by the method for assessing the efficacy of a colloid stabilizer given above. Once this has been determined, smears can be treated in the same way as sections; an example of how to react isolated cells in suspension (for assaying glucose 6-phosphate dehydrogenase activity) is given below.

Quantitative cytochemical studies have been made on cervical smears used in tests for malignancy (Millett *et al.*, 1980; Ibrahim *et al.*, 1983; Bitensky *et al.*, 1984); on blood cells (Stuart and Simpson, 1970); and on isolated mastocytoma cells (Pitsillides *et al.*, 1988; Chayen *et al.*, 1990).

Method for assaying activity in cells in suspension

The following gives an example of an assay done on isolated cells in suspension. The cells were mastocytoma cells and it had been demonstrated that they required a 30% solution of Polypep 5115 to give adequate stabilization. It was required to assay glucose 6-phosphate dehydrogenase activity in these cells.

Reaction medium

A 0.315% solution of purified neotetrazolium chloride in a 0.05 M glycylglycine buffer (pH 8.0) containing 30% (w/v) of Polypep 5115 (Sigma), NADP (3 mM), glucose 6-phosphate (5 mM) and phenazine methosulphate (0.7 mM). After these additions, the pH should be pH 8.0, or adjusted to this pH if necessary.

The reaction medium is equilibrated to 37 °C. Then 0.1 ml of a suspension of the cells (about 2×10^7 cells/ml) is added and mixed gently with the reaction medium at 37 °C.

Centrifugation

1. After a suitable time of reaction, the reaction medium with cells is centrifuged lightly (e.g. 850–900 rpm) for 5 min to ensure gentle sedimentation of the cells.
2. The supernatant is removed.
3. The cells are washed by resuspending them in a 1% aqueous solution of poly-vinyl alcohol (PVA grade G04/140) in 0.05 M glycylglycine buffer at pH 8.0.

4. The cells are immediately centrifuged again (e.g. at 900–1000 rpm) for 7 min, and the supernatant is removed.
5. The cells are then fixed in formol–calcium for 15 min.
6. The cells are sedimented again by centrifugation (e.g. 1000 rpm for 5 min).
7. They are resuspended in a 1% solution of PVA and allowed to settle out without centrifugation.
8. Slides are coated with a smear of 5% PVA, used very much as glycerine-albumin is used in conventional histology, which is allowed to dry (or become tacky). The cells are deposited on such slides and allowed to dry. They can then be mounted, for example in Farrants' medium or in a 1% solution of PVA.

IONS FOR STABILIZING SECTIONS

Although tissue stabilizers, such as PVA, can protect sections and isolated cells under most conditions, there are occasions when additional stabilization is necessary. This may occur, for example, as in the reaction for Na^+-K^+-ATPase activity, which is very labile and where the sections must be preincubated in one medium before being subjected to the full reaction medium. Under such circumstances, additional stabilization may be afforded by including a tissue-stabilizing ion in the medium. Acetate ions, even at close to neutral pH values, are well known to 'harden' protoplasm. Calcium ions have long been known to cause 'gelation' of protoplasm and immobilization of fatty acids and phospholipids. Either may be helpful, depending on the circumstances: for example, calcium ions cannot be used to stabilize tissue or sections for reactions in which calcium ions may also play a part, either as inhibitors or activators of an enzymatic process.

TO PREPARE A STOCK SOLUTION OF POLYVINYL ALCOHOL (PVA)

Heat the buffer solution on a heated stirrer (e.g. MAG/30H, Voss Instruments Ltd) to not more than 60 °C. Add a portion (about 15%) of the PVA and mix until all is dissolved. Add further portions of the PVA until the total amount of PVA is dissolved.

When preparing a large volume, which may take several hours, it is best to add a small amount of PVA and then leave it to dissolve, adding some more later and continuing this process at a leisurely pace; the solution can be prepared intermittently while doing other laboratory work. Allow the solution to cool and then make it up to the final volume with buffer.

Note: It is imperative that the PVA solution is never heated above 60 °C.

3

Methods of Quantification

To a surprisingly large extent every histochemist deals with quantification. For example, very few will report a reaction as being either positive or negative; most will modify the reaction with some adjectival word or phrase: 'strongly' or 'weakly' positive, 'intense' reaction, etc. This already implies some assessment of the amount of the reaction; this assessment may be arbitrary ('a strong reaction') or it may be relative, e.g. to the reaction found in other tissues or under other experimental or physiological conditions. Particularly because the quantitative aspect is not openly recognized, the errors involved are not even considered. One facet of this problem was lucidly discussed by Gomori (1952) who gave the example of certain amphibians and fish which change colour very rapidly: the animal does not contain more or less pigment per unit area—it only changes the state of contraction or expansion of its pigment-containing cells. So too in histochemistry: the same amount of a tissue component or enzyme can produce a few intense granules of reaction, or an apparently strong diffuse stain; the tendency will be to say that only the latter shows a strongly positive reaction, i.e. only the latter contains much of the component studied. In fact the same actual amount of tissue component, and even of the reaction product, may be present in both cases but dispersed differently.

Various procedures have been tried for quantifying histochemical reactions. In the early studies, a simple photocell was mounted above, or in the place of, the eyepiece. As will be discussed later, and as was emphasized by Gomori (1952), this can cause considerable error owing to the optically inhomogeneous distribution of most histochemical reactions: only very weak reactions of fairly evenly dispersed 'stain' can be measured adequately by such procedures. Others, recognizing the potential errors inherent in such optical methods, preferred to elute the coloured reaction product from the sections; the amount of coloured reaction product was then measured by conventional spectrophotometry (e.g. Defendi and Pearson, 1955; Jardetsky and Glick, 1956). However, such techniques lose much of the special value of quantitative histochemistry: they do not allow one to differentiate between the activities of the various cell types that may constitute that section of the tissue. The development of scanning and integrating microdensitometry has made this possible and has extended biochemistry down to the level of the individual cell (Chayen, 1984).

ELUTION TECHNIQUES

The histochemist who lacks sophisticated microphotometric apparatus may still wish to quantify the amount of the histochemical reaction present in each section and so relate the amount of stain, assessed visually in the different parts, with the actual activity of an enzyme, or the actual amount of a particular substance, present in that section. Various valuable methods have been tried, notably by Defendi and Pearson (1955) and by Jardetsky and Glick (1956). The elegant micromethods developed largely by Glick (1962, 1963) and by Lowry (e.g. 1964) and others (Matschinsky *et al.*, 1968; Giacobini, 1969) from the microbiochemical techniques of Lindestrøm-Lang and Holter at the Carlsberg Laboratories in Copenhagen belong more to the realm of biochemistry than histochemistry; they are basically disruptive analytical chemical methods.

The newer elution techniques for quantitative histochemistry (Altman, 1972a,b) were derived from the procedure used in Professor Brachet's Laboratory in Brussels in 1951. Sections were treated with toluidine blue to stain the nucleic acids (as done by Davidson and Waymouth, 1944); the dye was then extracted from the section by dilute acid and estimated in a spectrophotometer. This gave a reasonable estimation of the amount of nucleic acids in the section. In a similar fashion, Defendi and Pearson (1955) performed a histochemical tetrazolium reaction for succinate dehydrogenase and then extracted the formazan from each section; the amount of enzyme activity was proportional to the amount of formazan extracted, which was measured in a spectrophotometer. Consequently, in the procedure which we advocate, dehydrogenase enzymes in the section are allowed to act on their specific substrate in the presence of a tetrazole and of specific cofactor (if required); on accepting hydrogen from the dehydrogenation process the tetrazole is reduced to a coloured formazan which is precipitated at, or close to, the site at which it accepted the hydrogen. The localization of this activity can be examined by inspection of the section. [*Note*: phase-contrast microscopy is often helpful in determining the localization of the formazan in an otherwise unstained section.] By means of various elution procedures (Altman, 1969a,b; 1972a,b) the formazans of all commonly used tetrazolium salts can be eluted and measured quantitatively, even those that bind strongly to the tissue and so yield the best localization. It is noteworthy that Altman (1972b) has shown that the same order of enzyme activity can be measured by these procedures as by conventional biochemical methods, so that correctly performed enzyme histochemistry (including the use of colloid stabilizers and intermediate hydrogen carriers) may be as precise as conventional biochemistry and involves no loss or inhibition.

Elution method

Four serial sections are reacted for the particular enzyme, e.g. glucose 6-phosphate dehydrogenase, with one of the tetrazolium salts as the final hydrogen-acceptor. After the reaction, the formazan-stained sections are rinsed in distilled water and left to dry in air at laboratory temperature. (If triphenyl tetrazolium is used as the hydrogen acceptor, it is advisable to rinse the sections very briefly in hot distilled water to avoid this formazan dissolving on prolonged immersion in cold water.) Each section is numbered and placed in a photographic enlarger. The enlarged image is projected onto white paper and the outline is traced with a pencil. The area included within this outline is then measured by means of a planimeter, and the magnification factor through the photographic enlarger at this particular setting is determined. Thus, knowing the area of the projected image (by planimetry) and the magnification of this image, the actual area of the section can be determined in μm^2.

Each section, on its glass support, is then cut out from the rest of the slide, to give a formazan-stained section on a piece of glass which is just larger than the section. This glass fragment, with its section, is dropped into a test-tube and the minimum but measured amount of the eluant, sufficient to cover the sample, is pipetted into the tube. Usually 0.5 ml is used.

Most formazans dissolve readily, with gentle shaking, into dimethyl-formamide (see Table 1) but some are elutable only at an alkaline pH. For these an alkaline, buffered dimethylformamide (DMF) is used. It is composed of 1 volume of buffer to 9 volumes of DMF. The composition of the buffer (Altman, 1969a) is as follows:

0.1 M sodium hydroxide	95 ml
0.1 M glycine + 0.1 M sodium chloride	5 ml

(Keep this in a stoppered bottle at $+4\,°C$. The pH must be at least 11.5 and should be checked before use.)

In practice it is found that, for formazans that elute at an alkaline pH, the sections should be rinsed for a few minutes with this buffer alone before eluting in the 1 : 9 buffer : DMF solution. (Gentle warming to 40 °C in a water bath further assists this elution.) Moreover, in such studies, control sections should be incubated in the presence of the tetrazole alone and then subjected to this elution procedure to measure how much unreacted tetrazole is adsorbed onto the section. This is necessary only when the alkaline DMF is used because the alkalinity converts unreacted tetrazole to formazan. The difference between the amount of formazan extracted from these control sections and from the test sections

Table 1. Elution and quantification of various formazans (after Altman, 1972a,b)

Formazan	Solvent	Absorption maximum (nm)	An extinction of 1.0 is given by the following weights of the formazan (μg) in 1 ml solvent	Molecular weight of formazan
MTT (m)	Dimethylformamide	550	21.0	335
INT (m)	Alkaline dimethylformamide	645	5.5	472
TNBT (d)	Alkaline dimethylformamide	680	11.0	839
NBT (d)	Alkaline dimethylformamide	675	9.4	749
NT (d)	Dimethylformamide	530	30.0	599
BT (d)	Dimethylformamide	575	92.0	659
TV (m)	Dimethylformamide	510	27.0	351
TT (m)	Dimethylformamide	485	21.0	301

M, monotetrazole; d, ditetrazole; TNBT, tetranitroblue tetrazolium; NBT, nitroblue tetrazolium; NT, neotetrazolium; BT, blue tetrazolium; TV, tetrazolium violet; TT, triphenyl tetrazolium. No common names are generally used for the other tetrazoles.

(+ substrate + cofactor, etc.) is the amount of tetrazole reduced to formazan by enzymatic dehydrogenation of substrate and reduction of the tetrazole.

The dissolved formazan is measured in a microcell in a conventional spectrophotometer at the wavelength at which it absorbs maximally. The weight of formazan equivalent to the recorded absorption can be calculated readily from the data in Table 1. It is usual to measure four or five test sections (and an equal number of controls, where required) and to express the results as the mean value, with standard deviation.

The weight of formazan produced by the dehydrogenation reaction (or the amount of hydrogen, as calculated below) must then be related to the mass of the tissue. Different experimental conditions may call for different parameters of tissue mass. For example, if the water content of a tissue doubles during the experiment, a 10 μm section will have half the dry mass of tissue at the end of the experiment. For this reason, the best parameter is probably the nucleic acid content of tissue. This has the additional advantage that, if the polypeptide is used as the section stabilizer, it can be estimated very simply on the same sections after the formazan has been eluted (Butcher, 1968, 1971a,b). Generally, however, it is sufficient to relate the amount of formazan to unit area, as measured by planimetry. This would be entirely satisfactory if it were certain that the thickness of the sections were constant; then this calculation would be equivalent to the amount of formazan per unit volume. It has been found that the thickness of the sections cut by any one worker can be remarkably uniform but that the sections expand or contract according to the rate at which they are cut (Butcher, 1971a). Uniformity can be obtained by using the automatic motor-driven section-cutting device as is fitted to the 'automatic' Brights cryostat.

Calculation of the amount of hydrogen liberated

For any selected tetrazole, the amount of hydrogen liberated by the reaction in the section is directly proportional to the amount of formazan produced. But 1 g (or 1 μg) of one formazan does not represent the same amount of hydrogen as does the same weight of another formazan. The reasons are as follows: when the formazan is formed, each tetrazole ring accepts two atoms of hydrogen. Consequently monotetrazoles (like MTT) accept two atoms but ditetrazoles (such as neotetrazolium or nitroblue tetrazolium) require four atoms, i.e. two molecules of hydrogen for each molecule of the formazan. Hence 1 mole of a monoformazan represents 1 mole of hydrogen whereas 1 mole of a diformazan is equivalent to 2 moles of hydrogen. Thus, to derive the amount of hydrogen (in micromoles) from a weight of a monotetrazole, divide the weight (in micrograms) by the molecular weight of the formazan (Table 1); for a ditetrazole divide by the molecular weight but multiply by 2 (for full details see Altman, 1972b).

PRECISE QUANTIFICATION: SCANNING AND INTEGRATING MICRODENSITOMETRY

Histochemistry has always been poorly regarded by workers in the more precise biological sciences, mainly because of the following features:

1. The inactivation and the loss of activity caused by chemical fixation and the loss of soluble material during the reaction. These features have been corrected by the use of controlled chilling and sectioning methods, and by the use of colloid stabilizers in the reaction medium.
2. The largely arbitrary nature of the reaction mixtures and the lack of specificity of the reactions. The methods given in this book, supplemented by the original publications cited in the text, are designed to overcome these objections.
3. The lack of quantification. One of the major advances in histochemistry, or quantitative cytochemistry, of recent years has been the development of scanning and integrating microdensitometry, which completely eliminates this objection.

Background to microdensitometry

The early work on microscopic photometry was done on reactions that were known to be stoichiometric, such as the Feulgen nucleal reaction for DNA (as reviewed by Leuchtenberger, 1954) or the modified Sakaguchi reaction for arginine. To a large extent these investigations were rewarding because care was taken to ensure that the colour was fairly homogeneously distributed throughout each nucleus and the intensity of the stain, or chromophore, was low. However, even given such ideal conditions, the results were open to some doubt. The inadequacy of simple microscopic photometry (discussed by Chayen and Denby, 1968) becomes particularly serious when neither condition pertains, as occurs with most histochemical reactions. There is a vast literature dealing with the errors of using simple photometry for measuring a reasonably intense chromophore that is dispersed in an optically inhomogeneous fashion (e.g. Walker, 1958; Walker and Richards, 1959; Pollister and Ornstein, 1959; Chayen and Denby, 1968; Wied and Bahr, 1970; Bitensky, 1980). This problem has been overcome by the development of scanning and integrating microdensitometry.

Basis of scanning and integrating microdensitometry

Suppose the activity, represented by an amount of coloured reaction product evenly dispersed within the cell (Table 2), results in an 80% transmission of the light incident upon it. This transmission value can be converted to extinction (E) by use of the Beer–Lambert law which states that $E = \log_{10}(I_0/I)$, where I_0 is the intensity of the illuminating light (or clear field; i.e. 100% transmission) and I is the intensity of the light transmitted through the coloured object (or cell):

$$E = \log_{10}(100/80) = 0.097$$

Let us now suppose that in another cell, of similar size and activity, the coloured reaction product is concentrated into half the projected area of the cell (Table 2). Since the extinction is directly proportional to the concentration, the extinction of this amount of chromophore (E) will be $2 \times 0.097 = 0.194$. In this case the intensity of the transmitted light (I_a) from this half of the cell will be:

$$0.194 = \log(100/I_a)$$
$$I_a = 64\%$$

Since the other half of the cell, without any reaction product, will transmit 100% (I_b) of the light incident upon it, the total amount of light transmitted over the whole field of measurement (i.e. the whole cell) will be $(I_a + I_b)/2$ which is $(64 + 100)/2$ or 82%.

This means that if a simple photometer is used to measure the amount of absorption (i.e. reaction product) in this cell, in which all the colour occurs in one half of the total area of the cell, the error will be $82 - 80/80 \times 100$ or 2.5%. But the concentration of a chromophore is related to the extinction, not the percentage transmission. An area or cell which shows an 82% transmission will have an extinction, as calculated by the Beer–Lambert equation, of $\log(100/82)$ or 0.086 instead of the value of 0.097 that it would have if the same amount

Table 2. Errors caused by inhomogeneous distribution of chromophore

	Area occupied by the chromophore	Transmission, $T(\%)$	Extinction, E
Effects with low extinction values			
	A	80	0.097
	$A/2$	82	0.086
	$9A/10$	91.08	0.041
Effects with high extinction values			
	A	40	0.4
	$A/2$	58	0.24
	$9A/10$	90.001	0.046

of chromophore was evenly dispersed throughout the cell. Thus the error is now 11.3%; the second cell would seem to have significantly less activity than the first.

The error is greatly magnified if the area occupied by the chromophore is only one-tenth of the projected area of the cell, as it may be when lysosomes are stained. By the same type of calculation, the measured extinction would be 0.041 instead of 0.097, as it would be if all the chromophore (or stain) had been redispersed to fill the whole cell uniformly. The extinction value of 0.041 would imply that the second cell had less than half the activity that it actually possesses (Table 2).

So far, we have been considering errors in measuring low levels of activity, corresponding to an extinction of 0.097. Because of the logarithmic relationship between percentage transmission and extinction, the errors increase greatly at higher extinction values, as will occur in 'well stained' preparations. Consequently, in the examples discussed above, if the overall extinction would be 0.4 (40% of transmission) when the chromophore was uniformly dispersed, the transmission through the cell in which the chromophore occupied half of the projected area of the cell would be $(15.9 + 100)/2$ or 58%; the extinction would be 0.24. This would be an error of nearly 50%, as against 11.3% in the previous example. If the chromophore occupied only one-tenth of the projected area of the cell, that one-tenth of the cell would transmit 0.01% of the light while the other nine-tenths would still be transmitting 100% of the light. The overall transmission would be $(0.01 + 100)/10$, or 90.001%, and the measured extinction would be 0.046. This would imply that this cell had almost ten times less chromophore than it actually contained.

Scanning and integrating microdensitometry

This problem is overcome very simply by means of scanning and integrating microdensitometry (Deeley, 1955). The basis of the procedure is as follows.

The Beer–Lambert law states that

$$E = \log_{10}(I_0/I) = kcl$$

where E is the extinction (or absorbance), k is the absorption coefficient (or absorptivity) of the chromophore, c is the concentration and l is the path length through the chromophore.

In microdensitometric studies we are normally interested in the mass (M) of the chromophore (representing the amount of active moieties or of enzymatic activities) rather than the concentration. Since concentration is equal to mass per unit volume, and volume is area (A) multiplied by length (l), we can substitute for c:

$$c = \frac{M}{Al}$$

so that

$$E = k.(M/Al).l = \frac{kM}{A}$$

Consequently

$$M = \frac{EA}{k}$$

If the chromophore were uniformly dispersed throughout the area (A) to be measured (e.g. throughout a cell of area A), a simple measurement of the extinction (or absorbance) of the cell would give us a value proportional to its mass in that area. But this pertains only very rarely: generally the chromophore is dispersed very inhomogeneously. The alternative would be to measure the extinction, and the area, of each optical inhomogeneity, and to summate the results so that

$$M = \frac{E_1 A_1 + E_2 A_2 + E_3 A_3 + \ldots + E_n A_n}{k}$$

The scanning and integrating microdensitometer overcomes this practically clumsy operation by dividing the selected area to be measured (A) into a very large number of small regions of area a. If each region is optically homogeneous, the measured extinction in that region will be measured accurately. The instrument summates (integrates) all the individual values of E, so that

$$M = \frac{(E_1 + E_2 + E_3 + \ldots + E_n).a}{k}$$

Size of scanning spot

It should be apparent that the size of the scanning spot used in a given study can be critical. Suppose you wish to measure the activity of a mitochondrial enzyme, where the chromophore is staining the mitochondria. The chromophore (stain or reaction product) will be localized in small packets, each corresponding to a mitochondrion, and each of perhaps 0.5 μm diameter. Consequently, should you be so unwise as to use a scanning spot of 1.0 μm diameter, then when the spot is located over one mitochondrion half the field (the scanning spot) will be occupied by the chromophore and half will be occupied by clear, uncoloured field (with 100% transmission). This will entail an error of the type discussed in relation to Table 2; the exact degree of error will depend on the intensity with which the mitochondrion is stained. Clearly therefore, for such studies, the size of the scanning spot must not exceed 0.5 μm diameter. Similarly, if lysosomal activities are to be measured, the diameter of the scanning spot must not exceed 0.25 μm.

The advantage of using a scanning spot of diameter 0.2 or 0.25 μm is that this is equivalent to the limit of resolution of the normal light microscope so that any material within this spot will be optically homogeneous.

Errors caused by using the incorrect size of scanning spot have been demonstrated and discussed by Bitensky *et al.* (1973) and by Bitensky (1980).

The Vickers M85A microdensitometer

The standard instrument that fulfils all the optical requirements for precise measurement of a coloured chromophore, whether this is a natural chromophore such as cytochrome P_{450} (Altman *et al.*, 1975) or the result of a cytochemical

Figure 8. Vickers M85A Scanning Cytophotometer (Reproduced by permission of Vickers Instruments).

chromogenic reaction, is the Vickers M85 and the newer M85A scanning and integrating microdensitometer (Figure 8). Basically this is a spectrophotometer built around a microscope, with two sources of light, one for conventional microscopy and the other for measurement. The light for the former comes from a lamp sited below (and behind) the microscope; it is focused by relatively conventional Köhler illumination through the condenser onto the specimen. Nomarski interference can be inserted to facilitate inspection of the histology in relatively uncoloured specimens. The section, or isolated cells, are examined by conventional optics. The light for measurement comes from a lamp in the main housing of the instrument, above the microscope. It passes through a monochromator, to select light of a specific wavelength. As in conventional spectrophotometry, the bandwidth can be varied to give greater or less spectral resolution (i.e. to decrease the amount of light of wavelengths close to that which has been selected). The instrument has a further control, not found in conventional spectrophotometers. This is to control the size of the scanning spot (a in the equation above).

The spot of light, of the selected wavelength, then passes through the objective onto the specimen. It is made to traverse a raster pattern; the absorption of each point in this pattern is measured separately by the photomultiplier set below the condenser. The instrument then summates all the measured absorptions (or integrates the responses—hence the name of the scanning and integrating microdensitometer). But the instrument has an additional advantage. A disc of light, visible to the operator but not to the measuring photomultiplier, is inserted onto the specimen. The size of this disc can be varied considerably so that, for example, it can encompass a single cell or a single nucleus. This disc of light, the optical mask, is monitored by another photomultiplier which instructs the instrument to record only those measurements that are made from within the mask. In this way, the operator optically dissects out the feature to be measured, optically isolating it from the rest of the tissue as far as measurement is concerned.

The exact procedures to be followed in making such measurements depend on the instrument available. For example, they are somewhat different with the older Vickers M85 than with the newer M85A microdensitometer. The operator is recommended to use the procedures in the handbook available with the instrument.

Assay by microdensitometry

Calibration

The microdensitometer records a value for the summated absorptions, but this value—the relative density or relative absorption—is somewhat arbitrary in that it depends on the response of that particular measuring photomultiplier. This response can be altered by changing the voltage applied to the photomultiplier. Consequently, although any one microdensitometer will record regularly a particular value for a given specimen, another microdensitometer may well record (regularly) a different relative absorption for that specimen. This can be corrected by calibrating the response of the instrument. Such calibration is necessary in any case, to ensure that the instrument is adjusted to give linearity of response. In earlier times, when the sole use of microdensitometers was to measure the amount of dye in Feulgen-stained nuclei, it was customary to set microdensitometers so that they did *not* record linearly over low absorption values: this ensured that only well stained structures were measured. However, nowadays, it is essential that the instrument does record linearly. (If it does not, it may be necessary to remonstrate with the manufacturers: it is a simple matter to adjust it so that it gives a linear response.)

The instrument is calibrated by constructing a response/extinction graph. The instrument is set up as for normal measurement, with 100% transmission in clear field. Then a known amount of extinction is superimposed onto the field. Ideally this is done by inserting a neutral density filter of known extinction, but it can be done electronically by imposing a defined density (extinction) on the density meter of the instrument. The instrument is allowed to scan the field containing this imposed extinction and the relative absorption is recorded on the digital meter. This procedure is repeated for several neutral density filters of various known extinction values (or by different settings on the density meter). From these readings of relative absorptions (as recorded by the instrument) for known extinction values, a straight-line graph should be obtained. From such graphs, any relative absorption recorded by this instrument can be compared with relative absorption values obtained with any other instrument.

It is often helpful to convert relative absorption values, recorded by an instrument, into absolute units of extinction. This is particularly true for most quantitative cytochemical assays. Consequently, when a series of measurements is made, another reading of the amount of relative absorption is recorded, under exactly the same conditions, for a clear field containing a defined extinction (e.g. a neutral density filter of 0.5, or an electronically superimposed extinction of 0.5 or 0.7). As the operator has already shown that the instrument is recording linearly, this allows all the relative absorption values to be converted to absolute extinction. For example, suppose that a neutral density filter of 0.5 produces a relative absorption of 600. Then a value, in the specimen, of 300 would be equal to $(300/600) \times 0.5 = 0.5 \times 0.5$, or an extinction of 0.25. This implies that the mean extinction in the selected field (e.g. in a single cell) is 0.25. It has become customary to record this value as the mean integrated extinction (MIE) multiplied by 100 to avoid excessive decimal points. This value would therefore be recorded as an MIE $\times$ 100 of 25.

Calculation in absolute terms

The amount of a material, or of an enzymatic activity, can be calculated in absolute terms if (a) the stoichiometry of the material, or activity, and the final coloured end-product is known, and (b) the extinction coefficient (or absorptivity, k) in sections has been determined. (*Note*: the extinction coefficient of many chromophores, or coloured end-products, may be very different when these are precipitated in sections from when they are in solution: the extinction coefficient in sections must be determined (as was done by Butcher, 1972; Butcher and Altman, 1973). Then, knowing the extinction (E, namely MIE) as discussed above and the extinction coefficient k, the mass (M) of the chromophore in a defined area (A; for example, the area occupied by one cell) can be calculated from the formula (discussed earlier):

$$M = \frac{EA}{k}$$

Example: To calculate the activity of a dehydrogenase enzyme Suppose that the relative absorption, due to the formazan of neotetrazolium chloride, within a mask of area A is 374. Suppose that the relative absorption for a clear field of the same area, including a neutral density filter of 1.0, was 900. Then, as discussed previously,

$$\text{MIE} = \frac{374}{900} = 0.42$$

The area A is measured. Suppose it is 708 μm^2 (or $708 \times 10^{-8} cm^2$). It has been shown (e.g. Butcher and Altman, 1973) that, at the wavelength at which this specimen was measured (585 nm), the molar extinction coefficient (k) of the formazan is 7400 (for one gramme-molécular weight per litre).

Then:
$$M = \frac{708 \times 10^{-8} \times 0.42}{1} \times \frac{10^{-3}}{7400}$$

$$= 40.2 \times 10^{-14} \text{ moles formazan}$$

It was shown by Butcher and Altman (1973) that when the formazan of neotetrazolium chloride is measured at 585 nm, one mole of the dye is equivalent to one mole of hydrogen. Consequently 40.2×10^{-14} moles formazan are equivalent to 40.2×10^{-14} moles hydrogen, or 402 fmoles hydrogen.

POLARIZED LIGHT MICROSCOPY

General concepts

A single ray of light, moving directly from the source to the detector, is in fact oscillating in the form of a wave. Even a very narrow beam of light consists of many rays, each oscillating in its own direction perpendicularly to that of the beam.

The absorption and refraction of light passing through a material are produced by the interaction of the oscillating electric field of the wave of light with electrically charged particles in the material. In most materials the electromagnetic field produced by these charged particles is quantitatively the same in all directions, but in some materials this is not so. Consequently the speed of light in one or other direction, after it has passed through the material, is retarded. This cannot be detected by normal illumination, in which the light is vibrating in all directions. It can be detected with plane-polarized light, namely with light that vibrates in one direction only, all the other light having been filtered out of the beam by the polarizer.

The speed with which plane-polarized light moves through a material is related to the refractive index of that material:

$$\text{Refractive index} = \frac{\text{Velocity of light in air}}{\text{Velocity in the material}}$$

For most materials the refractive index is the same in all directions of the material irrespective of the direction of the light. Such materials are isotropic. However, in some structures, typically certain types of crystals, the refractive index measured along one axis (e.g. the long axis) is different from that measured when the light vibrates along another axis (e.g. the short axis). That is, light travels more quickly when vibrating along one axis than across the other axis. Such material is called anisotropic and shows the property of birefringence (i.e. it has two refractive indices). Consequently birefringence is defined as the difference between the refractive index measured with light parallel to the long axis $(n_{\parallel})$ and that measured with the light vibrating perpendicularly to the long axis $(n_{\perp})$: $n_{\parallel} - n_{\perp}$.

However, what we can actually measure with the polarizing light microscope is the *optical path difference* (opd) in the material, namely the interaction of both refractive indices that causes the change in the velocity of light. This will be influenced by the thickness (t) of the material:

$$\text{opd} = (n_{\parallel} - n_{\perp})t$$

It follows therefore that, although birefringence is a physical constant for a given material, a thicker specimen will appear brighter than a thinner specimen because what we actually observe is the optical path difference (opd), not the birefringence $(n_{\parallel} - n_{\perp})$.

Types of birefringence

There are three types of natural birefringence that may be relevant to histochemists.

Intrinsic or crystalline birefringence

This is determined by the chemical constitution of the material so that it is relatively little influenced by the medium in which the material is immersed. It is found particularly in crystals such as urates, which may occur in tissues particularly in gout (Watts *et al.*, 1971). It has also been used for demonstrating abnormal deposits of apatite or hydroxyapatite (Kent *et al.*, 1983).

Form or textural birefringence

Structures that contain oriented elongate submicroscopic particles or micelles may show this form of birefringence even though the particles themselves may lack intrinsic birefringence. The basis for this type of birefringence has been discussed by Chayen and Denby (1968; also Chayen, 1983). It depends on the different velocity of the light as it runs along the direction of the oriented fibrils as against the velocity when the light traverses the array of fibrils or the mounting medium. In contrast to intrinsic birefringence, form birefringence is markedly affected by the refractive index of the material in which the structures are mounted. Consequently measurement of the birefringence when the section is mounted sequentially in two media of very different refractive indices will differentiate this form of birefringence from intrinsic birefringence.

Strain or flow birefringence

Some materials that themselves lack birefringence become birefringent when they are made to flow along a restricted pathway so that they become oriented. To the histochemist this form of birefringence may be relevant only with regard to the microscope objectives, because the glass used in making these optical components must be free from strain. If it is not, the optical components of the microscope will show strain birefringence so that the microscope field, under crossed polars, will never become completely black.

Apparatus

The polarizing microscope should have a rotatable stage, and a slot in the microscope tube, at 45°, to accommodate a quartz red plate or a compensator. It should have a rotatable polarizer below the condenser and a rotatable analyser in the main tube of the microscope. Ideally the objectives should be of strain free glass.

Characteristic features in polarized light microscopy

Habit This includes shape and dimensions, particularly of crystals.

Extinction angle Most birefringent structures will extinguish (appear black) when the long axis is set to the N–S (north–south; 12 to 6 o'clock) or E–W position. This is particularly important for form birefringence because it is possible to have structures that show bright when viewed in the N–S position because they have been sectioned transversely (as demonstrated by Dunham *et al.*, 1988).

Some crystals do not extinguish at the N–S position but at a characteristic angle from N–S. For example, crystals of monosodium urate monohydrate, found characteristically in gout, extinguish at 12° from the N–S position (Watts *et al.*, 1971).

Sign of birefringence This is a major parameter since it defines whether the optically slow axis is along the long or the short geometric axis of the structure or crystal. To determine this, the structure is positioned with the long axis at 45°, at the N–E angle. The quartz red plate is inserted into the microscope tube making the whole field appear red. Then the retardation of light (R) caused by the plate will be either more or less retarded by the structure that is being examined. Then

$$R_t = R_p + R_s$$

where R_t is the total retardation, R_p is that due to the plate, and R_s is the retardation induced by the specimen. If the structure becomes blue it means that it has increased the retardation; that is, it is positively birefringent. This is typical of calcium pyrophosphate dihydrate crystals. If it becomes yellow, R_s is negative (as is found with crystals of monosodium urate monohydrate, characteristic of gout). This procedure was used to demonstrate the orientation of the collagen in the surface lamina of cartilage (Dunham *et al.*, 1988).

Refractive index This, and the symmetry of the birefringence, has been valuable in differentiating different types of crystalline deposits (Watts *et al.*, 1971).

Applications of polarized light microscopy

Probably the most striking demonstration of the practical use of quantitative polarized light measurements has been their application to the preservation of the myocardium (form birefringence) both during prolonged operation (e.g. Cankovic-Darracott *et al.*, 1977) and in the donor heart before and after transplantation (Darracott-Cankovic *et al.*, 1987). In a study on 172 donor hearts the quantitative birefringence measurements of the myocardial contractility correlated with the clinical outcome of transplantation at a level of $P < 0.001$ (Darracott-Cankovic *et al.*, 1989).

It has been used to demonstrate and define crystals of various urates in the muscle of patients with gout who had been treated with allopurinol (Watts *et al.*, 1971). It was also used to define crystals of apatite close to the site of the fracture of osteoporotic subjects (Kent *et al.*, 1983). Polarized light microscopy was used by Flint (1982) in defining starch, crystalline fats and cellulose in food products.

Induced birefringence

Sweat *et al.* (1964) showed that Sirius red F3BA was a good stain for collagen and could be used with polarization microscopy. Junqueira *et al.* (1979a) showed that this dye could be used for the quantitative estimation of collagen in solution. They then showed that not only was it an excellent stain for collagen but it enhanced the birefringence of collagen, so increasing the specificity of the reaction when assessed by polarized light microscopy (Junqueira *et al.*, 1979b).

Modis's (1974) studies on connective tissue showed that polarized light microscopy was of considerable value in enhancing the effect of staining such tissues. This concept of induced birefringence was found to be of particular value in measuring changes in the integrity of the glycosaminoglycans of cartilage and bone. For example, Kent et al. (1983), in studying fractured necks of femur in osteoporotic humans, found that, although the total amount of glycosamino-glycans measured with Alcian blue was the same as in normal individuals, their orientation (measured by quantitative polarized light microscopy) was considerably decreased. Similar decreases in the orientation of the proteoglycans, measured by this method of induced birefringence of Alcian blue staining, were found during the development of natural murine osteoarthritis (Dunham et al., 1990).

4

Common Histological Stains

RAPID DIAGNOSTIC SECTIONS AND STORAGE OF TISSUE

Histochemistry has a number of special advantages. On the one hand it allows one to relate chemical activity and function to histological structure; on the other hand it gives more chemical specificity to the histology or histopathology. For the whole range of interests that lie between these extremes, it is important to be able to subject sections to some preferred histological stain. But it is quite obvious that the end-product of the histological staining method should be a section that resembles, as closely as possible, one prepared by the conventional histological procedures, which include fixation, dehydration and embedding in paraffin wax. To achieve this end, some of the conventional histological techniques have had to be modified in a seemingly irrational way. However, these modifications have not only produced histological preparations which, by their conventional appearance, make the histology clearer for the histochemist; they have had the additional advantage of opening new possibilities for the histopathologist. For example, it is now possible to cut a frozen section for rapid surgical diagnosis; to have it stained within a few minutes of taking the specimen; and yet to have a preparation that easily rivals one which would take several hours or even a few days by conventional paraffin-wax histology. The pathologist can use such sections for diagnosis just as if they were conventional preparations, but he can also have additional sections cut for such histochemical procedures as may aid his work. Another drawback to the use of cryostat sections has been the difficulty of storing the frozen blocks over a period of some years, because in a routine pathology laboratory they must be kept in case it becomes necessary to refer to them at some later stage. This has been overcome by the procedure by which the frozen tissue can subsequently be fixed and embedded in paraffin wax; such tissue yields sections in which the histology is indistinguishable from that of similar pieces of tissue which have never been chilled but have been subjected only to fixation and embedding.

SPECIAL METHODS OF PREPARING TISSUE

Most tissues can be chilled fresh in hexane at $-70\,°C$ as described in Chapter 1 (p. 4). Muscle may benefit by being treated with 5% PVA for 15–30 min before it is chilled. (Sometimes the addition of small amounts of carnosine to the PVA can improve the histology of the muscle.) Pretreatment with PVA (5%, with or without 1–2% calcium chloride) is very advisable when delicate tissue is to be chilled; retina and necrotic malignant tissue typically benefit from such pretreatment.

Contrary to some belief, tissue which has been fixed in formalin can be chilled and sectioned just as effectively as fresh tissue. Certain precautions should be taken: the temperature of the chilling bath must be below $-65\,°C$ and the temperature of the cabinet must be $-30\,°C$ with the knife really well chilled by prolonged contact with solid carbon dioxide. The slides must be well albumenized (i.e. smeared with egg albumen and left to dry for 30 min before use).

FIXATIVES

Picric–formalin fixative (Bitensky *et al.*, 1963)

Formaldehyde (40% technical)	10 ml
Absolute ethanol	56 ml
Sodium chloride	0.18 g
Picric acid (wet)	0.15 g
Distilled water to make up to	100 ml

Note: the picric acid is wet. It is placed on a filter paper to remove excess moisture and this picric acid is weighed. The reason that picric acid is always kept wet is that, when dry, it can become explosive. For this reason, the mouth of the bottle in which the stock is kept should be checked frequently to ensure that no picric acid has dried around the stopper. If it has, it should be cleaned with wet tissue paper.

Picro-acetic acid fixative (Bitensky *et al.*, 1963)

As the picric–formalin, but with the addition of 5% acetic acid.

Acetic–ethanol fixative

Either:	
Glacial acetic acid	10 ml
Absolute ethanol	30 ml
Mix just before use	

or:

Glacial acetic acid	5 ml
Absolute ethanol	95 ml

Mix just before required

Formol–calcium

Formaldehyde (40% technical)	10 ml
Calcium chloride (CaCl$_2$.2H$_2$O)	5 g
Distilled water to make up to	100 ml

Especially if this is to be kept as a stock solution, it is advisable to add marble chips.

Formol–saline

Formaldehyde (40% technical)	10 ml
Sodium chloride	0.85 g
Distilled water to make up to	100 ml

Heidenhain's Susa (for cryostat sections)

Mercuric chloride	4.0 g
Sodium chloride	0.5 g
Trichloroacetic acid	2.0 g
Acetic acid (glacial)	4.0 ml
Formaldehyde (40% technical)	20 ml
Distilled water	100 ml

STAINING PROCEDURES

Toluidine blue

Especially when using cryostat sections, it is helpful (or even necessary) to have a rapid staining method for checking how well the sectioning is proceeding. It is also valuable to take every fifth or tenth section for histological staining to disclose the histology of sections being used for histochemical procedures. This can be done very simply by dipping a section, on the slide, into a solution of toluidine blue (as below). Such stained sections can be made permanent if required.

1. Dry, unfixed sections can be used. Sections of difficult tissues, such as bone, may benefit from prolonged drying, even overnight, to enhance adhesion of the section to the slide.

2. Immerse in a 0.1% solution of toluidine blue in 0.1 M acetate buffer, pH 6.2, for up to 10 s. At this stage, the section may be over-stained, to compensate for loss of stain during rinsing.
3. Rinse in distilled water.
4. Blot dry, pressing gently to remove moisture.
5. Immerse in absolute alcohol (or industrial spirit) for 30 s.
6. Remove and dry.
7. Mount in Euparal or Styrolite.

Formol–calcium haematoxylin and eosin method

1. The cryostat section is fixed in formol–calcium for 5 min.
2. Filter Harris's haematoxylin into a Coplin staining jar; stain for 5 min.
3. Wash in tap water until the section is blue.
4. Immerse in acid alcohol (1% HCl in 70% alcohol) as required (e.g. about 7 s).
5. Wash in tap water (or alkaline water; see below) until the histological structures are sufficiently differentiated. This is best checked under a microscope. (*Note*: it may be necessary to put the section back into the acid alcohol if it is insufficiently differentiated. Then 'blue' again.)
6. Immerse in a 1% solution of eosin, e.g. for 1 min.
7. Wash in running tap water until the nuclei are clearly visible as blue structures in a pink cytoplasm (or red, according to taste).
8. Dehydrate sequentially in 70%, 90% and absolute alcohol.
9. Clear in xylene.
10. Mount in Styrolite or similar mountant.

Normal haematoxylin–eosin method for cryostat sections

This is done at room temperature.

1. For some tissues, the best preparations are obtained if the sections are taken directly from the cryostat cabinet (at about $-25\,°C$) and immersed immediately into the fixative. For other tissues, better results will be obtained if the sections are allowed to dry at room temperature for several minutes before they are fixed.
2. Immerse in the picric–formalin fixative (5 min).
3. Wash in 70% alcohol (10 s).
4. Leave in absolute alcohol for 1 min.
5. Stain in Ehrlich's haematoxylin for 20 min.
6. Rinse in tap water.
7. Differentiate in acid–alcohol (1% concentrated hydrochloric acid in 70% alcohol) according to taste (e.g. up to 10 s).
8. 'Blue' in alkaline water (5% sodium bicarbonate solution) for about 30 s.

9. Stain in a 0.5% aqueous solution of eosin for 40–60 s, according to taste.
10. Rinse briefly in tap water.
11. Dehydrate through a graded series of alcohols, clear in xylol and mount in Styrolite.

Modified haematoxylin–eosin method for cryostat sections

This procedure may be useful if the normal method does not give results comparable to those obtainable in paraffin-embedded sections in which the dehydration and embedding have caused some degree of shrinkage, with concomitant increased clarity of histological features.

1. Use either sections taken directly from the cryostat cabinet or sections that have been allowed to dry, depending on the tissue (as discussed above).
2. Immerse in the picric–formalin fixative (5 min).
3. Wash in 70% alcohol (10 s).
4. Leave in the periodic acid solution for 5 min. (Steps 4–8 help to convert the histological appearance to that seen in more conventional fixed and embedded sections.)
5. Wash in 70% alcohol (10 s).
6. Immerse in reducing rinse for 5 min.
7. Wash in 70% alcohol (10 s).
8. Leave in absolute alcohol for 1 min.
9. Stain in Ehrlich's haematoxylin for 20 min.
10. Rinse in tap water.
11. Differentiate in acid–alcohol (1% concentrated hydrochloric acid in 70% alcohol) according to taste (e.g. up to 10 s).
12. 'Blue' in alkaline water (5% sodium bicarbonate solution) for about 30 s.
13. Stain in a 0.5% aqueous solution of eosin for 40–60 s, according to taste.
14. Rinse briefly in tap water.
15. Dehydrate through a graded series of alcohols, clear in xylol and mount in a suitable non-aqueous mountant such as DePeX or Styrolite.

Extra solutions required for this method

Periodic acid solution Dissolve 0.4 g periodic acid in 15 ml distilled water. To this add 35 ml absolute alcohol in which is dissolved 0.135 g hydrated crystalline sodium acetate. The final solution should be stored in the dark.

Reducing rinse To 20 ml distilled water add 1 g potassium iodide and 1 g sodium thiosulphate. Add, stirring continuously, 30 ml absolute alcohol and then 0.5 ml 2M hydrochloric acid. A precipitate of sulphur will form; this is allowed to settle.

Haematoxylin–eosin method for autoradiographs

1. Cryostat sections should be fixed in acetic–ethanol (1 volume glacial acetic acid : 3 volumes absolute alcohol) for 30 min.
2. Dry the sections overnight.
3. Perform the autoradiography.
4. After suitable exposure, and photographic fixation of the emulsion, stain in freshly filtered Ehrlich's haematoxylin for, for example, 45 min. The precise time will depend on the thickness of the photographic emulsion.
5. 'Blue' in tap water (or alkaline water; see above).
6. If required, that is if the section is too deeply stained, the section may be immersed in acidified water (a drop of hydrochloric acid in tap water).
7. Stain in a 0.05% aqueous solution of eosin for, for example, 30 s.
8. Rinse in tap water.
9. Dry in air.
10. Mount directly into Styrolite or an equivalent mountant.

Van Gieson stain for collagen

1. Take the fresh section from the cryostat cabinet; dry it in air for 5 min.
2. Fix either in picrate–formalin for 15 min or in 1 : 3 acetic–alcohol for 10 min.
3. Rinse in tap water.
4. Stain in Mayer's haemalum for 3 min.
5. Rinse briefly in tap water.
6. Differentiate very briefly (if necessary) in 1% acid alcohol (as above). (*Note*: further differentiation occurs during step 10 owing to the picric acid in the Van Gieson solution.)
7. Wash in tap water.
8. 'Blue' in 5% aqueous solution of sodium bicarbonate.
9. Wash in distilled water.
10. Stain with the Van Gieson solution for 0.5–1 min.
11. Blot dry.
12. Dehydrate in absolute ethanol, clear in xylol and mount in Styrolite.

Result

Nuclei—blue or black
Collagen fibres—red
Muscle and other tissue—yellow

Solutions required for this method

Mayer's haemalum Dissolve 1 g haematoxylin in 1 litre distilled water (heat if necessary). Add 50 g ammonium alum; shake to dissolve. Then add 0.2 g sodium iodate, 1 g citric acid and 50 g chloral hydrate.

Van Gieson's solution

1% aqueous solution of acid fuchsin	10 ml
Saturated aqueous solution of picric acid	90 ml
Distilled water	100 ml
Boil for 3 min	

Gordon and Sweet's silver impregnation method for reticulin

1. Take the fresh section from the cryostat cabinet and dry it thoroughly in air.
2. Fix in picric–formalin for 5 min.
3. Rinse in distilled water.
4. To oxidize the tissue, treat for 1–5 min in the acidified permanganate solution.
5. Rinse in distilled water.
6. Bleach until white with 1% oxalic acid solution.
7. Wash well in several changes of distilled water.
8. Mordant by immersing for 2–15 min in 2% aqueous iron alum.
9. Wash well in several changes of distilled water.
10. Treat for 5–8 s in Wilder's silver bath.
11. Rinse thoroughly in distilled water.
12. Reduce in formol–calcium (neutral 10% formaldehyde) for 10–30 s.
13. Rinse in tap water.
14. 'Fix' the silver in 5% solution of sodium thiosulphate (2–5 min).
15. Wash well in tap water.
16. Dehydrate through a graded series of alcohols, clear in xylol and mount in Styrolite.

Result

Reticulin fibres—jet black
Collagen fibres—golden brown

Solutions required for this method

Acidified permanganate solution

0.5% potassium permanganate in water	47.5 ml
3% sulphuric acid	2.5 ml

Wilder's silver bath To 5 ml of a 10% solution of silver nitrate in water add ammonia ('880') drop by drop. A precipitate will form. Continue to add the ammonia, drop by drop, until the precipitate is just redissolved. Then add 5 ml of a 3% solution of sodium hydroxide. Again add ammonia, drop by drop, until the solution just becomes clear. Add distilled water to make the final volume 50 ml.

This solution should be kept in a dark bottle. It should not be stored because it may become explosive on prolonged storage.

The phosphotungstic acid–haematoxylin method

(We are indebted to Dr S. Darracott for this procedure, as applied routinely to cryostat sections.)

1. Take the section from the cryostat cabinet. Dry the section in air.
2. Fix in picric–formalin (5 min).
3. Rinse in water.
4. Treat with Lugol's iodine (2–3 min).
5. Rinse in distilled water.
6. Treat with a 5% solution of sodium thiosulphate (2–3 min).
7. Rinse in distilled water.
8. Immerse in acidified potassium permanganate (2–3 min).
9. Rinse in tap water.
10. Bleach in 1% oxalic acid (until white).
11. Wash for 5 min in running water.
12. Leave in the PTAH (phosphotungstic acid–haematoxylin) solution (overnight).
13. Shake off the excess fluid but do not wash.
14. Place in 70% alcohol (2 min).
15. Dehydrate; clear in xylol; mount in Styrolite.

Result

This reaction is particularly good for identifying isolated myoblasts and for cross-striations in striated muscle. It is also good for cellular detail generally.

Note: it may not be widely appreciated that isolated myoblasts or smooth muscle can be identified, even in sections of paraffin-embedded tissue which have been stained with haematoxylin and eosin, and mounted in Styrolite. To do this, polarized light is used: muscle appears white against a darker background. This does not require a complete polarized light microscope: sheets of polaroid below the condenser and in or above the eypiece often suffice. Difficulty may be met if other mounting media are used. For details of polarized light see p. 29.

Solutions required for this method

Lugol's iodine (see p. 111).

Acidified potassium permanganate As for Gordon and Sweet method (p. 41).

PTAH solution Dissolve 0.1 g haematoxylin in distilled water (up to 50 ml). Also dissolve 2.0 g phosphotungstic acid in distilled water (again up to 50 ml may be used). Mix. Make up the final volume to 100 ml. Leave to ripen for 3 months.
   Rapid ripening can be achieved by adding 0.017 g potassium permanganate to the solution.

Iron haematoxylin

1. Fix dried cryostat sections in picric–formalin (5 min).
2. Wash with water.
3. Immerse in iron alum solution (5% ferric ammonium sulphate) for 30–45 min at 56 °C. (Alternatively this stage can be done at room temperature, but it then takes 12–24 h.) This step mordants the tissue.
4. Wash well in distilled water.
5. Stain in the haematoxylin solution at 56 °C for 30–45 min (or at room temperature for 12–24 h).
6. Wash well in distilled water.
7. Differentiate in a 2% solution of ferric ammonium sulphate to taste, i.e. until the particular structures to be studied are clearly defined.
8. Wash in running water for 5 min.
9. Dehydrate through a graded series of alcohols; clear in xylol and mount in Styrolite. (Some structures can be seen more clearly if the section is dehydrated and mounted in Euparal.)

The haematoxylin solution

Haematoxylin	0.5 g
Absolute ethanol	10 ml
Distilled water	90 ml

Allow to ripen for 4–5 weeks. Alternatively add 0.1 g sodium iodate to ripen immediately.

The methyl violet method for the demonstration of amyloid

1. Dry a fresh cryostat section in air.
2. Immerse in a 1% aqueous solution of methyl violet (2–5 min).

3. Rinse in distilled water.
4. Differentiate in 0.5–1% acetic acid until the amyloid is pink and the rest of the tissue remains blue–violet.
5. Wash well in distilled water.
6. Mount in Farrants' medium.

Aqueous mounting medium: Z5 (Zaman and Chayen, 1981)

Polyvinyl alcohol (e.g. G18/140 grade: Wacker Chemical Co.)	12 g
Glycerin	20 g
Lactic acid	20 g
Sodium acetate : acetic acid buffer (0.4 M) up to	100 ml

The pH should be 6.5

To store 'frozen' blocks in paraffin wax

After fresh sections for histology and histochemistry have been cut from correctly chilled tissue, the block of tissue can be treated as follows:

1. Remove the tissue from the blockholder and put it into a refrigerator at + 4 °C. Leave for 1 h.
2. Immerse in formol–saline at room temperature. Fixation may take at least 24 h.
3. Dehydrate, clear and embed in paraffin wax by whatever method is preferred for normally fixed tissue.

These paraffin blocks can be treated in exactly the same way as those of tissue treated by conventional histological procedures. When required, they can be sectioned, dewaxed and stained by the usual histological methods, and should yield preparations indistinguishable from tissues which have never been chilled.

5

Analysis of Chemical Components of Cells and Tissues

GENERAL ANALYSIS

Histochemistry should be able to analyse the chemical composition of parts of cells. As an academic exercise this has been of interest in studies on cell growth and differentiation: for example, McLeish (1959) measured the increase in DNA and in arginine (and hence in histone-like protein) in nuclei of growing plant cells; Sandritter and co-workers measured similar changes in normal and malignant cells growing in tissue culture (e.g. Sandritter and Krygier, 1959; Sandritter and Fischer, 1962; Sandritter and Kleinhans, 1964) and Richards (1960) correlated changes in amounts of DNA and of protein in nuclei during cell division.

As a practical, applied problem this becomes more complex. The pathologist may see abnormal matter either inside or outside cells and be concerned to know of what it is composed: is it a mass of dead cells (which will contain nucleic acids and proteins); is it an abnormal protein complex, e.g. 'fibrinoid'; is cholesterol present; or does it contain calcium? Generally the histologist applies a number of tests, selected largely at random. However, in this part we will suggest that this type of problem can be investigated by a rational sequence of tests, very much as the analytical chemist applies the sequence of 'group separation tests' to decide the chemical composition of an inorganic sample.

Material

Sections of unfixed tissue. Serial sections will be required so that the tests can be applied, in sequence, to the same matter.

Stage 1: Determination of the major classes of compound present (Figure 9)

In this stage, the major classes of compound present in the material are to be determined:

POSSIBLE MATERIAL   CHARACTERISTIC TESTS

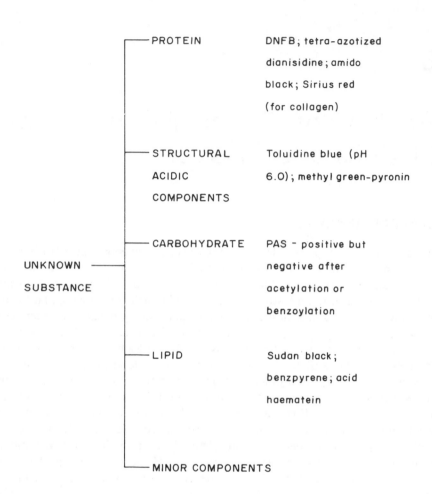

Figure 9.

1a   Test for **protein** by, for example, the DNFB or the tetra-azotized dianisidine method (or by a number of such methods used separately). If these are positive, the nature of the protein will be examined in stage 2.

1b   Test for **structural acidic groups** such as nucleic acids or glycosaminoglycans. At this stage, all that is required is to stain with toluidine blue (at pH 6.0) and with the methyl green–pyronin mixture (at pH 4.2). If these are positive, tests in stage 3 will have to be done to decide on the nature of the acidic material present.

1c   Test for **carbohydrate** by the use of a controlled periodic acid–Schiff test. If this is positive, tests in stage 4 will have to be used.

1d   Investigate the material for the presence of **lipid**. Sudan black is a good general indicator of lipid although benzpyrene (which must be used with caution, and only by a skilled worker, because it is carcinogenic) is better both as regards sensitivity and because it does not involve the use of a fat solvent. If the test is positive, apply the methods of stage 5.

1e   When all these have been tested (1a–d inclusive), and when the major components have been defined (stages 2–5 inclusive, below) **minor** components can be examined as described in stage 6.

Stage 2 (Figure 10)

Proceed with this stage only if a positive reaction is obtained with one of the characterizing tests for protein shown in Figure 9. If the tissue is negative to all these, it is unlikely to contain protein. If positive, it is necessary to determine whether the protein is predominantly basic, acidic or neutral. [But it must be remembered that the apparently basic or acidic nature of the 'protein' may be due to material conjugated to the protein; this will have to be removed before the true nature of the protein component of the conjugate can be revealed.]

2a   **It is basic if it stains with acidic dyes** (such as fast green) **at a high pH**. [*Note*: above pH 7 most $-NH_2$ groups present in proteins are relatively unionized so they do not bind acidic dyes; the more basic groups remain ionized and so stain with fast green.]

2b   **It is acidic if it stains with basic dyes** (such as toluidine blue) **at low pH values**. [*Note*: below pH 6 the carboxyl groups present in proteins are relatively unionized so that if the material stains at pH 4.2 it is likely to contain conjugated acidic matter such as nucleic acid or acidic glycosaminoglycans.]

2c   **It is a neutral protein if it shows neither of these staining characteristics**, i.e. it may stain with both the acidic dye (fast green) and the basic dye (toluidine blue) at about neutral pH but staining with the acidic dye is suppressed at more alkaline pH values and staining with the basic dye is decreased at acidic pH values. (Methods of estimating the pK of protein by these procedures are well discussed by Levine (1939).)

    If the protein appears to be basic, this is probably due to its containing a high proportion of basic amino acids. Therefore:

2d   (i) **Test for arginine** by the modified Sakaguchi reaction.
(ii) **Test for lysine**, which can be assumed to be present provided all the following criteria are met:

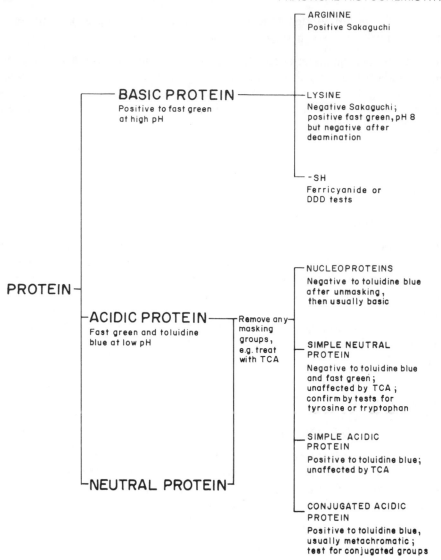

Figure 10.

1. The protein stains strongly with fast green at pH 8.
2. Such staining is negative after deamination.
3. The Sakaguchi reaction (for arginine) is either negative or suitably weak.

(iii) **Test for −SH groups** by the ferricyanide or the DDD reaction.

2e If the protein appears to be acidic this could be because it contains a high proportion of acidic amino acids like glutamic acid (whose ionization is suppressed at pH 4.2); however, it is more probable that the protein is a neutral (or even a basic) protein which is conjugated to an acidic compound. Hence both neutral and 'acidic' proteins can be tested together, as follows.

Treat the section with a 5% aqueous solution of trichloroacetic acid (TCA) at 90 °C for 15 min. This will remove the nucleic acids. Then:

Simple neutral protein will stain with fast green and with toluidine blue only close to neutrality (as before); it will be unaffected by the TCA. It is advisable to prove that such material is proteinaceous by testing for tyrosine or for tryptophan, as described later in this chapter.

Simple acidic protein will stain positively with basic dyes, like toluidine blue, but not appreciably with acidic dyes (like fast green) at pH values of between 4.2 and 6 as before; their staining will not have been affected by the TCA (although some intensification may be observed because of the coagulating effect of the TCA and the unmasking of acidic side-chains due to the denaturation of the protein).

2f Nucleoproteins will have stained strongly and orthochromatically (blue) with toluidine blue at pH 4.2 and with the methyl green–pyronin method before the TCA treatment. After TCA extraction these reactions will be abolished and the residual protein will stain either as a neutral or as a basic protein (e g the latter will be stained by fast green at pH 8). If it behaves as a basic protein, it may be subjected to the tests for basic amino acids detailed above.

2g Other conjugated acidic proteins may be either lipoprotein or protein conjugated with acidic polysaccharide. The former will behave as a simple protein and will also stain for lipid (see stage 5). Protein conjugated to acidic polysaccharide will stain positively—and usually metachromatically (purple to red)—with toluidine blue and this property will not have been lost as a result of the TCA treatment. It should then be tested for the acidic conjugate as in stage 4 (for polysaccharide) or stage 3 (for other acidic groups such as sulphate or phosphate).

Stage 3 (Figure 11)

Acidic substances have been shown to be present (as discussed in stage 1). In this stage, their nature is determined:

3a Stain with methyl green–pyronin. If this produces a green or blue-green colour, **test for DNA** by the Feulgen reaction or by Kurnick's methyl green method (or by ultraviolet microscopy, if available). If these tests yield apparently positive results, use controlled digestion with crystalline deoxyribonuclease to confirm that DNA is present.

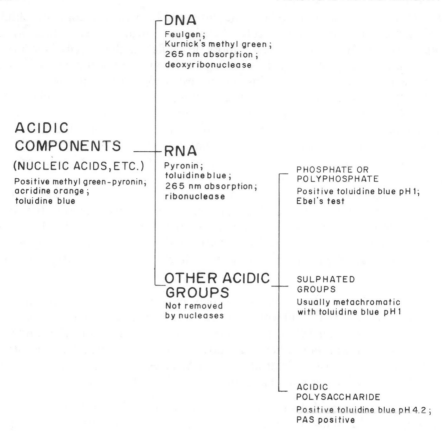

Figure 11.

3b If DNA is not present, or if it is not the sole constituent, all that is known is that acidic matter is present, i.e. it stains at pH 4.2 with basic dyes like pyronin (in the methyl green–pyronin test) or toluidine blue.

3c Test for RNA by observing whether the basophilia is decreased after controlled digestion with ribonuclease. (If ultraviolet microscopy is available, the maximal absorption in the 265 nm region, due to the purines and pyrimidines, is helpful in proving that nucleic acid is present.)

3d If the basophilia is not removed by digestion with ribonuclease and deoxyribonuclease, tests for **other acidic groups** must be used. These tests should also be applied where nucleases are not available. *Note*: RNA can usually be removed almost as effectively as by ribonuclease by immersing the section in 1 M hydrochloric acid at 60 °C for 6 min.

3e Stain with toluidine blue at pH 1. Only **phosphate** or **sulphate** is likely to remain ionized at this pH and so to be capable of staining.

3f Apply Ebel's test for phosphate and polyphosphate. If this is negative, the acidic group is probably sulphate. Usually such substances stain metachromatically (pink or red) with toluidine blue.

3g If the material stains with toluidine blue at pH 4.2 but the staining is abolished at pH 1, it is probably due to acidic groups of the type found in **acidic** (but not sulphated) **polysaccharides**. The ability of such groups to stain with basic dyes is lost after decarboxylation; they should be positive with the periodic acid–Schiff method.

Stage 4 (Figure 12)

The material is positive with the alcoholic periodic acid–Schiff (PAS) test, but is negative (or the reaction is markedly depressed) after benzoylation. Hence it contains **carbohydrate**.

Is it removed by rinsing in water or in dilute acid (e.g. 0.1 M trichloroacetic acid)? If it is lost after this treatment it is either free carbohydrate or a simple polymer such as glycogen or starch.

4a (i) Test with Lugol's iodine for **starch**.

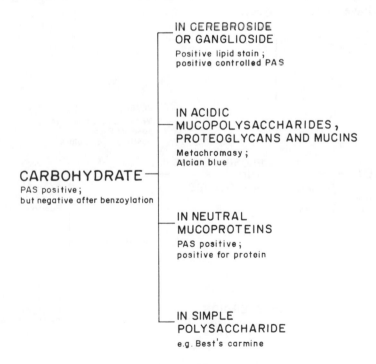

Figure 12.

(ii) Test with Best's carmine for **glycogen**. Or use the alcoholic PAS test both with and without prior treatment with diastase or with α and β amylases.

4b　If it is not removed by dilute acid it may be linked to protein. Hence apply a test for protein (stage 2).

4c　Does the carbohydrate material react metachromatically with toluidine blue? If so, it may be an **acidic polysaccharide**, also possibly linked with protein. Test with Alcian blue for acidic mucopolysaccharides or glycosaminoglycans to confirm this.

4d　The carbohydrate material may comprise part of a **cerebroside** or **ganglioside** complex. These will be positive with lipid stains (stage 5). Moreover, the presence of unsaturated fatty acids in such complexes may give an apparent PAS reaction which will not be abolished by benzoylation but which will require oxidation (e.g. by bromination; this will 'saturate' the fatty acids) before treatment with periodic acid.

Stage 5 (Figure 13)

The material contains **lipid** if it stains with benzpyrene or with Sudan black (although the 'burnt' Sudan black method or some other unmasking technique may have to be employed).

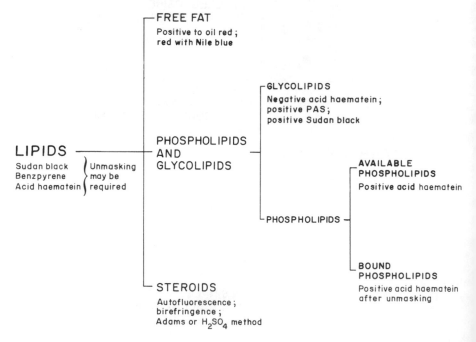

Figure 13.

5a   **Steroids** may not respond well to such lipid colourants. Consequently test for autofluorescence or for birefringence. Adam's reaction, or the sulphuric acid method, may help to demonstrate steroids.

5b   **Free fat** will take up lipid colourants very readily, even those with lower affinity for lipids than is possessed by Sudan black (which also stains phospholipids which are not as 'fatty' as free triglycerides). Hence test with oil red. Moreover neutral triglycerides should yield a red colour with Nile blue.

5c   Test with the acid haematein method. If this is positive (but negative after bromination), the material contains **phospholipid**. The extent to which the phospholipid is masked can be tested by the unmasking procedure (below).

5d   If the material stains with benzpyrene or Sudan black (i.e. it contains lipid) but is negative to oil red (hence not freely available triglyceride) and to the acid haematein test, even after unmasking procedures (i.e. it does not contain phospholipid) and it gives a positive PAS reaction (which is lost after acetylation or benzoylation) then it contains a **glycolipid**.

Stage 6 (Figure 14)

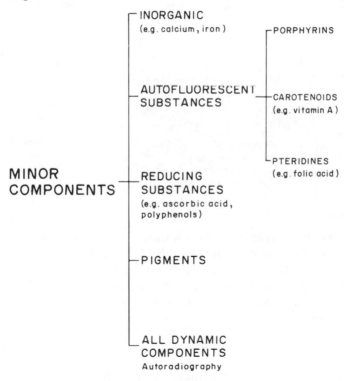

Figure 14.

PRACTICAL METHODS

STAGE 2: REACTIONS FOR PROTEIN

DINITROFLUOROBENZENE METHOD (see Danielli, 1953)

Rationale

Dinitrofluorobenzene (DNFB) reacts with amino acid residues and groups such as tyrosine, $-NH_2$ and $-SH$ (Sanger, 1945). The colour produced is too weak for histochemistry. It is intensified by synthesizing an azo dye from the DNFB–protein complex inside the section.

Free NH₂ group   DNFB links to the   One of the NO₂
in a protein   NH₂ group, eliminating   groups is
                                   HF which must be   reduced by
                                   removed by neutralization   dithionite
                                   with sodium bicarbonate

+ nitrous
acid

The newly formed NH₂
group is converted
into the azo bond which   azo dye
can then couple the
phenol or amine to
produce an azo dye

Application

Since the DNFB will react with groups which must occur in protein, it can be used as a general protein stain. It can probably be made specific for a particular group by blocking the other groups which can react. For example, Danielli (1953) suggested that the DNFB will react with tyrosine alone if $-NH_2$ groups are blocked by prior treatment with nitrous acid, and $-SH$ groups by treatment with iodoacetamide.

Method

The sections need not be fixed for this procedure. [If fixation is required (to free some of the reactive groups which otherwise might be masked, e.g. by nucleic acid), fix them in 5% acetic acid in absolute ethanol for 10 min. Take the sections through graded concentrations of alcohol to water before proceeding to 1.]

1. Place sections for 2 h in the DNFB solution, in a Coplin jar. Stir occasionally to maintain the saturated condition of the bicarbonate solution.
2. Wash in 60% alcohol (to remove adsorbed DNFB).
3. Place in 20% solution of sodium dithionite at 37 °C for 10 min. [*Note*: the solution should be in a Coplin jar and at 37 °C before the sections are immersed.]
4. Wash in distilled water at room temperature.

Stages 5–8 inclusive should be done in an ice bath and the solutions should be stood in the bath to cool before they are needed.

5. Place the slides in the nitrous acid solution for 5 min.
6. Pass the slides through three consecutive washes, 1 min in each, of acidified distilled water.
7. Transfer the slides to the H-acid solution; leave for 15 min.
8. Wash in distilled water.
9. Remove from the ice bath. Dehydrate. Mount in Euparal or Styrollte.

Result

The reactive groups in the protein, and hence protein generally, stain red-purple.

Solutions required for this method

DNFB solution

Dinitrofluorobenzene   0.5 ml
65% alcohol saturated with sodium bicarbonate   30 ml
Add excess solid sodium bicarbonate to ensure that the alcohol is fully saturated.

Sodium dithionite solution

20% in distilled water
[*Note*: dissolve in warm water (e.g. at 40 °C).]

Nitrous acid

2% solution of sodium nitrite   1 volume
0.1 M hydrochloric acid   1 volume
Mix just before use

H-acid solution

H-acid (8-amino-1-naphthol-3,6-disulphonic acid)   0.25 g
1% sodium bicarbonate solution   50 ml

TETRA-AZOTIZED DIANISIDINE METHOD (see Danielli, 1953)

Rationale

Diazonium hydroxides react with histidine, with tryptophan and with tyrosine (also with many phenols), yielding a coloured complex. This colour can be intensified by coupling another phenol or amine onto the unreacted diazo group of the diazonium hydroxide. Of the various diazonium hydroxides that are available, probably the most suitable is the tetra-azotized dianisidine.

phenolic side group
(such as tyrosine)
on peptide chain of
protein

tetra-azotized dianisidine
linked to the phenolic
group of the protein
(usually weakly coloured)

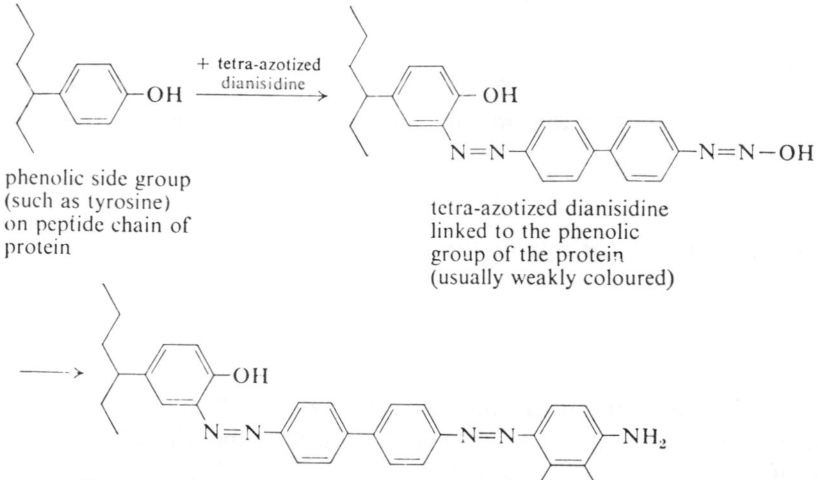

The colour is intensified by adding a synthetic amine or phenol which couples with the free diazo bond. There is no reason why the terminal $-NH_2$ should not be converted into a diazo group by treatment with nitrous acid, and be allowed to bind yet another molecule of the synthetic amine, as was done in the DNFB method. This would intensify the colour still further, if required

Application

Since the tetra-azotized dianisidine reacts with tyrosine as well as with histidine and tryptophan, it acts as a useful general stain for protein. It can be made specific for one or other of these amino acids by pretreating the sections with substances that block the reactive groups of the other amino acids. According to Danielli (1953) pretreatment with dinitrofluorobenzene (DNFB) will block tyrosine and probably histidine so that any reaction with tetra-azotized dianisidine would then be due to tryptophan alone. Pretreatment with performic acid should eliminate tryptophan and leave only histidine and tyrosine available for the tetra-azotized dianisidine reaction. Benzoylation should block all three amino acids. The advantage of this, and of the DNFB reaction (which becomes coloured only after the azo dye is synthesized on to the protein-bound DNFB), is that they are stoichiometric reactions which can be measured quantitatively.

Method

The sections need not be fixed for this procedure. [Should fixation be required (e.g. to free more of the potentially reactive groups), fix the cryostat sections for 10 min in acetic–alcohol (5 : 95 v/v or 1 : 3 v/v). Take the sections through graded concentrations of alcohol to water before proceeding to step 1.]

In an ice bath (with precooled solutions):

1. Place slide in freshly prepared cooled solution of tetra-azotized dianisidine for 6 min.
2. Transfer to another freshly prepared solution of tetra-azotized dianisidine for a further 6 min.
   Note: this solution should be prepared during the first 6 min reaction. The reason for fresh solutions is that the tetra-azotized dianisidine decomposes very rapidly so that one can be certain of good activity for only about 6 min.
3. Wash well in 2% cooled barbiturate solution.
4. Wash well in cooled distilled water.
   Note: it is essential to remove adsorbed, unreacted tetra-azotized dianisidine because it will also be intensified in step 5 and yield a strong but spurious stain. The reacted compound is held by true chemical linkage and cannot be washed out of the protein.
5. Stain for 15 min in cold H-acid solution in a Coplin jar.
6. Wash in cold distilled water.

Remove from ice bath.

7. Dehydrate and mount in Euparal or Styrolite.

Result

Red or brown-red.

Solutions required for this method

Barbiturate solution

Sodium diethyl barbiturate (Veronal)   6 g
Make up to 300 ml with distilled water

Tetra-azotized dianisidine solution

Grind 0.15 g tetra-azotized dianisidine with cooled barbiturate solution in a
cooled mortar until dissolved. Make up to 100 ml with barbiturate solution.

H-acid

H-acid   0.5 g
Sodium bicarbonate   1.0 g
Make up to 100 ml with distilled water

The bicarbonate is a convenient way of achieving an alkaline pH.

Pretreatment with performic acid

Solution
30% (100 vol) hydrogen peroxide   4 ml
Conc. sulphuric acid   0.5 ml
98% formic acid   40 ml

The performic acid should be vigorously stirred before use to remove gas and
the peroxide must be fresh.
    Place the section in this solution for 45 min at room temperature.

BAKER'S METHOD FOR **TYROSINE** (Baker, 1956)

Rationale

The phenolic amino acid, tyrosine, is a useful marker for most proteins (but
not for collagen). The most used histochemical procedure for this amino acid
has been a modification of the chemical Millon's test (e.g. Bensley and Gersh,
1933). Baker developed the following technique from the very sensitive
biochemical Folin reaction (Folin and Marenzi, 1929) and it yields a fairly strong
stain in 10 μm sections. In this reaction (Gibbs, 1927; Baker, 1956) the phenol

is converted into a nitrosophenol by the sodium nitrite and the mercury links to the nitrogen atom of this nitroso group to form a red compound.

Method

Fresh cryostat sections may be fixed in acetic alcohol (5% glacial acetic acid in absolute alcohol) for 10 min and then taken through graded concentrations of alcohol to water.

1. Put 25 ml of mercuric sulphate solution into a 50 ml beaker.
2. Add 2.5 ml sodium nitrite solution.
3. Put the section in the beaker and heat gently until the mercuric sulphate–sodium nitrite solution is just boiling.
4. Remove the sections. Wash them well in three changes of distilled water.
5. Either mount in glycerin or glycerin-jelly or dehydrate and mount in Euparal or Styrolite.

Result

Phenols, especially tyrosine and hence most proteins, stain red, pink or yellow-red.

Solutions required for this method

Mercuric sulphate solution

Add 10 ml concentrated sulphuric acid to 90 ml distilled water. Then add 10 g mercuric sulphate (preferably MAR grade). Heat to dissolve. Cool. Make up to 200 ml with distilled water. This solution is stable.

Sodium nitrite solution

Sodium nitrite   0.25 g
Make up to 100 ml with distilled water

QUANTITATIVE STAINING FOR *TOTAL PROTEIN* BY AMIDO BLACK

Rationale

Many studies, reviewed by Schauenstein *et al.* (1980), have established that amido black B can be used to assay quantitatively the amount of protein in isolated cells. It has been used for measuring protein content in analytical biochemistry (Schaffner and Weissmann, 1973; Bradford, 1976). Schauenstein *et al.* (1980) demonstrated the stoichiometry of staining with this dye in isolated cells, confirming their results by conventional biochemical assays. The precise mechanism by which this dye reacts stoichiometrically with 'protein' is still unclear but there seems little room for doubt that it can be used to give a good measure of 'total protein'. A pH of 5.5 is required for the formation of stable dye–protein complexes.

Two procedures have been given by Schauenstein *et al.* (1980). One is based on an aqueous solution of the dye, in which the molar extinction coefficient of the insolubilized dye–protein complex, at 620 nm, is about 96 000. The other, in which the dye is dissolved in ethanolic trichloroacetic acid, gives extinction coefficients two or three times greater. With solid films of protein, the stoichiometric reaction with this dye is maximal after about 10 min with the aqueous medium; with the ethanolic trichloroacetic acid procedure it takes up to 20 min.

Method

1. Fix in ethanol : ether (1 : 1) for 15 min.
2. Stain in aqueous or ethanolic TCA amido black solution at 20 °C for 15 min.
3. Wash in water, pH 5.5, for 3 min.
4. Three washes, each of 5 min, in water, pH 5.5.
5. Rinse in acetone.
6. Dry in air.
7. Mount either in glycerin or, after xylol, in a suitable mountant.
8. Measure at 620 nm.

Solutions required for this method

Aqueous dye solution

Amido black 10 B	200 mg
Water, adjusted to pH 5.5	100 ml

Ethanolic dye solution

Amido black 10 B	150 mg
10% trichloroacetic acid	50 ml
Absolute ethanol	50 ml

Comments

Schauenstein *et al.* (1980) recommended that the water used in all stages should be at pH 5.5 but should not be buffered. It seems possible that the ethanolic dye solution, containing trichloroacetic acid, may retain some soluble protein that might be lost in the aqueous dye solution.

More recently, Schauenstein *et al.* (1983) found it necessary to wash the stained sections for 3 days, with gentle stirring, in a mixture of 9 volumes of 95% ethanol and one volume of 2% trichloroacetic acid. They then dehydrated the sections with ethanol, cleared them with xylene and mounted them in a xylene-compatible mountant.

PICROSIRIUS RED METHOD FOR **COLLAGEN**

Rationale

Sirius red was first used as a more permanent substitute for acid fuchsin in the Van Gieson procedure for staining collagen (Sweat *et al.*, 1964). It was then recognized (Constantine and Mowry, 1968) that it enhanced the birefringence of the collagen. Junqueira *et al.* (1979a) then showed that it could be used to assay collagen in solution; it appeared to be highly specific and quantitative. These authors then proceeded to apply the procedure to tissue sections (Junqueira *et al.*, 1979b) and showed that, especially if the enhanced birefringence was used as a final discriminant for collagen, this method was highly specific. There was even some suggestion that collagens type I, II and III might be distinguishable by virtue of the colours observable with polarized light (Junqueira *et al.*, 1978, 1982).

According to Junqueira *et al.* (1979b), Sirius supra red F3BA is a long molecule (about 46 Å long), containing six sulphonic acid groups. Although it may bind to basic proteins generally, this non-specific binding is largely eliminated in the presence of picric acid; such non-specific binding will not, in general, produce enhanced birefringence except in collagen. It is suggested that, at the low pH used in this procedure (optimal between pH 2 and 3), the dye binds to the amino groups of lysine and hydroxylysine, and to the guanidine group of arginine. They showed that the concentration of picric acid is not critical provided that it was above 0.35 g/ml; that at least 1 mg/ml of the dye was required for maximal staining; and that a reaction time of 1 h was necessary for the reaction with collagen to be complete. In some tissues, enhanced staining is obtained when the non-collagenous ground substances (e.g. proteoglycans) are removed; this may be indicative of their interaction with collagen. The bound dye can be eluted from sections by treating for 30 min, at 37 °C, with 0.1 M sodium hydroxide and measured in a spectrophotometer; its absorption maximum is 540 nm. From such studies it seems that 126 molecules of Sirius red bind to each molecule of collagen. From their figures, it appears to have an extinction coefficient, under these conditions, of about 20 000.

Method

Unfixed sections may be used or cryostat sections may be fixed for 15 min in the picro-formalin fixative (p. 36). Junqueira *et al.* (1979b) fixed tissues in Bouin's fluid for 24 h and embedded them in paraffin wax.

With cryostat sections of unfixed cartilage or bone, it is often helpful to dry the sections at 37 °C overnight to promote their adhesion to the slide, before fixation or exposing them to the staining procedure. The use of a strong adherent, rather than simply glycerine–albumin, smeared on the slide before picking up the sections, is also recommended.

Then:

1. Dip the section in distilled water for 10 s to expand the collagen.
2. Wipe off excess water.
3. Immerse in the staining solution for 1 h.
4. Drain off excess stain.
5. Immerse in 0.01 M HCl for 2 (or 5) min.
6. Dehydrate rapidly in two changes of absolute alcohol.
7. Clear in xylene and mount in Styrolite.
8. Measure at 540 nm.

Result

Collagen stains red and shows strong red birefringence when viewed between crossed polars.

Comment

Enhanced staining can be obtained by removal of proteoglycan-like material before this staining schedule. For this, Junqueira et al. (1979b) used a 0.5% solution of papain (Papain IV F VIII from Difco Labs) in a 0.02 M phosphate buffer, pH 4.7, containing 5 mM sodium bisulphite and 0.5 mM EDTA at 37 °C for 90 min.

Solutions required for this method

Sirius red solution

Sirius red F3B200	0.1 g
Saturated aqueous picric acid solution	100 ml

The pH should be about pH 2. The Sirius red F3BA or F3B200 can be obtained from Bayer or Mobay Chemical Corp., New Jersey, USA.

Note: Picric acid is normally stored under water. It is perfectly safe as long as the neck and mouth of the bottle are kept clean and free of *dried* picric acid (which can be explosive, for example when grinding the stopper into the bottle). If it is to be weighed, it is placed on a filter paper to remove excess moisture and weighed on another filter paper. A 'saturated' solution will contain about 1.2 g in 100 ml of water.

Hydrochloric acid

0.01 M hydrochloric acid

FAST GREEN METHOD FOR BASIC PROTEIN (Alfert and Geschwind, 1953)

Rationale

At pH 8 the main groups which are sufficiently ionized to bind acidic dyes are the guanidine groups of arginine and the ϵ-amino groups of lysine (e.g. see Klotz, 1950). They are usually bound to other acidic groups in the tissue, particularly to the nucleic acids. Hence these have to be removed to disclose fully the basic proteins. This is best done by treatment with the relevant nuclease (described on pp. 89 and 93).

The basic groups of the protein tend to bind the acidic ions of buffer solutions and these can therefore interfere with the binding of the dye. Consequently Alfert and Geschwind suggested that the dye should be dissolved in distilled water and the pH of the dye solution brought to pH 8 by judicious addition of alkali. However, we find that McIlvaine's buffer is a more convenient way of achieving the required pH, and the phosphate and citrate ions do not seem to produce this competitive inhibition of the binding of the dye.

Method

Fresh cryostat sections should be fixed (e.g. for 15 min) either in acetic alcohol (5 : 95 v/v) or in absolute alcohol (to avoid possible removal of histones by acetic acid) and then taken through graded concentrations of alcohol to water. (Alfert and Geschwind suggested fixation in 10% neutral formalin.)

1. Remove nucleic acids by the use of the relevant nuclease (as on pp. 89 and 93).
2. Wash well in distilled water.
3. Stain in Fast green for 30 min at room temperature.
4. Wash briefly in distilled water.
5. Transfer the section directly to 95% alcohol.
6. Dehydrate and mount in Euparal or, after xylene, in Styrolite.

Result

Basic protein stains green or green-blue. Histones should be relatively strongly coloured.

Solutions required for this method

Fast green FCF

0.1% in citrate–phosphate buffer, pH 8.0

Buffer (McIlvaine's)

0.1 M citric acid   2.75 ml
(0.48 g in 25 ml)
0.2 M disodium phosphate   97.25 ml
(7.1 g Na_2HPO_4 anhydrous in 250 ml)

DEAMINATION METHOD

Terminal $-NH_2$ and ϵ-amino groups are 'destroyed' by this procedure.

Mix equal volumes of 0.1 M hydrochloric acid and 0.1 M sodium nitrite (6.9 g in 1000 ml). Leave sections in this freshly prepared solution for 10 min at room temperature. Transfer to fresh solution for a further 10 min.

ASSESSMENT OF 'pK' VALUES (see Levine, 1939; Klotz, 1950)

Rationale

For basic or acidic groups to be capable of asserting their basic or acidic properties, it is necessary that they should exist in the ionized form. For example, the carboxyl group $-COOH$ is not actually acidic unless it exists in the ionized (charged) form $-COO^-$. Similarly the amino group $-NH_2$ is not basic unless it is ionized to $-\overset{+}{N}H_3$. The ionization of acidic groups becomes progressively suppressed below their pK value (i.e. at more acid pH values) and that of basic groups above their pK value. (pK may be defined as that pH value at which the group is 50% ionized.) In histochemistry the pK value cannot be estimated accurately but its assessment is often valuable. In part this difficulty of achieving an accurate estimate is due to the fact that ions in the solution, especially those of buffers, may affect the ability of dyes to react with the charged groups. But another difficulty is that of ascribing the same precise pK values for groups when they occur in tissue as when they are studied in isolated chemical systems. For all that, the results may give a helpful approximation. In proteins the groups which can contribute to the acidic or basic properties of the molecule are as shown in the formulae below, where the pK value is that of isolated, purified amino acids (Fieser and Fieser, 1953).

From these values, which are only approximate, it would seem that protein should not be able to bind basic (positively charged) dyes at pH values much below pH 4 (at which value half the ionization of the $-COOH$ groups of glutamic acid may be suppressed); nor should it bind acidic dyes (negatively charged dyes) appreciably at or above pH 8 unless it contains much lysine, arginine or cysteine (or possibly tyrosine).

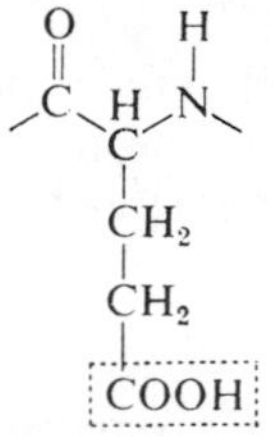

glutamate
residue
The carboxyl group
has a pK of 4

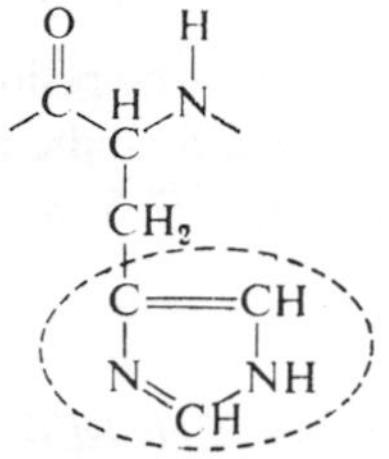

histidine
residue
The imidazole group
has a pK of 6–7

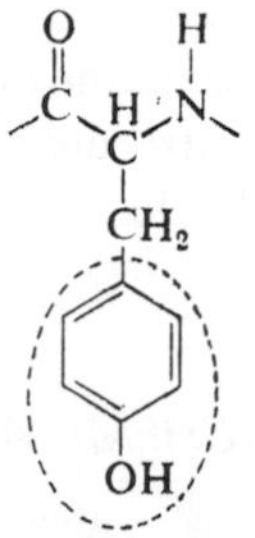

tyrosine
residue
The phenolic group
has a pK of 10

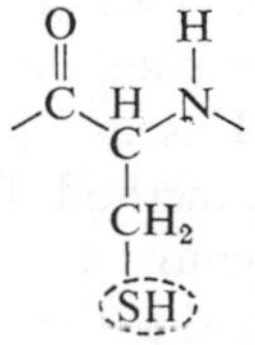

cysteine
residue
The sulphydryl group
has a pK of 9–11

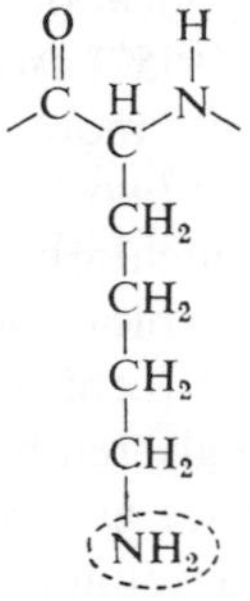

lysine
residue
The ε-amino group
has a pK of 9–11

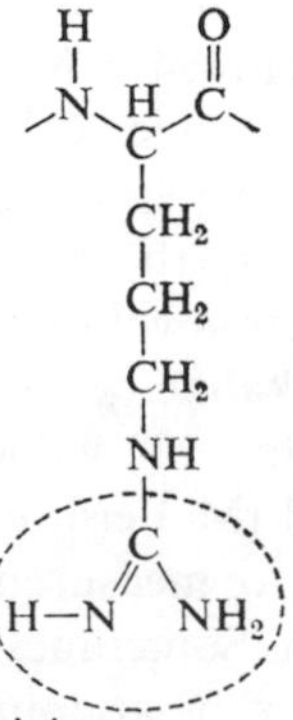

arginine
residue
The guanidine grouping
has a pK of 12–13

The same considerations apply to any acidic and basic groups in tissue sections, not only to those in protein. A rough guide to the acidity of a region of a cell or tissue can be obtained by staining with acidic and with basic dyes over a range of pH values. However, greater precision is obtained if dilute solutions (10^{-4} M) of the dye in very dilute buffer solutions (e.g. McIlvaine's citrate–phosphate buffer diluted ten times) are used for long periods (e.g. for 36 h). (It is difficult to talk of the molar concentration of dyestuffs because they are impure mixtures. We follow Levine in using molarity as a guide to the weight of dye to be used.) The sections are then washed thoroughly but briefly in distilled water. They are blotted dry and put straight into tertiary butanol, after which they may be mounted in Euparal or Styrolite.

The basic dyes can be toluidine blue or methylene blue; the acidic dyes can be Fast green, although other workers have used Ponceau 2R or orange G (but both of these are 'levelling dyes' and this may complicate the effect—see Baker, 1958).

Some of the amino acids, shown as the residues attached to the peptide chain, which can contribute to the basic and acidic properties of a protein are shown on p. 65. The terminal $-NH_2$ (pK 7.5–8.5) and $-COOH$ (pK 3) are not shown.

THE HISTOCHEMICAL SAKAGUCHI REACTION FOR *ARGININE*

Rationale

Biochemically the amino acid is estimated by Sakaguchi's (1925) method in which α-naphthol is linked to the guanido-amino group of arginine at a very alkaline pH. Colour is then produced by the addition of hypobromite or hypochlorite. The specificity of the histochemical application of this reaction was investigated fully by Baker (1947). McLeish *et al.* (1957) used the dichloronaphthol because this gives a stronger colour and reacts more rapidly, and used hypochlorite because it is easy to obtain. There are two hazards to the method: first that, after producing the colour, the hypochlorite must be removed because its prolonged action will reduce the final colour; and secondly that the mountant should be alkaline because the colour formed is a pH indicator, being straw coloured at low pH values but strongly red-brown at higher pH values. The final method (McLeish *et al.*, 1957; see also McLeish, 1959) can be used for the quantitative measurement of the amount of arginine available to be stained in the section. Sometimes it may be necessary to use special fixation to unmask and yet retain the arginine in tissue sections. This is the reason why La Cour *et al.* (1958) and Howe (1959) used Lewitsky's fluid; McLeish (1959) found formalin sufficient for his plant tissues.

Method

Cryostat sections should be fixed either in Lewitsky's fluid (15 min) or in 10% neutral formalin (McLeish and Sherratt, 1958; McLeish, 1959). (For this reaction the tissue itself may be fixed in either of these fluids and may be embedded in paraffin wax before it is sectioned.) Then proceed as follows:

1. Pass the section up the alcohol series to absolute alcohol.
2. Dip the whole slide into celloidin solution. Shake off excess; drain and air-dry.

3. Take the section through graded concentrations of alcohol to water.
4. Treat it in freshly prepared reaction medium for 6 min.
5. Rinse rapidly in 5% solution of urea [i.e. to stop the reaction. *Note*: prolonged treatment with urea will decolourize the reaction product.]
6. Place in 1% solution of sodium hydroxide for 5 min.
7. Mount in alkaline glycerol [the alkalinity is needed to retain the colour].

Result

Orange-red colour denotes arginine.

Solutions required for this method

Lewitsky's fluid

10% formalin (i.e. 10% of the 40% formaldehyde)   1 volume
1% chromic acid (chromium trioxide)   1 volume
Mix just before use
It is advisable to wash the sections well after fixing in this fluid (e.g. 5 min in each of three baths of distilled water).

Celloidin solution

Ether (diethyl ether)   50 ml
Absolute ethanol   50 ml
Celloidin   1 g
This is used to protect the tissue against the subsequent treatments, which tend to remove the section from the slide.

Reaction medium

1% aqueous solution of sodium hydroxide   30 ml
1% solution of 2,4-dichloro-α-naphthol in 70% ethanol   0.6 ml
'Milton' (proprietary brand of sodium hypochlorite)   1.2 ml
Mix immediately before it is to be applied to the sections.
[*Note*: the 'Milton' should be from a freshly opened bottle as its efficiency for this reaction becomes reduced after a week or two of having been opened.]

Urea solution

5%, aqueous

Sodium hydroxide solution

1%, aqueous

Alkaline glycerol

Glycerol	9 ml
10% aqueous solution of sodium hydroxide	1 ml

Note: In their description of a histochemical Sakaguchi method, Carver *et al.* (1953) recommended dehydration into tertiary butanol; then into aniline oil for 3 min; washing in xylene for 10 s; and making the preparation permanent by mounting in the following mountant:

Aniline	0.1 ml
Xylene	100 ml

Add one volume of this to four volumes of Permount.

METHODS FOR **THIOL** AND **DISULPHIDE** GROUPS

Background

Thiol (−SH) groups are of very great importance in cellular biochemistry and it behoves the histochemist to be able to measure the cellular concentration of free thiol groups and to be able to relate this to the content of disulphide (−S−S−) groups: the relationship between these may indicate the redox (reduction–oxidation) state of the cells. Yet this has proved remarkably difficult to do. Part of the difficulty lies in the fact that these groups may be in equilibrium so that some −S−S− groups can readily be converted to −SH during the reaction (e.g. with the DDD reagent: see below). Another major difficulty, that has caused confusing results in the past, is related to the demonstration of disulphide groups. These are inert, but can be demonstrated by reducing them to −SH, after which they can be reacted as for free −SH groups. However, the procedures for reducing the disulphide groups adequately can also cause disruption of the cells or the sections: what were once insoluble cross-linked proteins are now bereft of their linkages and may become soluble. The result has been that, on occasion, the amount of −SH groups (measured in intact cells or sections) has been found to be in excess of total −SH plus −S−S− (measured after reduction of −S−S− in disrupted specimens where much of the protein-bound −SH and −S−S− has been rendered soluble and so lost from the specimen).

A further difficulty, found also in biochemical estimations (Habeeb, 1973), is that some of the −SH groups, not linked as disulphides, may be 'masked', possibly by hydrogen bonding; 8 M guanidine or urea may be added to the reaction medium to try to overcome this type of masking.

There are two possible methods for staining −SH groups quantitatively. Although each has been investigated in some detail it is still uncertain as to which will prove to have finally overcome these problems.

DDD METHOD FOR —SH GROUPS (e.g. Esterbauer, 1972, 1973; Nöhammer *et al.*, 1977)

Rationale

The reagent, 2,2′-dihydroxy-6,6′-dinaphthyl disulphide (DDD) was synthesized by Barrnett and Seligman (1952), who examined its ability to respond specifically to protein-bound —SH groups (Barrnett and Seligman, 1954). DDD is a disulphide that reacts with —SH groups by exchanging hydrogen, splitting the disulphide group of the DDD. One half becomes linked to the tissue —SH group, forming a stable —S—S— bond; the other (2-hydroxy-6-naphthyl sulphide) is liberated and can be extracted with acetone or ether, which also removes excess DDD. Provided that the —S—S— linkage is formed to a fixed tissue component, it will be bound within the tissue. Although it is uncoloured, it can be converted to an intensely coloured azo dye by coupling with a diazonium salt such as Fast blue B. The reaction is reversible and therefore readily influenced by the Law of Mass Action.

According to Nöhammer *et al.* (1977) the reaction consists of two stages: the fast reaction equilibrates after about 7 h and represents the reaction with free or reactive protein —SH groups; the slow reaction, which reaches equilibrium only after 14 days, may be indicative of the total reactive protein-sulphur, including —S—S— groups. Thus, even after a 7 h reaction, some of the reaction will be due to unmasked —S—S— and a correction factor for this may be required. Esterbauer (1972, 1973) calculated the molar extinction coefficient of the DDD–Fast blue B complex, linked to protein-SH, to be 19 000. Using this value, Nöhammer et al. (1977) obtained similar values for protein-SH content in various cells measured microdensitometrically and by a conventional biochemical procedure.

Method

1. Sections or smears may be fixed in ether : ethanol (1 : 1 v/v) for 15 min.
2. Stain in Coplin jars for 7 h in the DDD solution.
3. Wash in acetone for 3 min.
4. Wash in acetate buffer, pH 4.0, for 3 min.
5. Wash in acetone for 3 min, three times.
6. Immerse in the Fast blue B solution, in a Coplin jar, for 15 min.
7. Wash in tap water for 30 min.
8. Dry in air.
9. Mount in Farrants' or other water-miscible medium.

Measurement: at 560 nm.

Calculation

The Beer–Lambert law gives the relationship:

$$c = E/kl$$

where c is the concentration in moles per litre; E is the extinction (or mean integrated extinction when measured by scanning and integrating microdensitometry); k, the molar extinction coefficient, is taken to be 19 000 for the coupled dye, as measured by Esterbauer (1972, 1973); and l is the path length (in centimetres).

Since $c = $ mass/volume, and volume $=$ area $\times$ length, this equation becomes:

$$M = \frac{EA}{k}$$

where M is the mass and A is the area. Then

$$M = \frac{EA}{19\,000 \times 10^3}$$

where the factor of 10^3 converts molarity (i.e. per litre) to moles.

Solutions required for this method

DDD solution

Dissolve 100 mg DDD in 55 ml ethanol. Then make up to 100 ml with 45 ml veronal buffer, pH 8.6.

Fast blue B solution

Add 100 mg Fast blue B to 80 ml of a 6 M solution of urea in distilled water (36 g in 100 ml). Bring to room temperature. Then add 20 ml 0.05 M Tris buffer, pH 7.4.

PRUSSIAN BLUE REACTIONS FOR −SH AND −S−S− GROUPS

Rationale

The principle of this method is the fact that sulphydryl groups are such strong reducing agents that they can reduce ferricyanide to ferrocyanide. The latter reacts with ferric ions to produce an intense and insoluble precipitate, generally referred to as Prussian blue (but see Davidson (1937) concerning Turnbull's blue; also Adams (1956)). This principle has been used in the histochemical procedures of Chèvremont and Fréderic (1943) and of Gomori (1952, 1956). The reaction is as follows:

$$2[Fe(CN)_6]^{3-} + \quad 2Pr-SH \quad \rightarrow \quad 2[Fe(CN)_6]^{4-} + \quad Pr-S-S-Pr$$

| ferricyanide ions | protein-bound sulphydryl (e.g. cysteine) | ferrocyanide | cross-linkage between protein-bound sulphur atoms (e.g. cystine) |

$$3K_4[Fe(CN)_6] + \quad 4Fe^{3+} \quad \rightarrow \quad Fe_4[Fe(CN)_6]_3 \quad + \quad 12K^+$$

| potassium ferrocyanide | ferric ions | ferric ferrocyanide (Prussian blue) |

Thus 1 mol of ferrocyanide is generated per mole of −SH, and 3 mol are required to produce 1 mol of ferric ferrocyanide.

It is true that this reaction will be given by many strongly reducing groups (as discussed by Gomori, 1952). In practice this means that it requires to be controlled by testing a serial section in which the sulphydryl groups have been specifically blocked. It is also true that, at the low pH necessary for this reaction (about pH 2.4), hydrolysis of the ferricyanide may liberate cyanide ions which break disulphide bonds, so contributing to the reaction. This seems to be a general hazard of these techniques since the DDD reaction also detects −SH groups liberated from −S−S− during the reaction (Nöhammer *et al.*, 1977).

Method

Unfixed cryostat sections can be used directly. Otherwise the sections may be fixed in absolute ethanol, or in acetic ethanol (e.g. 5% acetic acid in absolute ethanol) in which case they must be taken down to water before use.

1. Immerse the sections in the reaction solution for 7 min.
2. Transfer to freshly prepared reaction solution for 7 min. If necessary, immerse in another freshly prepared reaction solution for a further 7 min. Paraffin sections may require yet another transfer to freshly prepared reaction solution for another 7 min. This should be sufficient to produce adequate staining. [*Note*: the reaction solution must be prepared just before use.]
3. Wash in three changes of distilled water.
4. Dry in air and mount in a synthetic resin such as Styrolite.

Result

—SH and other tissue-bound strongly reducing moieties stain blue.

Control

The specificity of this method depends on the use of an adequate and specific blockade of —SH groups. Either of the following blockades can be used; both have been useful in our experience.

Blockade 1

Treat sections for 24–48 h in a saturated aqueous solution of mercuric chloride. [Excess mercury may be removed by washing in Lugol's iodine, followed by treatment with a 5% solution of sodium thiosulphate, provided that the mercury which is blocking the —SH groups is not affected.] Then test with the ferricyanide method: —SH groups should not react after this blockade.

Blockade 2

Immerse sections for 2–3 days in a saturated solution of phenylmercuric chloride in butanol.

[*Note*: Casselman (1959) suggested that 0.1 M *N*-ethyl maleimide in phosphate buffer at pH 7.4 for 4 h at 37 °C or 0.1 M iodoacetic acid brought to pH 8.0, at 37 °C for 20 h, is a very effective blocking agent for sulphydryl groups.]

Note: A 'saturated solution' is a solution which contains undissolved mercuric chloride or phenylmercuric chloride; the solution is left to settle and the supernatant is used. The advantage of the use of phenylmercuric chloride is that there is less likelihood of it leaving globules of free mercury in the tissue section.

Measurement: at 675 nm.

Solutions required for these techniques

Solution A

Ferric chloride	1.35 g
Distilled water to	100 ml

Solution B

Potassium ferricyanide	0.1 g
Distilled water to	100 ml

This must be freshly prepared.

Reaction solution

Add 3 volumes of solution A to 1 volume of solution B just before application to the sections or smears. Adjust the pH to 2.4 with 1 M sodium acetate.

Lugol's iodine

Iodine	1 g
Potassium iodide	2 g
Distilled water to	100 ml

Thiosulphate solution

5% aqueous solution of sodium thiosulphate

DEMONSTRATION OF **TOTAL −SH** AND **−S−S−**

Rationale

There are two problems that have to be overcome for assaying the total −SH and −S−S− content. The first is to obtain a method that adequately reduces disulphides to sulphydryl groups; the second is to retain the unmasked sulphydryl groups within the sections or isolated cells once the cross-linkages have been severed by this reduction. The use of trichloroacetic acid in this procedure, subsequent to reduction with borohydride, is designed to diminish loss of material.

Method

1. Sections or smears may be fixed in ethanol or acetic ethanol (5 : 95) or in 10% (w/v) trichloroacetic acid.
2. They are immersed in the methanolic solution of sodium borohydride for 10 min.
3. Wash in 10% (w/v) of trichloroacetic acid.
4. Stain for −SH, by the Prussian blue method, as previously.
5. Wash briefly either in distilled water or in 10% trichloroacetic acid.
6. Dry in air and mount in a synthetic resin such as Styrolite.

Extra solutions required for this method

Trichloroacetic acid

Trichloroacetic acid   10 g
Distilled water to   100 ml

Sodium borohydride

Sodium borohydride is perfectly safe provided it is handled sensibly. It should be kept in a desiccator, with the vent opened slightly to allow any hydrogen that may be liberated to escape; it should be kept and used in a fume cupboard.

To 100 ml of absolute methanol, add a few drops of 0.5 M sodium hydroxide to bring the pH to 10.

Then weigh out approximately 1 g sodium borohydride. This may be done by weighing it into a weighing tube on a normal balance. It should be done fairly expeditiously so that an accuracy of only ±0.5% is sufficient.

Then, again in the fume cupboard, add the sodium borohydride to the alkaline methanol. It will fizz, with the liberation of hydrogen. The sections should be placed immediately in this solution. The solution should be discarded, down the drain, after use.

TO DEMONSTRATE −S−S− GROUPS

In general it is adequate to measure the amount of −SH and of total −SH and −S−S− groups by the Prussian blue (or DDD) method, as detailed above. Then the amount of −S−S− groups can be derived by subtraction of the first value from the second. However, if a separate assessment is required, the following schedule may be helpful.

Take three serial sections.

1. Block pre-existing −SH groups, for example by treatment with mercuric chloride (see above).

2. Remove free mercury by rinsing in Lugol's iodine and then in the thiosulphate solution (see above).
3. Rinse well in water.
4. Treat one of these sections with the ferricyanide method for sulphydryl groups (see above). Any positive stain should be due to non-specific reducing moieties.
5. Treat the remaining two sections with the borohydride method for total $-SH$ and $-S-S-$ groups (as detailed above).
6. Treat one of these slides with a sulphydryl blocking method (see above) to test the specificity of the final reaction.
7. Stain both slides by the ferricyanide method. The unblocked sections should be blue wherever disulphide bonds occurred. The proof that these were disulphides (but have now been converted to sulphydryl groups) is (a) that they did not occur at step 4, and (b) that the blocked section (step 6) does not show a blue colour at these sites.

METHOD FOR **GLUTATHIONE** (Smith *et al.*, 1979)

Rationale

Reduced glutathione is a small, rapidly diffusible molecule. It is likely to be lost during fixation and during normal chemical reactions for free thiols. It has been shown (Loveridge *et al.*, 1975) that, with suitable proportions of ferric ions to ferricyanide ions and at high concentrations, glutathione reduces the latter more specifically than other small reducing molecules (which were more active against a different proportionality of these reactants, related apparently to the redox potential of the reacting solution). Thus the aim of the reaction is to use such high concentrations of reactants that they trap the glutathione before it can leave the tissue or cells. Smith *et al.* (1979) validated the method quantitatively against conventional biochemical estimations.

Method

1. Flash-dried cryostat sections must be used. They should be removed from the cryostat immediately and left to dry at room temperature, preferably in a desiccator.
2. Immerse in the staining solution in a Coplin jar for 20 s.
3. Wash briefly in distilled water.
4. Dry in air.
5. Mount in Farrants' medium.

Result

Strong blue colour.

Solutions required for this method

Solution 1

33% (w/v) of ferric chloride in distilled water.

Solution 2

30% w/v potassium ferricyanide in distilled water.

Immediately before use, mix one volume of solution 1 with one volume of solution 2.

Measurement

The absorption curve of this precipitated Prussian blue dye has a maximum at 675 nm and a minimum (which includes tissue 'scatter') at 475 nm. Consequently measurements are made at each wavelength (i.e. A_{675} and A_{475}) and the amount of dye is expressed as the difference (i.e. $A_{675} - A_{475}$).

Comment

This method is not entirely specific for reduced glutathione. In particular ascorbate can cause interference (but see Smith *et al.*, 1979).

IMMUNOCYTOCHEMISTRY FOR DETECTING PROTEINS

Apart from the Sirius red procedure, the methods described above do not define the actual protein molecule but only the active or available side-groups of the protein. So, for example, a layer of keratin may be stained decisively by the ferricyanide method subsequent to reduction with borohydride, but it will be the concentration of sulphur-containing amino acids that will have been stained, not the keratin itself.

Immunocytochemistry purports to demonstrate the presence of a specified protein. In recent years, this subject has expanded very greatly, with a vast literature. Fundamentally it involves the isolation of the pure protein or peptide to be localized, and the production of a specific antibody to it. Then the section is exposed to the antibody, which should react solely with the antigen. The localization of the antigen in the section can be visualized either directly, for example by autoradiography provided that the antibody has been suitably labelled with a radioactive marker, or indirectly by reaction with an antiglobulin antibody (or protein A) labelled with some visualizing marker such as a fluorochrome, or an enzyme, or a heavy metal (particularly for electron microscopy). Useful references for this subject are Sternberger (1979), Bullock and Petruzs (1982) and Cuello (1983).

This is too large a subject to be reviewed here, but certain points that are pertinent to quantification or to histochemistry may be noted. As regards the former, relatively few quantitative, validatory studies have been made. In one (Gau and Chard, 1976), sections from a tissue of known content of a hormone were studied by an immunoperoxidase technique. Despite careful calibration, the immunocytochemical results did not agree with the known content of the hormone, defined by suitable extraction and immunoassay. There may be many reasons why the results were discrepant; the reader is recommended the review by van der Sluis and Boer (1986) for a survey of controls that are required to validate an immunocytochemical analysis. As those authors said, 'In many instances difficulties in establishing specificity of immunocytochemical staining seem to have been brushed aside too easily by the allure of a successful localization'. As regards the histochemical aspects of immunocyto-chemistry, it must be remembered that, however precise the localization of the antibody–antigen reaction may be, the localization of an enzyme-linked antibody is as good—or as poor—as the localization of the histochemical reaction product by which the localization of the antibody is perceived. Remarkably little attention seems to have been given to this aspect.

One other point merits recognition, namely that, although immunocyto-chemistry may demonstrate the presence of a protein, it gives no information as regards whether that protein is present in an active or inactive form. So, for example, most cells may contain ornithine decarboxylase. Immunocytochemistry may be able to demonstrate its presence. But this enzyme can change in activity very rapidly (in a matter of minutes) from being totally inert to being highly active. Other enzymes, such as glucose 6-phosphate dehydrogenase, can change from a relatively slowly acting to a rapidly acting enzyme; both forms will have the same immunoactivity.

STAGE 3: REACTIONS FOR NUCLEIC ACIDS AND POLYPHOSPHATE

THE FEULGEN REACTION FOR *DEOXYRIBONUCLEIC ACID* (DNA)

Background

The Feulgen nucleal reaction (to distinguish it from the Feulgen plasmal reaction) was developed by Feulgen in his biochemical studies on nucleic acids (Feulgen and Rossenbeck, 1924). As the 'Feulgen reaction for DNA', it has been very widely used; being apparently stoichiometric for DNA, it was one of the first procedures to be measured by cytophotometry (e.g. Leuchtenberger, 1954).

It became established as a valuable tool when it was used to show the constancy of DNA per nucleus. For example, biochemical analyses of a mass of liver defined the amount of DNA per unit weight of liver; nuclear counts gave the number of nuclei per unit weight of liver; the amount of DNA, per nucleus, based on this type of calculation was not equivalent to twice, or four times, the concentration in the spermatozoa. Cytophotometry showed that the amount of DNA in each nucleus of a non-growing population was indeed twice, or four times, or even eight times, the amount found in the spermatozoa: the biochemical assessments had been misled by the fact that the sample contained diploid, tetraploid and even octoploid nuclei (Swift, 1953; Leuchtenberger, 1958).

Problems with the stoichiometry of the reaction were highlighted by studies (Walker and Richards, 1959) which showed that the stoichiometry varied considerably with the pH at which the staining reaction occurred. It was also realized by many workers that the optimal length of the acid hydrolysis, then normally in 1 M hydrochloric acid at 60 °C, differed for different fixatives. These problems have been settled by the use of low-temperature hydrolysis in strong acid (which decreases the breakdown of the backbone of the DNA: Jordanov, 1963) and of the buffered Schiff reagent (Fukuda *et al.*, 1978).

In recent years, two relatively new applications of this reaction have rendered it of increasing value. The first is the finding that what appears to be metabolically active DNA, and also newly synthesized DNA, show different degrees of lability to the low-temperature hydrolysis. Thus Millett *et al.* (1982) were able to distinguish cytological smears obtained from patients with cancer from those from other conditions on the basis of the very labile DNA demonstrable in hydrolysis-profile graphs. The second is the method for determining the amount of DNA synthesis in a population, based on Feulgen histograms (Coulton *et al.*, 1981). The DNA-synthesis index, based on such calculations, correlated well with the more conventional index based on the uptake of tritiated thymidine (Henderson *et al.*, 1981).

Rationale

Basically the Feulgen reaction involves a reaction between an aldehyde (or ketone) and Schiff's reagent. This is a normal procedure in organic chemistry. In the native state, neither DNA nor ribonucleic acid (RNA) contain free aldehydic (or ketonic) groups, but it is well established that the deoxyribose moieties of DNA (but not the ribose moieties of RNA) can be induced, by treatment with acid, to disclose such groups. The precise chemical mechanism is still unclear but may be related to the ability of the phosphate to migrate to different sites in the sugars (Brown and Todd, 1955). Consequently the lack of one oxygen atom in the deoxyribose moiety would be critical in allowing the ring form of the sugar to be opened up to give a straight-chain molecule with a terminal exposed aldehyde group.

Treatment with acid causes the removal of purines, which exposes the deoxyribose associated with the purines. Thus, at best, the Feulgen reaction is stoichiometric with the number of purine nucleotides in the DNA molecule. One of the advantages of the cold acid hydrolysis (5 M HCl at 20 or 25 °C) is that it tends to preserve the apurinic acid (DNA minus purines) by being less damaging to the DNA backbone than higher temperatures (Jordanov, 1963).

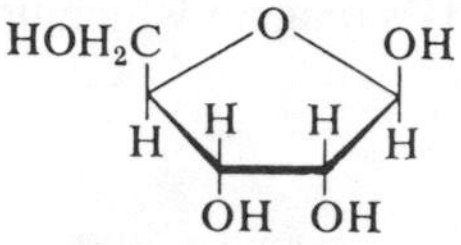

D-Ribofuranose
The ring form of ribose
as found in RNA

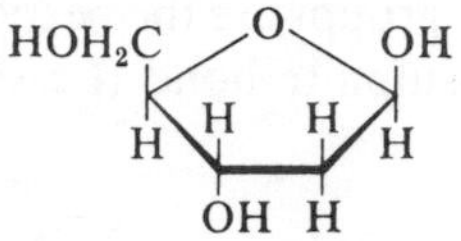

D-2-Deoxyribofuranose
The ring form of deoxyribose
as found in DNA

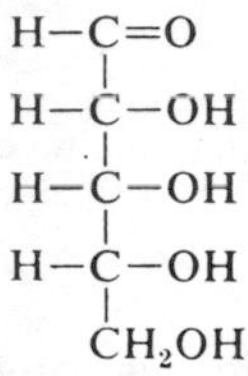

$$
\begin{array}{l}
H-C=O \\
H-C-OH \\
H-C-OH \\
H-C-OH \\
CH_2OH
\end{array}
$$

Straight-chain form of
ribose

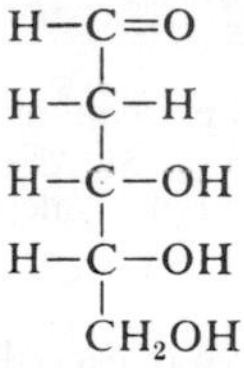

$$
\begin{array}{l}
H-C=O \\
H-C-H \\
H-C-OH \\
H-C-OH \\
CH_2OH
\end{array}
$$

Straight-chain form of
deoxyribose

It will be seen that, in the straight-chain form, both ribose and deoxyribose have a terminal aldehyde group:

$$H-C=O$$

Another chemical difference between ribose and deoxyribose is the fact that only the latter can form levulinic acid when treated with dilute acid:

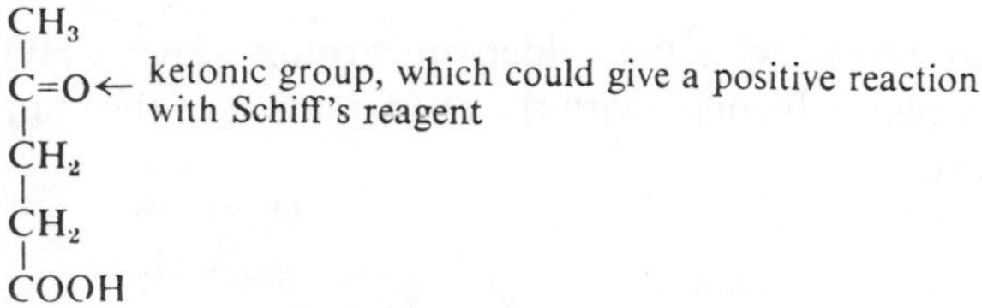

$$
\begin{array}{l}
CH_3 \\
C=O \leftarrow \text{ ketonic group, which could give a positive reaction} \\
\quad\quad\quad \text{with Schiff's reagent} \\
CH_2 \\
CH_2 \\
COOH
\end{array}
$$

Essentially the Feulgen nucleal reaction involves the following:

1. There are no aldehyde (or ketones) at the beginning of the procedure.

2. The acidic treatment releases purines, converting the DNA essentially to apurinic acid and allowing the exposure of the aldehyde groups. Incidentally it also removes the RNA, but anyway this would not colour with the Schiff's reagent. At elevated temperatures (e.g. 60 °C), acidic treatment will cause a variable degree of breakage of the backbone of the apurinic acid, with loss of DNA from the cells; some fixatives can enhance this loss, while others will protect the DNA from hydrolysis, so that longer hydrolysis times will be required.

3. The aldehyde groups are then exposed to Schiff's reagent which links to them through a bisulphite bond (Lessler, 1953):

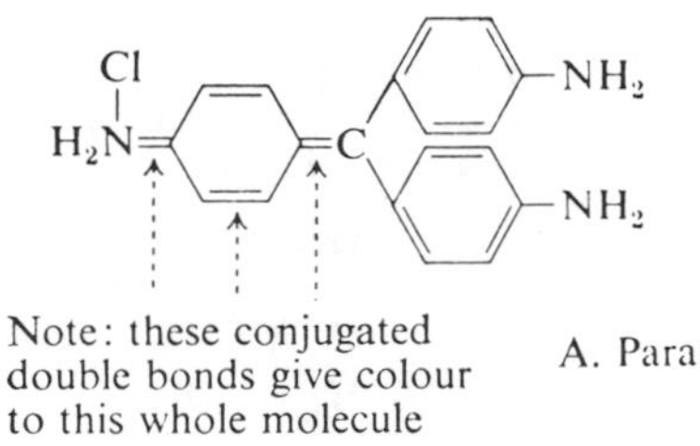

Note: these conjugated double bonds give colour to this whole molecule

A. Pararosaniline (main constituent of basic fuchsin)

When this dye is dissolved in sulphurous acid it is converted to the leuco-form (colourless because the conjugated double bonds have been lost) and it forms a complex with the sulphurous acid:

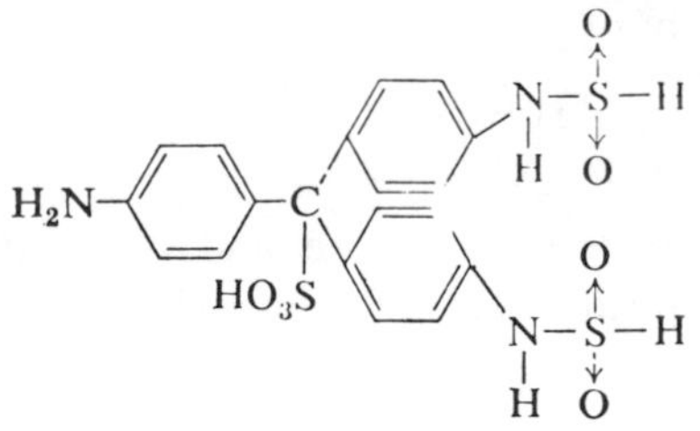

When this reacts with two aldehyde groups (2R—CHO), a coloured addition complex is formed (with the re-formation of the conjugated double bond system).

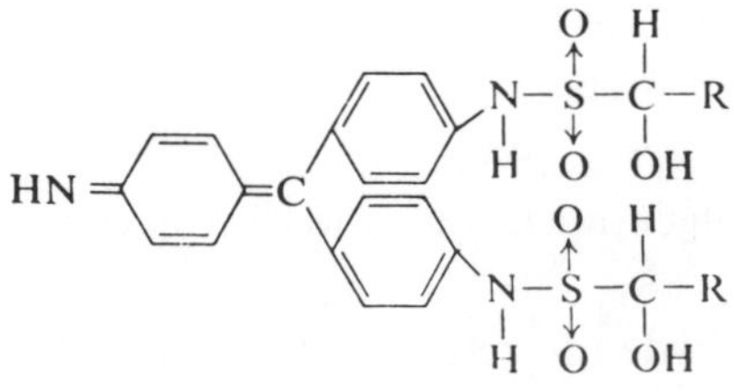

4. After treatment with Schiff's reagent the unbound reagent must be washed out of the section. If this is not done it will oxidize very readily, i.e. the leuco-form (B) will be converted back into the pararosaniline form (A). This is not a Schiff complex and is spurious. [*Note*: for this reason it is nonsensical to wash the sections with tap water or with water to which an oxidant has been added because these will produce the coloured pararosaniline form from any adsorbed leuco-basic fuchsin. Yet some workers actually recommend that this be done. It will certainly increase the amount of colour in the section but no-one will be able to assess how much of this colour is a true Feulgen reaction, and how much is due to adsorbed reoxidized fuchsin.]

Consequently the section is washed in SO$_2$-water. This will couple any free pararosaniline (A) to produce the leuco-form (B) and will wash it out of the section.

5. Particularly for quantitative studies, it should be noted that the Schiff's reagent loses sulphur dioxide on being kept for any length of time and its pH rises. The stoichiometry of the reaction depends critically on the pH at which the reaction is done.

Precautions

The Feulgen nucleal reaction depends on the production of aldehyde groups as a consequence of the acidic treatment. The presence of a colour after the reaction has been performed correctly shows that such groups became available during this procedure; it is not yet proof that they required this particular treatment (depending on whether the hot or cold hydrolysis has been used) for their unmasking. For example, plasmals are characterized by exposing aldehyde groups after mild hydrolysis. Hence it is necessary to use one serial section as a control; this is placed in 1 M hydrochloric acid at room temperature for 15 min and then exposed to the Schiff's reagent. It is the difference between the test and control section that demonstrates the presence of DNA. In general, such a control will be negative or only weakly coloured relative to the test section so that such controls need not be done invariably. However, if the control is strongly positive, it may be necessary either to remove the interfering material, which is likely to be lipid in nature and therefore extractable with lipid solvents, or to block these aldehydic (or ketonic) groups by prior treatment with a blocking agent such as dimedone (see below).

Methods

Since the most widely used method is that which involves treatment with 1 M hydrochloric acid at 60 °C, we have retained this procedure. However, especially for quantitative studies, we strongly recommend the cold hydrolysis procedure.

Hot hydrolysis method

1. When dealing with cryostat sections, fix in acetic ethanol (1 : 3 v/v of glacial acetic acid : absolute ethanol) for 10 min.
2. Take through graded concentrations of alcohol to water.
3. Rinse in 1 M hydrochloric acid.
4. Plunge into 1 M hydrochloric acid at 60 °C. For tissue fixed in acetic ethanol, maintain at this temperature for 6 min. (For tissue fixed with other fixatives, longer hydrolysis times may be required. For example, sections from formalin-fixed tissue may require 12 min; after chromic fixation, the required hydrolysis time may be as long as 18–22 min.)
5. Rinse briefly in 1 M hydrochloric acid at room temperature.
6. Immerse in the Schiff's reagent in the dark for 60 min. For meaningful stoichiometry, the pH of this reagent should be about pH 2.4 (Walker and Richards, 1959).
7. Rinse in three changes (2 min each) of freshly prepared SO_2-water.
8. Rinse in distilled water.
9. Dehydrate, clear in xylene and mount in Styrolite.

Result

DNA stains magenta.

Control

1. Fix serial section or specimen in 1 : 3 acetic ethanol for 10 min.
2. Take down to water.
3. Rinse in 1 M hydrochloric acid and leave at room temperature for 15 min.
4. Immerse in the Schiff's reagent in the dark for 60 min.
5. Rinse in three changes of freshly prepared SO_2-water.
6. Rinse in distilled water.
7. Dehydrate, clear in xylene and mount in Styrolite.

Result

Any stain produced by this control procedure is *not* due to DNA.

Additional control

Deoxyribonuclease can be used to make the results more rigorous (see later).

Solutions required for this method

Schiff's reagent

Dissolve 1 g basic fuchsin in 200 ml boiling distilled water. Shake and cool to 50 °C. Filter and add 30 ml 1 M hydrochloric acid and 3 g potassium metabisulphite. This solution is allowed to stand for 24 h in a dark-coloured, stoppered bottle. During this time the bisulphite decolorizes the fuchsin, i.e. it forms the colourless SO_2–pararosaniline complex (leuco-basic fuchsin). Impurities and other coloured matter are then removed by adsorption onto activated charcoal. Therefore, add 0.5 g of a decolorizing activated charcoal ['Norit', a commercial vegetable charcoal has been recommended—the properties of the charcoal used may be critical], and shake well for about 1 min. Filter rapidly (to avoid recolorizing the leuco-basic fuchsin) through a coarse filter paper (or glass wool). The solution should be clear and colourless; it can be stored for months in a well stoppered dark-coloured bottle.

This description of how to prepare Schiff's reagent is given primarily to demonstrate its constitution. It is often advisable to buy commercially prepared Schiff's reagent from a suitable source.

SO_2-water

1 M hydrochloric acid	25 ml
0.5% aqueous solution of potassium metabisulphite	50 ml
Mix immediately before use	

1 M hydrochloric acid

Analar hydrochloric acid (specific gravity 1.18)	89 ml
Distilled water	911 ml

Always add the acid to the water.

Cold hydrolysis method

This procedure follows that described by Sibatani and Naora (1952) and Leuchtenberger (1954) (see also De Cosse and Aiello, 1966), as given in detail by Fukuda et al. (1978). It has been of special use in investigating rates of synthesis of DNA (Coulton et al., 1980, 1981) and for discriminating between malignant and non-malignant cells (Millett et al., 1982).

Method

Fukuda et al. (1978) recommended fixation in absolute methanol. Acetic alcohol (1 : 3, or 5 : 95 glacial acetic acid to absolute ethanol) may be preferred.

1. Take the sections or smears down to water.
2. Rinse in distilled water.
3. Hydrolyse in 5 M HCl at 25 °C for 40 min. [*Note*: it is preferable to do a graded series of hydrolysis times, to define the optimal time.]
4. Rinse in 0.1 M HCl at 18 °C to remove traces of the strong acid.
5. Stain in the buffered dilute Schiff's reagent at 18 °C for at least 30 min.
6. Wash in three fresh baths of the buffered rinse solution, at 18 °C, for 2 min in each bath.
7. Rinse in distilled water.
8. Dehydrate, clear in xylene and mount in a suitable mountant such as Styrolite or Canada balsam (refractive index $n_{D_{20}}$ of 1.55).

Result and controls

These should be as for the hot hydrolysis method (above) but using the buffered Schiff's reagent.

Solutions required for this method

0.1 M hydrochloric acid

Analar hydrochloric acid (specific gravity 1.18)	8.9 ml
Distilled water to make up to	1000 ml

5 M hydrochloric acid

Analar hydrochloric acid (specific gravity 1.18)	445 ml
Distilled water to make up to	1000 ml

Always add the acid slowly to the water.

Glycine buffer (pH 2.3)

Solution A

Glycine	7.505 g
Sodium chloride	5.850 g
Distilled water to make up to	1000 ml

Solution B
0.1 M hydrochloric acid (as above)

Add 60.3 ml solution A to 39.7 ml solution B. Check that the final pH is 2.3. Adjust if necessary.

Buffered Schiff's reagent

Schiff's reagent (as for the hot hydrolysis method)   10 ml
Sodium metabisulphite solution (15% $NaHSO_3$)   10 ml
Glycine buffer at pH 2.3   80 ml

Rinse solution

15% sodium metabisulphite solution   10 ml
Glycine buffer (pH 2.3)   90 ml

DNA-SYNTHESIS INDEX

It is normal to measure the amount of DNA in, for example, 100 nuclei of the population being studied. The measurement is made by microdensitometry at a wavelength of 550 nm, and with a spot of 0.2 μm in the plane of the specimen. The results are usually expressed as a histogram, as in Figure 15. In this form of presentation all nuclei showing values of, for example, between 8.0 and 8.9 units are put together in one column; all those with values of between 9.0 and 9.9 units are put into the next column, and so on. More recently (Coulton *et al.*, 1981) it has been shown that such histograms can be converted into an index of the degree of DNA synthesis that was occurring in that population. One proviso must be noted: in comparing two or more populations, the same number of nuclei must have been measured in each sample.

Taking the population in Figure 15, the index of DNA synthesis is calculated as follows (and from Table 3).

From Figure 15 it can be seen that the $2c$ value was between 7 and 8 units of relative absorption (RA), as measured by the microdensitometer. Twenty

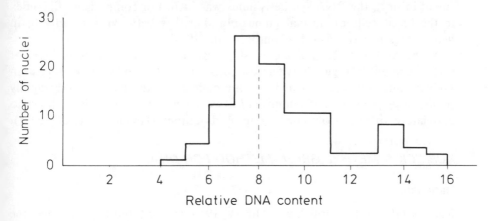

Figure 15.

Table 3. Method of calculating the DNA synthesis index from Figure 15

RA*	RA − 8	No. of nuclei in each category (n)	n(RA − 8)/8
8†	0	—	—
9	1	20	20/8
10	2	10	20/8
11	3	10	30/8
12	4	2	8/8
13	5	2	10/8
14	6	8	48/8
15	7	3	21/8
16	8	2	16/8
Total			173/8 = 21.6

*RA: relative absorption as indicated by the microdensitometer.
†Taken as the 2c value.

had values of between 8.1 and 9.0 units of relative absorption (for the purposes of this calculation, all these values are taken to be 9.0 units); these are put into a table (as in Table 3) and the synthesis index is calculated (as in the last column) by multiplying the difference from the 2c value (i.e. $9 - 8 = 1$) by the number of nuclei (20) that showed this value; this is then divided by the 2c value. The purpose of dividing by the 2c value is to give the degree of synthesis in this population irrespective of the 2c value for that species; it also eliminates slight changes in intensity of staining that can occur, for example on different days.

This procedure is repeated for each class of the histogram, for example for all nuclei showing values of between 9.1 and 10.0, or 10.1 and 11.0, etc. The results ($n(RA − 8)/8$, i.e. $n(RA − 2c)/2c$) are added together to give a final value for the population of nuclei of DNA values between 2c and 4c. In this case (basal cells in the regenerating region of the skin of a guinea-pig, close to the site of injury), the DNA-synthesis index was 21.6. For comparison, the index in the basal cells of normal guinea-pig skin was between 3.0 and 6.0; in hepatocytes it was 0.6–1.4 (Coulton et al., 1981).

An index of up to 2.0 indicates no significant degree of synthesis; an index of up to 6.0 might be suggestive of a slow mitotic index (e.g. of 5–10 per 1000 as in normal guinea-pig skin); and an index of around 20 is indicative of a fairly rapidly proliferating population (Coulton et al., 1981). These values have been shown to correlate with results obtained by autoradiography (Henderson et al., 1981).

KURNICK'S METHYL GREEN METHOD FOR DNA

Rationale

Kurnick (1950a, b; Kurnick and Mirsky, 1950) showed that triphenylmethane dyes which contain two charged amino groups stain polymerized nucleic acids selectively; if they have only one charged group (e.g. crystal violet) they do not show this selectivity:

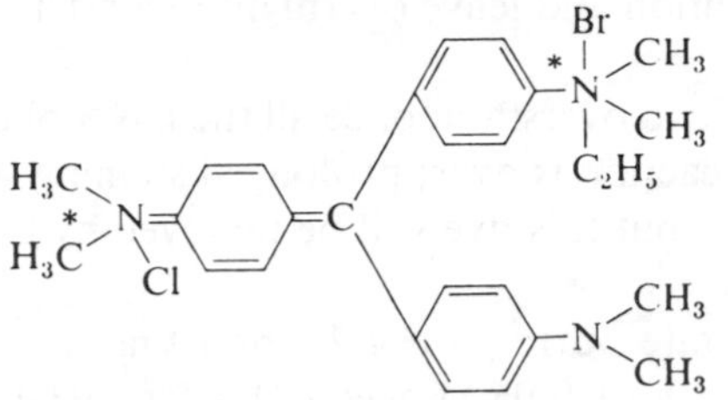

methyl green

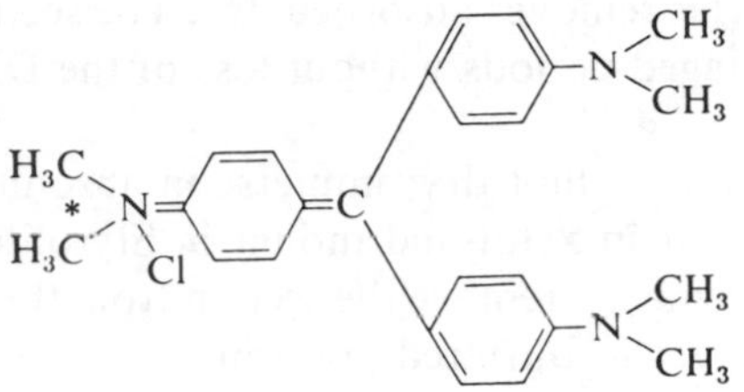

crystal violet

Kurnick suggested that the specificity shown by methyl green for polymerized DNA was due to the actual distance between these basic groups: they are spaced just sufficiently to fit on two phosphate moieties of the DNA, one above the other in the double helix.

Whether these ideas are correct or not, it has been shown that methyl green can be made to bind to DNA so strongly that it is not washed out even by dilute buffer; moreover, this binding is stoichiometric, so that one molecule of methyl green bound in the section is equivalent to a length of the DNA corresponding to 20 phosphate groups in the DNA molecule (see Kurnick, 1950b).

Method

Fresh cryostat sections can be used, but it must be realized that they will become fixed when they are immersed in the acetate buffer at pH 4.2 or by the pretreatment. Consequently it may be advisable to fix the fresh sections deliberately (rather than by the first coagulating fluid to which they are subjected). Absolute or 80% alcohol can be used for 10 min. Should this cause the chromatin to become too coagulated (and so unsuitable for quantitative measurement), then neutral formalin can be used (e.g. for 15 min).

For optically homogeneous nuclei, the sections should be treated with 10% sucrose for 15–30 min before they are fixed either in neutral formalin or in formol–sucrose (Ris and Mirsky, 1949).

1. Immerse sections in 0.1 M hydrochloric acid for 5 min (to split the DNA from histone-like protein).
2. Rinse in water.
3. Place the sections in 0.2 M acetate buffer, pH 4.2, for 5–10 min.
4. Transfer to the dye solution and leave overnight in a refrigerator at about +4 °C.

    [*Note*: it is impossible to overstain: once all the DNA phosphate groups have been stained the reaction is over; prolonged staining will allow more dye to become adsorbed but this dye will be removed by the dilute buffer used in the next step.]
5. Immerse in 0.05 M acetate buffer, pH 4.2, for 10 min.
6. Transfer to a second bath of 0.05 M acetate buffer, pH 4.2, for 10 min.

    [*Note*: the dilute buffer removes adsorbed dye. The sections can be left in this buffer for prolonged periods without loss of the DNA-bound dye (Chayen, 1952).]
7a. For permanent preparations: blot dry; immerse in absolute alcohol, and mount in Euparal or clear in xylol and mount in Styrolite.
7b. For quantitative measurements: remove the section from the 0.05 M acetate buffer and mount it direct in buffered glycerin.

One technical problem must be stressed. The phosphate groups of the DNA are not freely available in cells; they are usually linked to basic groups, such as the arginine and lysine moieties of nuclear histones (e.g. see Kurnick and Mirsky, 1950). Consequently some acidic treatment may be required to render the phosphate groups available to the dye. However, without such acidic treatment, the amount of methyl green staining may be related to the amount of active DNA.

Result

DNA stains green or green-blue.

[*Note*: some other polymeric molecules may also stain by this method. Also, it should be noted that DNA which is not highly polymerized (because of either disease processes or technical artefact) will not stain by this procedure.]

Control

The use of deoxyribonuclease (see below).

Solutions required for this method

Buffers

See Appendix.

Dye solution

[*Note*: methyl green very readily becomes degraded to methyl violet, which interferes with the specific reaction for DNA. Any methyl violet must be removed before the dye solution is used, as described below.]

A good brand of methyl green, with relatively little impurity, is strongly recommended. The degree of impurity will be shown by the amount of violet colour removed by chloroform (as described below).

1. Dissolve 0.25 g methyl green in 100 ml 0.2 M acetate buffer at pH 4.2.
2. Shake vigorously for several hours to ensure full solution.
3. Put the solution into a large separating funnel.
4. Add 100–200 ml chloroform. Shake well. Leave to separate and remove the chloroform (which will be heavily coloured by methyl violet). Add another 100–200 ml chloroform. Shake well, allow to separate, and remove the chloroform.
5. Repeat this procedure until the chloroform remains almost colourless. Leave the aqueous dye solution in an open vessel in the dark to allow any entrapped chloroform to evaporate.

This solution should be used shortly after it has been prepared. On the other hand, it can be stored in a stoppered dark coloured bottle, but before it is used it should be shaken with chloroform to ensure that there has been no excessive degradation to form methyl violet.

Buffered glycerine

Glycerine	4 volumes
0.2 M acetate buffer, pH 4.2	1 volume

THE USE OF DEOXYRIBONUCLEASE

Rationale

The final proof that a substance in sections is indeed DNA must be that it is removed specifically by deoxyribonuclease. But even the 'crystalline' enzyme may contain proteolytic activity; the removal of protein, to which may be attached an acidic substance, could simulate removal of DNA. To inhibit such proteolytic activity, 0.05 M cysteine or hydroxylamine may be added to the incubation medium (see Jacobson and Webb, 1952). Magnesium ions may be included to activate the deoxyribonuclease even though most tissues may contain enough of this element.

As a control for the specific action of the deoxyribonuclease, a serial section can be treated with the same incubation medium, for the same time, but without the addition of deoxyribonuclease to the solution. A more rigorous control is to use the same incubation medium, containing deoxyribonuclease, but also containing zinc to inhibit the action of the deoxyribonuclease. This ensures the presence of the enzyme, but in an inactive form.

Incubation solution

The final concentration of the deoxyribonuclease is to be 0.1%. First prepare a 0.05 M solution of cysteine hydrochloride (or hydroxylamine hydrochloride) in 0.05 M sodium acetate solution at pH 7. Dissolve enough of the deoxyribonuclease in this so that the final concentration of the enzyme will be 0.1% (e.g. dissolve 2 mg of the deoxyribonuclease in 1 ml of the cysteine–acetate buffer solution).

Adjust to pH 7 if necessary. Leave the solution to stand at room temperature (or at 37 °C) for 5–15 min. (This is to inactivate the proteolytic activity.)

Before it is to be used, dilute the solution with an equal volume of barbitone–HCl buffer at pH 7.

To the enzyme solution add 1% magnesium chloride.

Treatment

Add the complete medium to the section and leave it at room temperature. One hour should be sufficient to remove DNA from unfixed sections. Then stain with methyl green as above.

To confirm the specificity of the deoxyribonuclease

Should it be necessary to prove the specificity of the deoxyribonuclease used, the following procedure should be followed:

1. Take three serial sections.
2. Treat one with the deoxyribonuclease incubation medium for 1 h.
3. Treat the second for the same time with an exactly similar incubation medium to which has been added zinc sulphate (at a concentration of 0.1–0.01 M; 0.1 M sodium arsenate may be used instead of zinc sulphate). This should inhibit the deoxyribonuclease.
4. Leave the third section untreated (or treat it with the incubation medium lacking the enzyme entirely). Then stain all three slides together by Kurnick's methyl green method.

Result

DNA should stain green in the third slide; it should not be present in the first slide (having been removed by the active deoxyribonuclease). It should be present and stained green in the second slide (which has been subjected to the inactivated deoxyribonuclease).

THE METHYL GREEN–PYRONIN METHOD FOR **DNA AND RNA**

Rationale

As has been discussed (above) in connection with Kurnick's methyl green procedure, it is possible to obtain specific and quantitative staining of DNA by prolonged reaction with methyl green under controlled conditions. This specificity is said to be due to the two charged groups on the dye molecule both becoming linked to phosphate moieties in the DNA. The use of the mixture of the two basic dyes, methyl green and pyronin, to stain nuclei and cytoplasm differentially is of long standing; a possible mechanism has been suggested by Chayen (1952) and Brachet (1954). It seems that under the conditions of this reaction, and in particular with the relatively short periods of staining used, the more reactive, singly charged pyronin can successfully compete with methyl green except where the double charge on the methyl green gives it selective advantage, namely when it binds by both charges to an acidic polymer such as DNA. [Although it happens only infrequently, it should be noted that the methyl green can react with other acidic polymers.]

The reaction is done at pH 4.2 at what is about the isoelectric point of isolated nucleic acids (Kurnick and Mirsky, 1950). This has the advantages of having them in the charged condition and also in their least soluble state.

Method

Fresh cryostat sections can be used but it must be appreciated that they will be fixed by the acidic acetate buffer used in the staining mixture. Consequently the sections can be fixed in acetic alcohol, which is a good precipitant of nucleoprotein. The higher the concentration of the acetic acid used, the more the nucleic acids will be freed from the protein and hence the stronger the stain will be. Consequently 1 : 3 (v/v) acetic alcohol may be used, although 5% acetic acid in absolute ethanol is satisfactory. Fixation should be for 15 min.

1. If the sections have been fixed in an alcoholic fixative, rinse in 70% alcohol and transfer to acetate buffer.
2. Immerse in stain mixture for 30 min.
3. Rinse very briefly in acetate buffer.
4. Blot dry. This must be done carefully but firmly. (If any moisture is left, the methyl green will be lost from the sections at the next step.)
5. Immerse in absolute n-butanol for 2 min.
6. Mount in Euparal.

Solutions required for this method

Acetate buffer (pH 4.2)

Solution A. 0.2 M sodium acetate solution consists of 6.8 g CH$_2$.COONa.3H$_2$O made up to 250 ml in distilled water.

Solution B. 0.2 M acetic acid contains 2.85 ml of glacial acetic acid made up to 250 ml with distilled water.

Add 70 ml solution B to 30 ml solution A.

Stain mixture

First prepare a 0.25% (w/v) solution of methyl green in 0.2 M acetate buffer at pH 4.2. Extract with chloroform (as for Kurnick's methyl green method) until almost all the methyl violet impurity has been removed in the chloroform. Separate the aqueous methyl green solution from the chloroform and leave it to stand, uncovered, in the dark until the dissolved chloroform has evaporated.

Then add an equal volume of a 0.5% (w/v) solution of pyronin G in 0.2 M acetate buffer. Mix. This solution is stable for some weeks.

[*Note*: an alternative method for plant histochemistry is to use the phenolic mixture recommended by Darlington and La Cour (1947).]

Result

Nuclei stain green or blue; nucleoli and cytoplasm stain red.

[*Note*: mast cells stain a characteristic flame orange-red by this method. Plasma cells stain so intensely red that this procedure has been used to identify them.]

Interpretation

Very often, as a first approximation, the red colour is said to denote the presence of RNA and the blue or green colour that of DNA. But it must be emphasized that this can be used only as a rough approximation. In practice, when nuclei stain blue-green, this is probably a true indication of the presence of DNA; however, should the presence of DNA be in the least critical, this apparent localization should be confirmed by the use of deoxyribonuclease as discussed on p. 89. Alternatively, the use of a short hydrolysis with hydrochloric acid (see below) may be a useful and less costly way of strengthening the belief that this blue-green-staining matter is DNA. [*Note*: but remember that DNA stains red after acid hydrolysis.]

On the other hand, most studies on the histochemistry of nucleic acids are likely to be more concerned with whether there is more or less RNA in the cells. For example, it may be required to see whether cells are apparently synthesizing extra amounts of protein, whether they be potentially malignant cells or such cells as lymphocytes responding to produce antibody. Under these circumstances it may be gravely misleading to assume that 'pyroninophilia' equals RNA. The gravity of this assumption can best be emphasized by an example with which we were involved: histologically the section contained unusual cells which were mitotically active. When tested by the methyl green–pyronin method, these cells stained vividly red and so the suggestion given us was that this was a neoplastic growth. But, when they were tested either by ribonuclease or by the hydrochloric acid method, the ability of these cells to stain with pyronin was undiminished. Consequently the 'pyroninophilia' was not due to RNA. In fact, these cells were actively growing, but not malignant, bile duct cells in which the basophilia (to pyronin) was due to some acidic mucopolysaccharide matter. Hence, for precise histochemistry, the interpretation of pyronin-positive matter as RNA depends on the loss of this basophilia after treatment either with ribonuclease or with hydrochloric acid.

THE USE OF RIBONUCLEASE

Method

This enzyme acts optimally at about pH 7.7. Hence 1 mg of the crystalline preparation is dissolved in 1 ml 0.2 M veronal–HCl buffer at pH 7.7.

[*Note*: it is often far simpler to dissolve 1 mg of the enzyme in an M/40 aqueous solution of sodium bicarbonate; this is a convenient way of achieving a pH of 8.2, which is sufficiently close to the optimum.]

Then:

1. Treat one section with the ribonuclease solution for 1 h at room temperature.
2. Treat a control section for the same length of time either with the buffer (or bicarbonate) solution alone or with the ribonuclease solution to which has been added zinc sulphate (at a concentration of 0.1–0.01 M) to inactivate the enzyme.
3. Wash both sections and stain with methyl green–pyronin.

Result

Pyronin-positive matter, which is lost after treatment with ribonuclease but not after exposure to the control, can be taken to be RNA.

HYDROCHLORIC ACID METHOD FOR **RNA**

Rationale

Frequently a histochemist may not have ribonuclease readily available but may wish to place more credence on the methyl green–pyronin staining seen in a section than is reasonable by use of the mixture alone. This can be achieved by recourse to the fact that 1 M hydrochloric acid at 60 °C extracts all the RNA from sections in 5 min; it also causes sufficient alteration in DNA to make it stain with the pyronin, and not with the methyl green, component of the methyl green–pyronin mixture.

Method

Take two cryostat sections. Fix in acetic alcohol (either 1 : 3 or 5% as required).

1. Take both sections through graded concentrations of alcohol to 0.2 M acetate buffer at pH 4.2.
2. Immerse one in 1 M hydrochloric acid at 60 °C and maintain at 60 °C for 5 min.
3. Wash both in 0.2 M acetate buffer.
4. Stain both in the methyl green–pyronin mixture (see p. 92).
5. Blot dry; immerse in *n*-butyl alcohol for 2 min. Mount in Euparal (for details see the methyl green–pyronin method as given on p. 91).

Result

RNA stains red in the non-hydrolysed section but is absent from the hydrolysed section. The DNA should stain green-blue in the non-hydrolysed section but a reddish or purple colour in the hydrolysed section.

DIFFERENTIATION OF ACIDIC MOIETIES

Hyaluronate, heparin, chondroitin and keratan sulphates will stain strongly with pyronin. It may be helpful to know that the staining of these moieties can be inhibited by including relatively low concentrations (e.g. 0.05 M) of sodium chloride in the pyronin G solution (or methyl green–pyronin G solution). The concentration of sodium chloride may have to be increased to as much as 3.0 M before it inhibits the staining of RNA by pyronin G.

TOLUIDINE BLUE METHOD FOR **BASOPHILIA**

Rationale

Toluidine blue is a basic dye and so will stain acidic matter. The exact nature of that matter cannot be determined just by the fact that it stains with this dye; all that can be said is that it is acidic. The use of nucleases can show whether it is RNA or DNA. Another diagnostic feature is whether the dye stains orthochromatically, i.e. blue, or metachromatically, namely red or red-purple. (For the histochemical significance of metachromasia see below.)

Method

Fresh cryostat sections are recommended. Otherwise fix fresh sections in 1 : 3 (v/v) acetic alcohol (or 5% acetic acid in absolute ethanol), and take them through graded concentrations of alcohol to acetate buffer.

1. Stain in an aqueous solution of toluidine blue in acetate buffer for 30 min. [*Note*: 10 min may be sufficient.]
2. Wash in acetate buffer.
3. Blot dry.
4. Place in *n*-butanol for 2 min.
5. Mount in Euparal.

Solutions required for this method

Acetate buffers (0.1 M)

   Solution A. 2.85 ml glacial acetic acid, made up to 500 ml with distilled water.

   Solution B. 6.8 g sodium acetate made up to 500 ml with distilled water.

For pH 4.2 add 140 ml of A to 60 ml of B
For pH 5.0 add 60 ml of A to 140 ml of B
For pH 5.6 add 20 ml of A to 180 ml of B

Toluidine blue solution

Toluidine blue   0.1 g
Make up to 100 ml with the relevant buffer.

Result

Normal basophilia is shown by blue or purple-blue staining. (The exact hue depends on the composition of the mixture of dyes known collectively as 'toluidine blue'. For this reason it may be preferred to use azure B in place of toluidine blue.) Pink or red staining denotes metachromatic coloration (see below).

METACHROMATIC STAINING WITH TOLUIDINE BLUE

Rationale

There is still considerable uncertainty as to the exact nature of the phenomenon known as metachromasia (e.g. see Baker, 1958, chapter 13; Sylvén, 1954; R. Chayen and Roberts, 1955). The phenomenon itself seems to be quite simple: a pure basic dye in aqueous solutions absorbs light in a distinct manner and so appears to be coloured. [*Note*: the property of metachromasia is seen generally with basic rather than with acidic dyes but is not related, fundamentally, to the acidity of the dye.] When the dye reacts with tissue elements to produce a colour similar to that of the dye, it stains orthochromatically. However certain dyes, even when 'pure', may stain some tissue components orthochromatically but will produce a different colour with other components. As an example we may consider the effects obtained with toluidine blue. (This is rarely a simple 'pure' substance but is often a complex of different dyes of slightly different colours, as shown by Ball and Jackson (1954).) Even when it behaves as a single blue dye, it stains cells blue (orthochromatically) but proteoglycans ('acidic polysaccharides' generally) are stained red. This change of colour, especially when shown by a 'pure' dye, is called metachromasia; toluidine blue stains acidic polysaccharides metachromatically. The change of colour is usually, but not exclusively, towards the longer wavelength (blue to red); it can be shifted towards the shorter wavelength: for example, toluidine blue colours highly polymerized DNA green.

In elegant studies Sylvén (1954) showed that toluidine blue stains polymers metachromatically only when the acidic groups are packed closely together within the polymer: they should not be more than 0.5 nm apart. It may be a fair approximation to suggest that, when the dye reacts with an acidic polymeric molecule in the tissue section (by electrostatic interaction between the basic dye and the acidic group of the polymer), the dye shows its ordinary colour, that is, it stains orthochromatically. However, should these dye molecules come to be so closely packed along the polymer that they can interact with each other, the ionization of the dye becomes suppressed, giving rise to a change of colour. From the spectroscopic work of Michaelis (1947) there seems little doubt that the change in colour is due to the formation of dye micelles (dimers, trimers) probably associated side by side.

To summarize: basic dyes normally stain acidic substances orthochromatically. They stain metachromatically when the acidic substances are present as long polymeric molecules, with the acidic groups closely packed along the polymer. Consequently, acidic mucopolysaccharides typically are stained metachromatically but certain other acidic polymers may also show this type of coloration. The nature of the acidic groups can be tested by finding at what pH value their ionization is suppressed.

Method

Fresh cryostat sections should be used (because the state—for example the coagulation—of the cytoplasm can affect the metachromatic response markedly).

Alternatively the sections can be fixed in the picric–formalin fixative.

1. Immerse in a 0.1% solution of toluidine blue at the required pH. Stain for 30 min.
2. Rinse in buffer at the same pH.
3. Either blot dry or shake free of buffer; dry in air.
4. Place in *n*-butanol for 2 min.
5. Mount in Euparal.

Result

Orthochromatic coloration: blue or blue-purple (depending on the purity of the blue dye).

Metachromatic coloration: red, pink or red-purple.

Solutions required for this method

For pH values of 5.6, 5.0 or 4.2 use the acetate buffer recommended for the usual method for toluidine blue. For lower pH values use distilled water and add a few drops of 1 M hydrochloric acid to achieve the desired pH.

The dye is dissolved (0.1%) either in buffer or water. The pH of the dye-in-water solution should be adjusted to the required value after the dye has been dissolved, in case the dye itself should affect the pH.

[*Note*: Azure A can be used in place of toluidine blue. It has been claimed that Azure B (0.1–0.2 mg/ml in citrate buffer at pH 4) stains RNA metachromatically.]

Application

Nucleic acids in sections stain orthochromatically with toluidine blue at pH 4.2. At lower pH values their ability to bind the dye becomes progressively diminished.

Acidic polymers are stained metachromatically by toluidine blue at pH 5.6. If the acidity is due to carboxyl groups, this staining is abolished at lower pH values (e.g. 4.2). Of the acidic groups found commonly in biological material only sulphate groups (and polyphosphate) remain stainable at pH 1.0. Phosphate moieties of organic molecules become less stainable between pH values of 4.2 and 1.0.

EBEL'S TEST FOR **POLYPHOSPHATE**

Where a region of a section is stained metachromatically at low pH values, the possibility that the acidic groups (which are responsible for this reaction) may be phosphate can be tested by recourse to this test. Fundamentally it is based on the avidity of lead for phosphate. (This is used in the histochemical method for the demonstration of acid phosphatase.) The lead which becomes bound to phosphate can be visualized if it is converted into lead sulphide (black or brown).

Method (after Ebel, 1952)

Use unfixed cryostat sections.

1. Immerse in 10% lead nitrate in 0.05 M acetate buffer at pH 4.5.
2. Wash well with acetate buffer to remove excess lead nitrate.
3. Immerse in water which has been saturated with hydrogen sulphide. (Bubble the gas from a Kipp's apparatus or from a commercial cylinder through distilled water for 5 min. This *must* be done in a fume cupboard.)
4. Wash in distilled water.
5. Mount in Farrants' medium.

ACRIDINE ORANGE METHOD FOR **DNA AND RNA**

Rationale

The reaction of acridine orange with both nucleic acids has been much studied. It appears to be a complex situation (reference may be made to papers by Armstrong, 1956; De Bruyn *et al.*, 1953; Massart *et al.*, 1947; Schümmelfeder, 1958). Certainly this dye produces a different coloured fluorescence when it is bound to low polymer nucleic acids than when it is linked to high polymer DNA. But it also shows its 'high-polymer' colour when linked to other polymers, such as keratin. The exact colour seen depends on the excitation wavelength and on the eyepiece (barrier) filters used in the fluorescence equipment. The action of the dye is also very sensitive to the pH at which the staining reaction is done and to the fixative used.

Method (after Armstrong, 1956)

Fresh cryostat sections should be used or they may be fixed in acetic alcohol (1 : 3 v/v). Tissues can be fixed in calcium–formaldehyde, but fixatives containing heavy metals should not be used.

1. Immerse in acetate buffer, pH 4.2, for 5 min. [*Note*: this will also fix fresh cryostat sections.]
2. Stain for 15–30 min in acridine orange solution at pH 4.2

3. Wash in buffer at pH 4.2.
4. Mount in a drop of the buffer and ring the coverslip with wax.
5. Examine with a fluorescence microscope.

Results

(a) When excited with purple light, the DNA–acridine orange complex fluoresces green-yellow to bright yellow; the RNA–acridine orange complex gives a dull reddish brown to flame orange, depending on its concentration in the cells.

(b) With ultraviolet light (365 nm) DNA appears greenish yellow and RNA crimson red.

False positives: mast cell granules and cartilage matrix give red fluorescence. Vascular elastic elements fluoresce yellow and keratin is often green.

Solutions required for this buffer

Walpole's acetate buffer, pH 4.2

1 M sodium acetate (13.6 g/100 ml; $CH_3COONa.3H_2O$)   50 ml
1 M hydrochloric acid   35 ml

Acridine orange solution

1 : 10 000 (w/v) parts of acridine orange in Walpole's acetate buffer (pH 4.2).

STAGE 4: REACTIONS FOR POLYSACCHARIDES

Undoubtedly the diagnostic histochemical test for carbohydrate material is the periodic acid–Schiff (PAS) reaction. Even water-soluble carbohydrate matter may be demonstrated provided it is fixed in the tissue and an alcoholic PAS reaction is used. In plant histochemistry the recognition of the various carbohydrate-containing components depends largely on histological stains (see Table 4). In animal histology, the Alcian blue method is very widely relied on to demonstrate acidic mucopolysaccharide; Hale's colloidal iron method has also been much used.

PERIODIC ACID–SCHIFF (PAS) METHOD

Rationale

Periodic acid is a very strong oxidizing agent which will oxidize a vicinal diol (1,2-glycol), whether in the *cis* or *trans* form, to yield two aldehyde groups:

$$
\begin{array}{ccc}
\overset{\displaystyle HO}{\underset{\displaystyle H}{\overset{|}{\underset{|}{C}}}}-\overset{\displaystyle H}{\underset{\displaystyle OH}{\overset{|}{\underset{|}{C}}}}- & \longrightarrow & \overset{\displaystyle O}{\underset{\displaystyle H}{\overset{\|}{\underset{|}{C}}}}\;\overset{\displaystyle H}{\underset{\displaystyle O}{\overset{|}{\underset{\|}{C}}}}-
\end{array}
$$

$trans$-glycol   aldehydes

It should be noted that periodic acid will also oxidize α-amino aldehydes and ketones, α-amino alcohols such as serine, and also diamines, although the optimal pH for these oxidations is 7–8; pH 3–5 is optimal for the oxidation of α-glycols, hydroxyaldehydes, hydroxyketones and diketones (Dyer, 1956). But the oxidation is inhibited if any of the $-OH$ groups are substituted by any group other than these.

Periodic acid also oxidizes ethylenic linkages:

$$
-\overset{|}{C}=\overset{|}{C}-
$$

and this is of considerable consequence in histochemistry since these linkages found in unsaturated fatty acids, may give a strong and apparently specific reaction when so oxidized (see below under benzoylation). Moreover, sodium periodate, in particular, decomposes readily in sunlight to produce ozone which will oxidize many groups (see Dyer, 1956).

Consequently the object of this test is to use periodate at that concentration, and acting for a suitably short time, which will be most conducive to the

Table 4. Summary of the properties and colour reactions of cell-wall constituents (modified from Bonner, 1950, pp. 130, 131)

Constituent	Sign of birefringence	Ultraviolet absorption	Diagnostic colour reactions
Cellulose	Positive	Little	(a) Dichroic violet stain with chlorzinc iodine
			(b) Blue colour with KI–I$_2$ in conc. H$_2$SO$_4$
			(c) Dichroic stain with Congo red
			(d) Almost no stain with ruthenium red
Pectic substances	Statistically 0	Little	Stain red with ruthenium red
Lignin	0	Strong	Red reaction with phloroglucinol–HCl
Mannans, galactans, etc.	Usually birefringent	Little	No specific stains
Polyuronide hemicelluloses	0	Little	No specific stains
Cuticular substances and 'suberin'	Negative	Strong	Red or orange-red colour with scarlet H or Sudan III
Callose			Stained by resorcin blue (Johansen, 1940)

oxidation (to the aldehyde) of the $-CHOH-CHOH-$ groups of carbohydrates and not to allow over-oxidation.

The reaction sequence (Hotchkiss, 1948) is as follows:

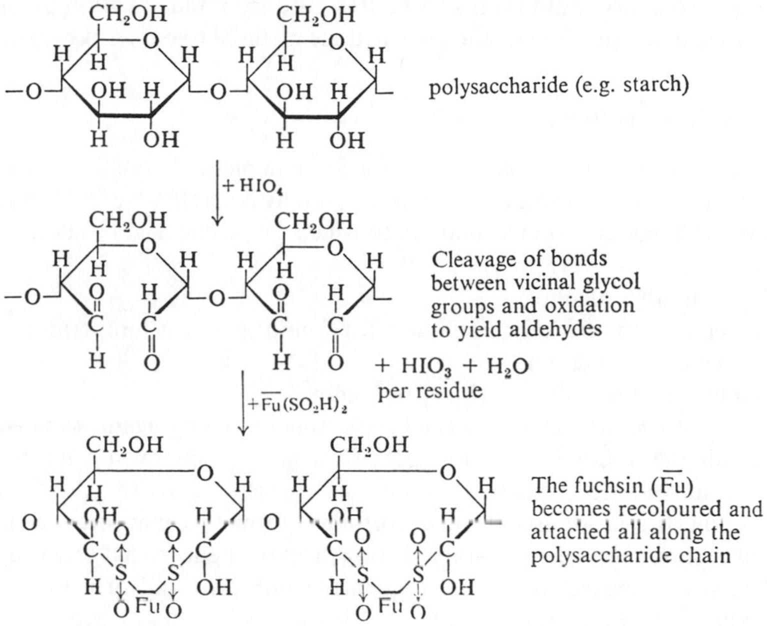

polysaccharide (e.g. starch)

$+ HIO_4$

Cleavage of bonds between vicinal glycol groups and oxidation to yield aldehydes

$+ HIO_3 + H_2O$ per residue

$+ \overline{Fu}(SO_2H)_2$

The fuchsin ($\overline{Fu}$) becomes recoloured and attached all along the polysaccharide chain

According to Hotchkiss (1948) a substance will give a positive result with this reaction if:

1. It contains the 1,2-glycol grouping in an unsubstituted form (except where $-OH$ is replaced by $-NH_2$ or alkylamino groups, and except for the oxidation products of the glycol grouping).
2. It does not diffuse away in the course of fixation.
3. Its oxidation product is not diffusible.
4. It is present in a sufficiently high concentration to give a detectable final colour.

From these criteria it follows that the sugar moiety of the nucleic acids will not be stained (because one of the 1,2-glycol groups is substituted); on the other hand those of cerebrosides and inositol-lipids will stain by this method.

Considerable controversy has been generated over the problem of how to fix polysaccharides in tissues. The much quoted work of Mancini (1948), which dealt with the diffusion of glycogen in a block of tissue during fixation and dehydration, need not be of any concern if fresh sections are used.

Polysaccharide material can then be fixed in the section, without diffusion from its site, by the picric–formalin fixative developed by Bitensky *et al.* (1962) to deal with the fixation of this type of material. More intense results can often be obtained if 5% acetic acid is added to the fixative, or if the sections are fixed in absolute alcohol; both these procedures produce intensification by partially solubilizing and redistributing the carbohydrate material (see Bitensky *et al.*, 1962).

Method (after Hotchkiss, 1948)

Fresh cryostat sections should be fixed for 5 min in picric–formalin fixative. [For an estimate of the total amount of glycogen, it may be preferable to fix in absolute alcohol or in picric–formalin solution to which 5% acetic acid has been added.]

1. Wash in 70% alcohol.
2. Immerse in the periodate solution for 5 min at room temperature.
3. Wash in 70% alcohol.
4. Immerse in the reducing rinse for 5 min.

   [*Note*: because periodate is a very strong oxidizing agent, even traces of it left in the section will recolour Schiff's reagent. Consequently it is necessary to remove all adsorbed traces of periodate with this reducing solution. If it is not used—and unfortunately some workers even recommend that it should be left out—there is no reason to believe that any staining of the tissue is indeed due to polysaccharide and not to adsorbed periodate.]
5. Wash in 70% alcohol.
6. Immerse in Schiff's reagent for 30 min.
7. Wash in three changes of SO_2-water.

   [*Note*: for the need for this please see under the Feulgen reaction, p. 81.]
8. Dehydrate (stain the nuclei with Ehrlich's haematoxylin if required); mount in Euparal or Styrolite.

Result

Red or purple-red indicates a positive PAS reaction.

Solutions required for this method

Picric–formalin fixative

40% formaldehyde (technical)	10 ml
Ethanol	56 ml
Sodium chloride	0.18 g
Picric acid	0.15 g
Distilled water to	100 ml

Note: Picric acid can be explosive when dry. It should be kept under water. The rim of the bottle should be kept clean. Remove some of the solid picric acid and place it on a filter paper to remove surplus moisture. Weigh 0.15 g of this moist picric acid. Replace the rest, under water, in the stock bottle.

Periodate solution

Dissolve 400 mg periodic acid in 15 ml distilled water. To this solution add 135 mg crystalline sodium acetate ($CH_3.COONa.3H_2O$) dissolved in 35 ml absolute ethanol. Prepare just before use.

Reducing rinse

Potassium iodide	1 g
Sodium thiosulphate	1 g
Distilled water	20 ml

When dissolved, add (while stirring the solution) 30 ml ethanol and then 0.5 ml 2 M hydrochloric acid. A precipitate of sulphur may form; this is left to settle out.

Controls

A positive reaction after this procedure could be due to:

1. Aldehydes or ketones present in the tissues and which have remained unaltered during the process.
2. Oxidizable moieties other than carbohydrate; the most obviously significant of these are the

$$-\overset{|}{C}=\overset{|}{C}-$$

groups in unsaturated fatty acids which are oxidized to aldehydes.
3. Carbohydrate matter, probably in a polysaccharide.

To prove that the positive reaction is due to carbohydrate (or more precisely to $-$CHOH$-$CHOH$-$ groups), a control section should be acetylated or benzoylated (see below); matter which is positive to the simple PAS reaction but negative to PAS after such treatment contains the $-$CHOH$-$CHOH$-$ grouping and almost certainly contains carbohydrate. The exact nature of the carbohydrate can then be examined by the other staining methods for polysaccharides (starch, glycogen, mucopolysaccharide) or it may be subjected to specific glycolytic enzymes.

To test whether some or all of the reaction is due to pre-existing aldehydes or ketones (1 above) another section is subjected to the Schiff's reagent either without prior treatment, or after 5 min in an M/20 alcoholic solution of hydrochloric acid at room temperature (in case plasmalogens are contributing to the response to Schiff's reagent). Should there be pre-existing aldehydes which confuse the true periodic acid–Schiff reaction, they can be blocked by treating the section with a saturated solution of dimedone in 5% acetic acid for 18 h at 60 °C; subsequent treatment with periodic acid does not uncouple this link between the dimedone and the aldehyde.

BENZOYLATION

Method

1. Fix for 8 min in absolute alcohol.
2. Change to fresh absolute alcohol for a further 8 min.
3. Place in dry acetonitrile in a desiccator containing phosphorus pentoxide for 3 min.
4. Transfer to fresh dry acetonitrile for a further 3 min.
5. Immerse in the benzoylation solution for 2 h.
6. Remove sections from desiccator.
7. Wash well in absolute alcohol.
8. Continue with PAS test.

Solution required for this method

Benzoylation solution

Acetonitrile	50 ml
Benzoyl chloride	4.2 ml
Dry pyridine	2.2 ml

Note: acetonitrile, benzoyl chloride and pyridine are harmful chemicals and should be handled in a fume cupboard. They should not be pipetted by mouth.

ACETYLATION

This is a useful technique in the study of polysaccharides. Occasionally a methylation–saponification method (e.g. Fisher and Lillie, 1954) may be helpful to remove sulphate groups; this procedure and acetylation have been considered by Materazzi and Ferretti (1970).

Method

1. Incubate in the acetylation solution at 60 °C for 18 h. (Occasionally 1 h may suffice—see Materazzi and Ferratti, 1970.)
2. Wash well in absolute alcohol.
3. Continue with the appropriate test.

Solution required for this method

Acetylation solution

Acetic anhydride 16 ml
Dry pyridine 24 ml
Note: preferably prepare this solution in a fume cupboard.

ALCIAN BLUE METHOD

Rationale

All the acid glycosaminoglycans (mucopolysaccharides) are negatively charged by virtue of the presence of sulphate ester groups or carboxyl groups of uronic acids, or both charged groups. Consequently they bind positively charged (cationic) dyes such as toluidine blue, Azure A and Alcian blue. Of these the last is used very extensively by histologists and histochemists; the many variants can be found in most textbooks of histochemistry. However, the precise use of Alcian blue is that developed by Scott and Dorling (1965) and reference should be made to this work. They applied the concept of the critical electrolyte concentration to differentiate between different glycosaminoglycans. They were able to show both in model experiments with chemically defined substances and in histological sections of selected tissues that most 'mucopolysaccharides' stained with 0.05% Alcian blue at pH 5.8 (protein interference was noted at pH 2.5); that goblet-cell mucin (mainly carboxyl groups) staining was reduced if the concentration of magnesium chloride was greater than 0.4 M; cartilage staining was lost at above 0.6 M magnesium chloride; mast cells no longer stained if the magnesium chloride concentration was above 0.75 M; and staining of corneal stroma (mainly chondroitin sulphate and keratan sulphate) was lost above 1 M. These results agreed with the known concentrations in these tissues of sialomucin, chondroitin sulphate, heparin and keratan sulphate respectively. However, they emphasized that staining with Alcian blue must be carried to equilibrium so that staining should continue for 8 h. Ideally it should also be shown that the binding of Alcian blue is reversible; this is done by soaking the stained section in the electrolyte solution alone, although destaining may require longer than staining. However, many workers have reported useful results using only 30 min staining even though this is not rigorously correct (e.g. Andersen *et al.*, 1970).

S'ome workers have met with difficulties with this method. Some of these difficulties may have arisen because the workers did not follow the method precisely. Other difficulties have been clarified by Scott and Mowry (1970). Many of the troubles arise from the fact that the contents of the bottle, sold under the name of 'Alcian blue', may not correlate well with the label. Scott and Mowry listed the following requirements:

1. Use Alcian blue 8GX (not GS, 5 or 7GX).
2. It should dissolve in water to give at least 5% solution.
3. A 1% (w/v) solution containing 2.0 M magnesium chloride and buffered to pH 5.7 with 0.025 M acetate buffer should not precipitate when fresh, although a fine precipitate may be seen after 24 h.
4. The spectrum and absorbencies should be as given by Scott (1970).
5. The sample should not be more than 3 years old.

With a dye that has these properties, good results should be obtainable.

Method (from Scott and Dorling, 1965)

1. Fixation: Scott and Dorling used tissues fixed in either neutral phosphate-buffered 4% formaldehyde or 95% ethanol at 4 °C. Cryostat sections should be fixed in either fixative for 5–10 min.
2. If fixed in ethanol, take sections through a graded concentration of alcohol to acetate buffer at pH 5.8.
3. Stain at room temperature in 50 ml of the dye solution in a Coplin jar for at least 8 h or overnight.
4. Rinse each section individually in acetate buffer, pH 5.8, containing 0.5 M magnesium chloride.
5. Transfer to fresh acetate buffer, containing 0.5 M magnesium chloride, in a Coplin jar.
6. Dehydrate through graded concentrations of alcohol (70%, 90%, absolute).
7. Clear in xylene; mount in Styrolite.

Result

The various acidic glycosaminoglycans or proteoglycans stain blue; this staining can be suppressed by the correct concentration of magnesium chloride (as below).

Solutions required for this method

Dissolve 0.05% Alcian blue in 0.025 M sodium acetate buffer at pH 5.8. To stain all acidic glycosaminoglycans, add 0.025 M magnesium chloride (Analar grade). This concentration of magnesium chloride can be varied if it is required to demonstrate only particular proteoglycans or glycosaminoglycans selectively. Use 0.5 M magnesium chloride to demonstrate chondroitin sulphate and keratan sulphate but to suppress staining with sialomucins, or 0.7 M magnesium chloride to stain only keratan sulphate.

The Alcian blue solution should be prepared on the day it is to be used. (*Note*: the concentrations of magnesium chloride to be included in the Alcian blue solution are given only as a guide. They may require modification for some tissues.)

INDUCED BIREFRINGENCE OF PROTEOGLYCANS

Although proteoglycans (glycosaminoglycans) apparently show no birefringence, their degree of orientation (molecular order) can be measured by staining with the appropriate concentration of magnesium chloride in the Alcian blue solution and inspecting the section under polarized light. As with collagen, maximum birefringence will be obtained when the proteoglycans are placed at the 45° position. Weak birefringence can be measured with a Brace–Köhler λ/30 compensator; stronger birefringence requires Sénarmont compensation.

This type of enquiry has been fairly extensively reviewed by Modis (1974). Its use with Alcian blue has allowed quantitative measurement of the orientation of proteoglycans, and the disorder imposed by disease both in bone (Kent *et al.*, 1983) and in articular cartilage (Dunham *et al.*, 1985).

DIAMINE METHODS FOR MUCOSUBSTANCES

Spicer (1965) developed a range of methods that seem to give valuable information concerning mucosubstances. In particular, a mixture of *N*,*N*-dimethyl-*m*-phenylenediamine with *N*,*N*-dimethyl-*p*-phenylenediamine (as the hydrochlorides), generally in a proportion of 6 : 1, and at a final diamine concentration of 0.07% at pH 3.4–4.0, may colour acid mucosubstances (sulphomucins and sialomucins) purple. The specificity may be enhanced if iron, as ferric chloride, is included in the dye solution. The 'low iron' diamine method, followed by Alcian blue, is said to stain sulphomucins and most sialomucins black; other sialomucins stain blue. A higher concentration of iron and of the diamines, again followed by staining with Alcian blue, is claimed to stain sulphomucins black and sialomucins blue. Gad and Sylvén (1969) tested the high concentrations of iron and diamines on model systems; their results indicated that this method stained sulphomucins provided that red, purple or violet reactions only were taken as truly positive.

To prepare the 'high concentration' dye solution, dissolve 120 mg *meta*-diamine and 20 mg *para*-diamine in 50 ml distilled water. Pour into the staining jar, which should already contain 1.4 ml 10% ferric chloride. Staining time, for paraffin sections, is 18 h. The sections can be 'counterstained' with Alcian blue (1% in 3% acetic acid, for 30 min). The sections can be dehydrated and mounted in the normal way. Gad and Sylvén (1969) used 1.4 ml of 37.2% ferric chloride and adjusted the final pH to 1.3–1.5.

HALE'S COLLOIDAL IRON METHOD (Hale, 1946)

Rationale

This method seems to depend on the adsorption of colloidal iron by tissue components (probably acidic groups); the iron is then stained by the Prussian blue method (see p. 71). There appears to be little reason for believing that this method is intrinsically specific for any chemical group but it (or modifications like those of Lillie and Mowry (1949) or Wigglesworth (1952)) has been used quite extensively as a histological stain for mucopolysaccharide (or for acidic polysaccharides or some types of 'mucin').

Method

1. Fix cryostat sections in the picric–formalin solution.
2. Take through graded concentrations of alcohol to water.
3. Flood the section (held horizontally in a staining dish) with the dialysed iron solution. Leave for 10 min.
4. Wash in water.
5. Immerse in acid–ferrocyanide solution for 10 min.
6. Wash in water. (The section may be counterstained with safranin at this stage.)
7. Dehydrate and mount in Euparal or Styrolite.

Result

Blue colour indicates 'mucins' or acidic polysaccharides.

Solutions required for this method

Dialysed iron

Pharmaceutical grade of dialysed iron (5% Fe_2O_3)   1 volume
2 M acetic acid   1 volume
   Should this not be suitable, or available, the dialysed iron can be prepared fresh (see Gomori, 1952).

Acidic ferrocyanide solution

1% solution of potassium ferrocyanide	85 ml
1 M hydrochloric acid	15 ml

BEST'S CARMINE METHOD FOR *GLYCOGEN*

Rationale

As Gomori (1952) stated, this is an empirical stain for glycogen. There is some doubt now whether it stains all glycogen; it may be sensitive to one form rather than to another (e.g. to the branched rather than to the long chain form of glycogen).

Method

This method, applied to cryostat sections in this way, has been developed and used routinely by Dr S. Darracott.

1. Fix cryostat sections in the picric–formalin solution (p. 36) for 5 min.
2. Wash in 70% alcohol.
3. Treat with periodate solution for 5 min.
4. Rinse in 70% alcohol.
5. Immerse the section in the reducing rinse for 5 min.
6. Take through graded concentrations of alcohol and stain with Ehrlich's haematoxylin for 15 min.
7. Wash in tap water (or in 5% sodium bicarbonate solution) to 'blue'.
8. Stain with Best's carmine for 10–15 min.
9. Wash in the differentiator (a few seconds should suffice).
10. Wash very briefly in absolute alcohol.
11. Clear in xylol and mount in Styrolite.

Result

Glycogen should stain red.

Solutions required for this method

Periodate and reducing rinse

As for the PAS reaction (p. 103).

Best's carmine

Stock solution

Carmine   2 g
Potassium carbonate   1 g
Potassium chloride   5 g
Distilled water   60 ml

Boil until the colour deepens (3–5 min). Cool. Add 20 ml concentrated ammonia (0.880).

Staining solution

Stock solution   12 ml
Concentrated ammonia   18 ml
Methanol   18 ml

Differentiator

Absolute ethanol   20 ml
Absolute methanol   10 ml
Distilled water   25 ml

[*Note*: since the sections are treated with periodate and with the reducing rinse, the results with Best's carmine are exactly analogous to those with PAS. Hence, in serial sections, those regions which are positive to the PAS reaction and are also stained red by Best's carmine must be considered to contain glycogen. (This can be proved more rigorously by extraction by diastase or amylase.) If such comparison is not required, steps 3, 4 and 5 can be left out.]

LUGOL'S IODINE

This procedure may be useful as a diagnostic test in polysaccharide histochemistry (e.g. Flint, 1982). Fresh cryostat sections may be used or they may be fixed in picric–formalin or in this fixative to which 5% acetic acid has been added (p. 36). Fixation with the acetic–picric–formalin fixative often discloses more of the polysaccharide to the iodine solution. The sections are stained with Lugol's iodine for various times, depending on the nature of the material. The sections are then rinsed and inspected wet; the wet sections may be mounted in Farrants' medium. Alternatively they may be blotted dry and mounted in Euparal.

Glycogen should stain brownish and native starch should be coloured deep blue.

To avoid solubilizing glycogen, Mancini (1948) dissolved the iodine in liquid paraffin.

Lugol's iodine

	Concentrated solution	Dilute solution
Iodine	1 g	0.3 g
Potassium iodide	2 g	6 g
Distilled water to	100 ml	100 ml

Note: sometimes it may be preferable to use the more dilute solution of iodine; at other times the concentrated solution may be found preferable.

THE USE OF DIASTASE AND AMYLASES

Suppose that tissue sections contain regions that are positive to the PAS reaction (and perhaps to Lugol's iodine and Best's carmine). The proof that such regions contain glycogen or starch is that they do not give these reactions after specific digestion with a diastase. The simplest way of treating with a diastase is to use saliva but 'at present, more appetizing and sanitary tests have replaced the time-honoured saliva test' (Gomori, 1952).

Method

Fix sections with the same fixative as is used for the PAS reaction. Wash in 70% alcohol and treat the sections for 3 h at 37 °C with a 0.5% solution of α-amylase in 0.004 M acetate buffer, pH 5.5. Control sections must be treated with the buffer alone to ensure that any loss of PAS-positive material is due to the specific action of the amylase and not mere solution of the stainable matter. The sections are then washed in 70% alcohol and subjected to the PAS reaction.

All glycogen and starch will be removed by such treatment. In special investigations it may be found helpful to distinguish between straight-chain and more complex, branched glycogens. To do this, the sections are treated similarly but a 0.5% solution of β-amylase in 0.004 M acetate buffer at pH 5.0 is used instead of the α-amylase. The β-amylase digests only straight-chain glycogen.

STAGE 5: REACTIONS FOR LIPIDS (Figure 13)

The histochemistry of lipids is a complex and rather confusing field of study. This is not surprising seeing that the biochemistry of these materials is not completely clear: knowledge of these materials was barely rudimentary before the work of Folch and colleagues (e.g. Folch *et al.*, 1951), which showed the inadequacy of the older extraction procedures, and which introduced the concept of proteolipids as well as lipoproteins (Folch and Lees, 1951), the former being insoluble in water whereas the latter are more hydrophilic. It is now known that there are many materials in which fatty acid, or a lipid component (such as sphingomyelin), is linked to a water-soluble moiety, as in the inositol–lipids. Consequently caution must be exercised over the demonstration of 'lipids'. For all that, there is still a tendency for histologists and histochemists to consider that the term 'lipid' is synonymous with 'fat'—by which is meant fat droplets or globules. These may range from large, microscopically obvious droplets or globules (most strikingly seen by phase-contrast or polarized light microscopic examination of fresh cryostat sections) down to minute droplets dispersed throughout the protoplasm and not resolvable by the optical microscope. Such minute fat droplets have sometimes, previously, been referred to as 'masked lipids', but this term should be avoided for this type of fat, because it is not 'masked' but merely finely dispersed (as discussed by Berg, 1951; Chayen *et al.*, 1959). The term 'masked lipid' should be restricted to those lipids which are actually masked by being covered by (or chemically bound to) some other material, such as protein or carbohydrate. It is this form of lipid which plays an essential role in the structure of cells (and not only in the membranes), and which does not react with lipid colorants or reagents until the masking molecule has been disjoined from the lipid. It is still not generally appreciated that a lipid–protein complex may contain a high proportion of lipid, and yet be water soluble: that is, it may not appear to be 'fatty'. Consequently, workers have reported methods as being capricious when in fact it may not be the method (assuming that it is done correctly) which is variable, but the degree to which the lipid is bound to its masking substance. Changes in this binding induced by physiological variations, trauma, drugs or hormones can provide important and sensitive indications of altered function of the tissue (Chayen, 1968a).

The tests for lipids are of two types:

1. Physical methods, which depend on the readiness of lipophilic molecules to partition between their solvent and the lipid in the section. These are therefore tests of the degree of 'fattiness' of the material.
2. Chemical methods, like the acid haematein test.

Of these, the physical methods are generally the more important so that, to a large extent, 'lipid' is characterized histochemically by its ability to concentrate dyes such as Sudan black or benzpyrene from semi-aqueous solutions in which these compounds are not very soluble.

SUDAN BLACK B METHOD

Rationale

Sudan black B is a fat-soluble colorant which is soluble in absolute alcohol but insoluble in water. It is a strong colorant, and has a high affinity for fatty material, including phospholipids (see Baker, 1958, pp. 300–1). It is used as a saturated solution in 70% ethanol in which it is still soluble (it is very much less soluble in 50% alcohol) and the section is exposed to this solution. The colorant partitions between the alcohol and the tissue lipids (in which it is much more soluble). Because of its intense colour, and of its strong affinity for all (or most) classes of lipids, the coloration of tissue components by Sudan black B is often taken as the histochemical criterion for the presence of lipids. The purpose of the various 'burnt Sudan black' methods is to increase the sensitivity of tissue lipids, including some masked lipids but also steroids (which, being solid, will stain only at higher temperatures, i.e. closer to their melting point). The drawback of the procedure is the fact that the section is exposed to 70% alcohol which may itself affect the lipids.

Method

Fresh cryostat sections must be used because most, if not all, fixatives can modify the nature of lipid–protein and lipid–carbohydrate complexes.

1. Place the section horizontally in a staining tray and filter the Sudan black solution through a funnel onto the section. Leave for 30 min (replenishing if necessary).
2. Remove any deposit of the Sudan black from the section by rinsing briefly in 70% ethanol. Transfer to 50% alcohol to judge if the deposit has been adequately removed.
3. Wash in 50% alcohol.
4. Wash in water.
5. Mount in Farrants' medium.

Result

All available lipids (other than steroids) should be coloured grey to blue-black.

Solution required for this method

A saturated solution of Sudan black B in 70% ethanol. This can be kept for prolonged periods, provided the solution is filtered onto the section.

[*Note*: Some workers prefer to use a 0.7% solution of Sudan black B in propylene glycol and to differentiate in 85% propylene glycol. Gomori's (1952) suggestion that the dye should be dissolved in 60% triethylphosphate is interesting but it seems not to be much used.]

MODIFIED SUDAN BLACK METHODS

The 'burnt Sudan black' method recommended by Burdon (1947) for use with microorganisms sometimes can be remarkably helpful when applied to tissue sections, especially if the presence of steroids is suspected. The procedure is the same as for the ordinary technique except that, when the solution of Sudan black B has covered the section, the alcoholic solution is set alight. When the alcohol has finished burning, the section is covered with fresh solution and the process is repeated a few times. Care must be taken to remove the heavy deposit of solid Sudan black with 70% alcohol.

Berenbaum (1958) has suggested other ways by which tissue lipids may be unmasked and coloured with Sudan black B.

OIL RED O METHOD FOR COLOURING FATS

Rationale

Oil red O is a strong colorant of fats but it requires a high concentration of fatty matter for it to be drawn out of its alcoholic solution into the section. Triglycerides ('fats') are 'fattier' than are phospholipids, which contain polar groups (i.e. hydrophilic and so lipophobic parts of the molecule); moreover the former occur more generally in highly concentrated globules of hydrophobic molecules whereas the latter may be associated with other more hydrophilic molecules. Consequently oil red O is an excellent colorant of fats but very poor for phospholipids.

Method

Fresh cryostat sections must be used.

1. Rinse the section in 60% isopropanol.
2. Immerse in oil red O solution for 10 min.
3. Rinse briefly in 60% isopropanol.
4. Wash in water; then either mount in Farrants' medium or proceed to step 5.
5. If required, stain in Mayer's haemalum or in Harris's haematoxylin for ½–1 min; wash in water; blue in 5% sodium bicarbonate for a few minutes; wash well in distilled water and mount in Farrants' medium.

Result

Fats are stained red or yellow-red, depending on their concentration.

Solutions required for this method

Stock solution of oil red O

A saturated solution (approx. 0.5%) of oil red O or oil red 4B in absolute isopropanol. This solution can be kept indefinitely.

'Staining' solution

Stock solution of oil red O   60 ml
Distilled water   40 ml
Leave to stand for 10 min and then filter before use.

[*Note*: this solution is over-saturated (to make the oil red O more likely to dissolve out of this solution and into the fat). Consequently it must be filtered just before use (cf. the Sudan black method).]

NILE BLUE METHOD FOR LIPIDS

Rationale

This technique has been investigated fully by the Oxford school of cytologists (reference should be made to Baker, 1958, pp. 301–3). An aqueous solution of Nile blue A contains the blue cation of the dye, the red spontaneous oxidation product of it (namely Nile red) and the orange-red imino-base of Nile blue. The Nile red dissolves in all fatty lipid components and acts as a general lipid colorant. Both the cation of the Nile blue and the imino-base can react with acidic groups, and when so linked they colour blue. Consequently phospholipids and fatty acids will stain blue (because these components of the Nile blue solution will overlay and mask any effect of Nile red) but so may other non-lipid acidic groups; the blue colour is specific for fatty acids and phospholipids only if the region so coloured has been proved to contain lipid. On the other hand, neutral fats will be coloured red. The difficulty in the past has been that the amount of Nile red present in any given solution may vary considerably. This is overcome either by adding Nile red (which can be obtained commercially) to a solution of Nile blue A, or by using commercial mixtures which have been reinforced with the red dye.

Method

Cryostat sections must be used.

1. Warm all solutions (i.e. 1% aqueous Nile blue, 1% acetic acid and water for washing) to either 37 °C or 60 °C before beginning the reaction.

2. Immerse the sections in a 1% aqueous solution of Nile blue at 37 °C or 60 °C for 5 min.
3. Wash the sections quickly in the warmed water.
4. Immerse for 30 s in warmed 1% acetic acid.
5. Wash in warmed water and mount in Farrants' medium.

Compare this section with a second section which is stained and differentiated the same as the first and then restained with a 0.02% aqueous solution of Nile blue at 37 °C or 60 °C for 5 min. Then wash it quickly in warmed water; differentiate for 30 s in warmed 1% acetic acid; wash in warmed water and mount in Farrants' medium.

Solutions required for this method

1% Nile blue

Nile blue	1 g
Distilled water to make up to	100 ml

0.02% Nile blue

Nile blue	20 mg
Distilled water to make up to	100 ml

BENZPYRENE METHOD FOR LIPIDS

Rationale

As has been discussed (in relation to the use of Sudan black), the coloration of lipid matter by lipophilic dyes depends on:

1. How soluble the dye is in its solvent as against its solubility in the lipid (i.e. the partition coefficient of the dye in the lipid and the solvent). It would be ideal if the dye could be 'dissolved' (in sufficient concentration) in a solvent in which it was virtually insoluble because then it would be more readily attracted to the tissue lipid. Moreover, a good solvent for the dye (such as absolute alcohol) is likely to be a solvent for the tissue lipid as well.
2. The intensity of the dye itself: a strongly coloured dye will be detected even in relatively low concentrations in the section.

The benzpyrene method (Berg, 1951) seems to meet both requirements better than any other technique. Benzpyrene is an intensely fluorescent molecule so that it meets requirement 2. Furthermore, it is possible to detect far less of a fluorescent dye than of a visible dye so that the sensitivity of this method should far exceed that of the others discussed in this section. The other special advantage of benzpyrene is that it can be dispersed in a 'hydrotropic solution' in water if this dispersion is stabilized by caffeine. It therefore approximates well to requirement 1.

In our experience it is indeed a very sensitive method for all types of lipids (except solid steroids). It can also be used as a stain for living cells (e.g. Dumonde *et al.*, 1965).

Method

Fresh cryostat sections should be used, although the technique works on fixed sections, even if treated with dichromate (as in the acid haematein procedure).

1. Cover the section with the benzpyrene reagent for 20 min.
2. Rinse well.
3. Mount in water and examine with near ultraviolet light (e.g. 365 nm) with a fluorescence microscope.

For cell suspensions

Add 0.6 ml of the benzpyrene reagent to about 1 ml of the cell suspension. Leave for 20 min. Make a smear or hanging-drop preparation (preferably in fresh suspending medium) and inspect with the fluorescence microscope.

Result

Lipid fluoresces silver to blue-white or even yellow, depending on the concentration of the benzpyrene in the lipid.

Control

Use the caffeine solution without added benzpyrene.

Solutions required for this method

To prepare the benzpyrene reagent (Berg, 1951):

1. Saturate distilled water with caffeine at room temperature for 2 days (about 1.5% caffeine).
2. Filter.

3. Add about 2 mg 3,4-benzpyrene (also called benzopyrene) to 100 ml of this caffeine solution.
4. Incubate at about 35 °C for 2 days.
5. Filter off the excess benzpyrene.
6. Dilute the solution with an equal volume of distilled water (to prevent the caffeine from crystallizing out of the solution).
7. Leave to stand for 2 h.
8. Filter.

This solution is now ready for use, although it can be kept for several months in a dark bottle. It should contain about 0.75 mg benzpyrene in 100 ml of solution.

Rapid method for preparing the benzpyrene reagent

1. Prepare a 1% solution of benzpyrene in acetone.
2. Add this, drop by drop, to 100 ml of a saturated solution of caffeine which must be stirred briskly. When a persistent turbidity is obtained, enough benzpyrene has been added. (0.1–0.2 ml has been recommended, but we have found rather more to be required.)
3. Dilute with an equal volume of distilled water.
4. Filter twice.

The solution is then ready for use.

Notes

1. *Great care should be taken in handling the concentrated solutions of benzpyrene because this substance is* **carcinogenic**.
2. In practice we prefer the slower method of preparing this reagent.

OTHER FLUORESCENT METHODS FOR LIPIDS

Rhodamine B (1 : 1000 or even 1 : 10 000 in water, or in some Ringer solution) has been much used as a vital fluorescent dye (fluorochrome), particularly for plant cells (see Strugger, 1938). It is quite a useful colorant of lipids. The **phosphine 3R** method of Volk and Popper (1944) is another useful fluorescence method for lipids.

ACID HAEMATEIN METHOD FOR **PHOSPHOLIPIDS**

Rationale

This method is specific for phospholipids (Chayen, 1968a) and can also yield valuable information regarding the state of these molecules inside the tissue.

Sections or smears are treated with dichromate first at room temperature and then at 60 °C. (It is possible that the prior treatment, at room temperature, can be left out.) Particularly at 60 °C the chromate reacts with the unsaturated bonds of the fatty acids of phospholipids (but not generally of free fatty acids alone or in triglycerides: Chayen, 1968a). The chromium becomes very tightly bound to the phospholipid. The section is then exposed to acidified haematein to form a metal–dye complex wherever chromium is bound in the section or isolated cells. The sections, or cells, are then immersed in a solution of borax–ferricyanide, which dissolves free haematein out of the specimen; the haematein is less readily removed from the phospholipids because, with these, it has become bound to the chromium which is linked to the phospholipids. The proof that this retention of the haematein is caused by its linkage to unsaturated bonds is proved by the use of control sections (or preparations of cells) in which all these bonds are saturated by bromination before the acid haematein reaction. The control preparations should not respond to the acid haematein reaction.

Calcium is added to the dichromate solution to render the lipids less soluble than they might be in the dichromate solution alone.

Some workers have been worried that the acid haematein method can give 'variable' results. In many, if not in all, cases these variations may have been due to various degrees of unmasking of the phospholipids either in life, or as a result of the way the tissue was treated. For example, the hot dichromate can cause some degree of unmasking; this may become exaggerated where the lipid–protein bonding was already weak. Another cause for concern was the occasional finding that lipid solvents enhanced, rather than eliminated, this reaction. It is now known that such solvents, acting for relatively short periods, may be the most effective unmasking agents available, showing up phospholipids which were too tightly bound (e.g. to protein) to be demonstrable by the acid haematein procedure alone. Prolonged treatment with methanol–chloroform, for up to 24 h, should fully unmask such tightly bound phospholipids and remove them from the sections (or isolated cells).

Method

It is preferable to use fresh cryostat sections for this method; alternatively, the cryostat sections may be stored in a desiccator in a refrigerator. If fixed tissue must be used, the tissue should be fixed in a chromic acid or dichromate fixative.

1. Immerse in dichromate–calcium solution in a Coplin jar at room temperature for 18 h. [This step can be eliminated, i.e. immerse directly into this solution at 60 °C as in step 2.]
2. Transfer the Coplin jar to an incubator at 60 °C; leave for 24 h.
3. Wash well in distilled water; use several changes to remove free dichromate.
4. Immerse the section in acidified, freshly prepared haematein for 5 h at 37 °C.

5. Rinse in distilled water.
6. Place the section in the borax–ferricyanide solution and leave for 18 h at 37 °C in the dark.
7. Rinse in distilled water.
8. Either take the section through graded concentrations of alcohol to xylene and mount in Styrolite, or shake off the water, dry the section in air, rinse in *n*-butanol and mount in Euparal.

Result

Phospholipids stain various shades of blue and black. Some types of lipid may stain deep brown. This is not a spurious reaction, as has been shown by studies with purified phospholipid material. The section may appear yellowish, but this is a general non-specific coloration.

Solutions required for this method

Dichromate–calcium solution

Potassium dichromate	5 g
Calcium chloride (anhydrous)	1 g
Distilled water to make up to	100 ml

Acidified haematein solution

1. To 0.05 g haematoxylin (BDH 'SS Reagent' grade) add 48 ml distilled water and 1 ml of a 1% solution of sodium iodate.
2. Heat until the solution boils. [*Note*: this oxidizes and so 'ripens' the haematein, as can be seen by the deepening colour.]
3. Cool and then add 1 ml glacial acetic acid. This solution should be prepared immediately before it is to be used.

Borax–ferricyanide solution

Potassium ferricyanide	0.25 g
Sodium tetraborate.10H$_2$O	0.25 g
Distilled water to make up to	100 ml

This solution should be kept in a dark bottle.

To unmask the phospholipids

To unmask the phospholipids (and so disclose the total concentration of these substances in the section), treat the sections with 1 : 2 (v/v) methanol–chloroform for 1–10 h at room temperature. Generally, optimal unmasking of the phospholipids occurs after between 3 and 5 h; prolonged treatment (up to 24 h) extracts the lipids. Stain with the acid haematein method.

Bromination as a control for the acid haematein method

1. Unmask phospholipid by treatment for 3–5 h with methanol–chloroform (1 : 2, v/v).
2. Immerse in a saturated solution of bromine in water for up to 24 h.
3. Rinse and test with the acid haematein method.

To prepare a saturated solution of bromine, add 1 ml liquid bromine to about 50 ml distilled water in a Coplin jar. Stir well. It is advisable to keep the jar closed whenever possible.

METHODS FOR **STEROIDS**

As was discussed by Cain (1950), because steroids are solid and often crystalline, they cannot be coloured by lipid dyes unless the reaction is done at a temperature at which the steroids melt (see the burnt Sudan black method, above). Sometimes they can be detected, without treatment, by their birefringence as assessed by polarized light microscopy (see p. 29). It has been claimed that their presence can be demonstrated by their fluorescence, as viewed in a fluorescence microscope. Steroids may show primary fluorescence, but their fluorescence may be greatly enhanced, sometimes with a striking change of colour, when subjected to concentrated sulphuric acid. This can be tested on cryostat sections provided that they are well dried in a vacuum desiccator and are not removed from the desiccating environment until the section has been covered with the concentrated sulphuric acid (Chayen *et al.*, 1966b). Another useful method for steroids is the microchemical Liebermann–Burchardt reaction. In this (Gomori, 1952, Schultz's method; also see Cain, 1950, for fuller discussion), sections are treated with a 2% solution of ferric alum for 24 h. They should then be rinsed and dried. After they are dry, they are mounted in a freshly prepared mixture of equal volumes of glacial acetic acid and concentrated sulphuric acid. The section must be inspected immediately: colours change from purple to dark blue or blue-green. The last colour indicates the presence of unsaturated steroid, such as cholesterol and its esters.

A potentially valuable and more truly histochemical technique for steroids is that described by Adams (1961). Methods for ketosteroids have been described by Adams *et al.* (1967).

METHODS FOR **CAROTENOIDS**

Probably the most useful way of demonstrating these substances is by their yellow to green autofluorescence. Although this fluorescence has often been said to be only transitory, it has been found to be remarkably permanent for vitamin A dissolved in lipid, and for the carotenoids of the retina, studied in cryostat sections (Chayen *et al.*, 1966b).

STAGE 6: MINOR COMPONENTS

IRON (see Casselman, 1959)

Free (reactive) iron, as in haemosiderin, can be demonstrated by one of the various modifications of Perls' procedure. Fresh cryostat sections can be used. It is necessary to avoid acidic fixatives, which may allow the iron to diffuse; Casselman preferred neutral formalin rather than alcohol.

Perls' method demonstrates only ferric, not ferrous, iron but it seems generally to be assumed that ferrous ions do not occur in tissues. For all that, the Turnbull blue technique can be used to test for ferrous iron.

Methods

Ferric iron (Perls' procedure: Culling, 1963)

Ferrocyanide ions are said to diffuse only slowly. Acid, which diffuses more rapidly, is added to ionize the iron that may be present in an insoluble form.
   Consequently:

1. Immerse the section in a freshly prepared, filtered 2% solution of potassium ferrocyanide for 5 min (to allow full entry of this reagent).
2. Transfer to the acidified solution of potassium ferrocyanide (equal volumes of the 2% potassium ferrocyanide solution and 0.2 M hydrochloric acid). Leave for 20 min.
3. Wash in distilled water. (Nuclei can then be stained with neutral red.)
4. Dehydrate and mount in Euparal.

Result

Reactive iron stains blue.

Ferrous iron (Turnbull blue method: Culling, 1963)

1. Immerse sections in water.
2. Immerse for 15 min in freshly prepared reagent of 1 : 1 20% potassium ferricyanide and 1% hydrochloric acid.
3. Wash, dehydrate, and mount in Euparal.

Masked iron (as in haemoglobin)

Treat the section for at least 30 min with 100 volume hydrogen peroxide to which has been added a few drops of a solution of sodium carbonate to render the peroxide slightly alkaline. Then stain for reactive iron as above.

CALCIUM

There are many problems in demonstrating calcium adequately. The use of alizarin compounds has been carefully considered by Puchter *et al.* (1969), who recommended the use of alizarin red S at pH 9. In contrast, Schäfer (1979) found this procedure to be inadequate. McGee-Russell (1955; 1958) recommended the use of nuclear fast red. Casselman (1959) discussed the various procedures then available, including a method for calcium oxalate; however, crystals of oxalates can best be demonstrated by polarized light microscopy (see p. 29).

Glyoxal-bis-(2-hydroxyanil) method for calcium

From the very detailed studies of Kashiwa and Atkinson (1963; Kashiwa and Marshall House, 1964) and of Schäfer (1979), it seems that the best currently available reagent for demonstrating calcium is glyoxal-bis-(2-hydroxyanil) or GBHA. The method of Kashiwa and Atkinson (1963), followed by Schäfer (1979) is presented below. Schäfer (1979) showed that it was semiquantitative when applied to cryostat sections.

Because of the very ready diffusion of calcium ions, it has been considered necessary to cut cryostat sections and float them out directly on to the reaction medium. However, if the flash-dry procedure described in Chapter 1 for picking the sections off the knife is followed, the section can be adhered to the slide and handled in the normal way.

For various modifications of the method, for exposing 'insoluble calcium' or for improving the specificity of the method, reference should be made to the publications of Kashiwa (cited above).

Method

Use fresh, unfixed cryostat sections.

1. Cut the section (e.g. at 20 μm) and cover with the GBHA solution in the cryostat (at about $-25\,^{\circ}$C). Leave for 2 min.
2. Wash, at room temperature, with 75% alcohol to remove excess stain.
3. Allow the section to dry at room temperature.
4. Immerse for 1 min in each of two baths of 96% ethanol and of absolute ethanol; clear in xylene.

5. Mount in a neutral medium such as Clearmount (E. Gurr); the stain fades
   in acidic mountants.

Result

Red coloration (generally red granulation).

Solution required for this method

0.1 g glyoxal-bis-(2-hydroxyanil) in 2 ml 3.4% NaOH in 75% ethanol. This
should be prepared just before it is to be used, and kept at about $-25\,°C$ (e.g.
in the cryostat).

The sodium hydroxide is required to ionize the GBHA, to enhance its
solubility.

Alternative procedure

1. Lay a slide, at a slight angle to the horizontal, in the cryostat. Add a small
   volume of GBHA solution (as above).
2. Quickly cut a 20 μm section and place it, by means of needles, onto the drop
   of GBHA solution in the cryostat at a temperature of less than $-20\,°C$. Leave
   for 2 min.
3. Remove the slide from the cryostat, taking care to keep the section on the
   slide.
4. Use blotting paper to remove excess reagent.
5. Wash with 75% ethanol.
6. Leave the section to dry onto the slide. Then dehydrate with two baths of
   96% ethanol, followed by two of absolute ethanol. Clear in xylene. Mount
   in Clearmount or other neutral mountant.

PIGMENTS

Many histochemical pathologists have paid considerable attention to tissue
pigments. The chemistry of these substances seems less important, and less
known, than is their histological terminology, and reference should be made
to Gomori (1952) or Barka and Anderson (1963).

REDUCING SUBSTANCES

Phenolic substances, especially polyphenols, are of some importance in
histochemistry since they include adrenaline and enterochromaffin substance
(Gomori, 1952). All phenols seem to react with ferric chloride to yield pale
colours which may be characteristic of the phenol. Diphenols are oxidized by

dichromate to form coloured quinones (the 'chromaffin' reaction): since dichromate is a mild fixative no other fixation is required so that the reaction and fixation should occur simultaneously in cryostat sections. However it may be advisable to try the reactions in full concentrations of either polyvinyl alcohol or Polypep 5115 which should retard the loss of these substances during the reaction.

The presence of many reducing substances can be observed in tissues by their reduction of a 1% aqueous solution of osmium tetroxide or of various solutions of silver nitrate. Ascorbic acid is so strong a reducing agent that it reduces acidified silver nitrate; others can act only on alkaline or ammoniacal solutions of silver nitrate (e.g. the 'argentaffin reaction' in which polyphenolic substances reduce ammoniacal silver nitrate to metallic silver).

Methods for ascorbate

The silver nitrate method on whole tissue (Chayen, 1953)

Tissue is fixed in 1 : 3 acetic ethanol which was shown to retain ascorbate within tissue. Fixation may be as long as 24 h.

It is then exposed to a 10% solution of silver nitrate in 10% acetic acid for at least 4 h. It is not washed but gently dehydrated, embedded and sectioned. After the wax has been removed with xylene, the sections are mounted in Styrolite. The presence of ascorbate is shown by the silver grains deposited within the sections.

The silver nitrate method applied to sections (Chayen, 1953)

The tissue should be fixed in either absolute ethanol or acetic ethanol (1 : 3). It is embedded and sectioned. The sections, still embedded in paraffin wax, are placed horizontally in the dark in a bath of 10% silver nitrate dissolved in 10% acetic acid. They should be left in this bath for between 4 and 7 h. The sections are dried (e.g. with filter paper); they are rinsed briefly in absolute alcohol; the wax is removed with xylene; and they are mounted (e.g. in Styrolite). The presence of ascorbate is denoted by fine silver grains. However, after the sections have been mounted for a few days, the staining becomes more diffuse and red.

Alternatively, cryostat sections can be used.

The ferric chloride : ferricyanide method (Loveridge et al., 1975)

Both glutathione and ascorbate will reduce ferricyanide to ferrocyanide to produce an intense dye. The avidity with which these react depends on the redox balance of the mixture of the two reagents. Thus it was shown that, at a ratio of 50 : 1 (ferric ions to ferricyanide ions), ascorbate reacted immediately, with little response from glutathione. These must be present in very high concentrations to trap the soluble ascorbate.

Cryostat sections are immersed in a solution containing 1.2 M ferric ions : 0.023 M ferricyanide ions. The reaction should be complete in 15 s (30 s may be used). The sections can then be rinsed briefly and mounted in an aqueous mountant (e.g. Z5 mountant). Alternatively they may be dehydrated and mounted in Styrolite because the dye is very insoluble.

Although both cysteine and glutathione will respond to this mixture of ferric ions and ferricyanide ions, their contribution to the amount of dye that is precipitated will be small. To demonstrate glutathione (with only a small contribution from ascorbate), the ratio of ferric ions to ferricyanide ions should be 6 : 4.5.

AUTOFLUORESCENT SUBSTANCES

Certain molecules, when they are irradiated with light of short wavelength, have the ability to absorb this light energy (the excitation light) and to transmit some of the absorbed energy as light of longer wavelength (transmitted light). This is what is meant by fluorescence. This should be of considerable value in histochemistry because a number of physiologically and pharmacologically significant molecules are capable of being rendered fluorescent (i.e. they show 'autofluorescence'); many of these molecules cannot be identified adequately by any other histochemical method. In view of this, it seems surprising that so little use has been made of primary fluorescence microscopy (namely the microscopic study of autofluorescent molecules; this is contrasted with secondary fluorescence, in which the fluorescence is induced by the use of fluorescent dyes or fluorochromes, such as benzpyrene). To a large extent the little reliance that has been placed on fluorescence microscopy has been caused by the fact that many factors, including certain fixatives, fat-solvents and heat, cause tissue to become intensely and apparently non-specifically autofluorescent. Some of this 'non-specific' autofluorescence may be caused by lipid peroxidation induced by the splitting of lipid–protein complexes by these agents. Other workers have been concerned that autofluorescence can be 'quenched' (i.e. the molecule ceases to emit light) by a wide range of agents, such as metals (used in some fixatives) and oxidative or reductive media in which the cells may be mounted.

Most of these objections are eliminated by the use of unfixed cryostat sections which can be mounted either dry in air, or in liquid paraffin (which can have a high concentration of oxygen), or in any medium which is especially suitable for the particular autofluorescent molecule to be studied. Some substances which can be identified by fluorescence microscopy are listed in Table 5. The application of this form of microscopy has been reviewed by Chayen and Denby (1968).

Table 5.

Substance	Fluorescence
Riboflavin	Yellow—in 1 M acetic acid it fluoresces intensely yellow-green when irradiated with ultraviolet light
Porphyrins	Red
Thiamine	White
Pteridines (including folic acid)	Generally blue-white, but other colours have been reported
Ceroid and lipofuscins	Yellow or brown

INDUCED FLUORESCENCE OF CATECHOLAMINES AND RELATED SUBSTANCES

Over the past few years much use has been made of formalin-induced fluorescence for detecting catecholamines and related compounds in cells (e.g. Eränkö, 1967). This appears to be a growing interest in certain specialized branches of histochemistry and reference should be made to the original papers by those wishing to utilize these very sensitive techniques. In general monoamines, kept at their original sites by freeze-drying or in cryostat sections (Hamberger and Norberg, 1964; Eränkö and Härkönen, 1965), are treated with slightly humid gaseous formaldehyde (from paraformaldehyde) at 80 °C for 1 h to convert them into a condensation product (tetrahydroisoquinoline derivative). This substance undergoes dehydrogenation, through the mediation of dry protein, to form a highly fluorescent dihydroisoquinoline derivative (Corrodi and Jonsson, 1967).

dihydroisoquinoline
very fluorescent

condensation
(can occur readily
in solution)

dehydrogenation
(occurs only in
relatively dry
films)

Corrodi and Jonsson showed that as little as 0.001% w/v of primary catecholamines or 5-hydroxytryptamine can be demonstrated in a 10 μm section. In investigations of an adrenergic nerve cell body of the superior cervical ganglion in the cat it has been calculated that the fluorescence method detected $0.4 \times 10^{-6} \mu g$ noradrenaline, and similar results have been obtained in other cells with dopamine. Particularly if quantitative microspectrofluorimetry is used, it is possible to distinguish between catecholamines and 5-hydroxytryptamine, the emission spectrum of the fluorescence of the former showing a maximum at 480 nm and of the latter at 525 nm. Spectral changes can also be induced by exposure of the fluorescent compounds to dry hydrogen chloride gas. Björklund et al. (1968) used such treatment and showed that the fluorescence excitation spectra of dopamine and noradrenaline became so different that the two substances could be clearly differentiated. Incorporating minute amounts of the dry hydrogen chloride gas into the formaldehyde vapour appears to increase the amount of fluorescence which is inducible, and so makes it possible to detect compounds whose fluorescence is too weak when exposed to formaldehyde alone (Björklund and Stenevi, 1970). For example, the fluorescence of tryptamine was 20–200 times greater with acid treatment than with the standard formaldehyde method. They suggested that the acid catalysed the condensation step.

The specificity of the induced fluorescence can be tested by treating the sections with low concentrations of sodium borohydride ($NaBH_4$) in alcoholic solution. This instantly reduces the fluorescent compounds to non-fluorescent molecules which become fluorescent again after renewed treatment with the formaldehyde gas.

There is doubt whether histamine can be demonstrated by these methods (Corrodi and Jonsson, 1967). However, it can be demonstrated by its reaction with o-phthalaldehyde, either in an anhydrous solvent or in gaseous form, to form a highly fluorescent compound (Shelley et al., 1968). The reagent can also be used as a supravital dye (Juhlin, 1967).

AUTORADIOGRAPHY*

Autoradiography allows the localization of radioactively labelled material in a specimen by means of the photographic action of the emitted ionizing particles. In all techniques the specimen is brought into contact with a photographic emulsion and stored in the dark for a certain time, the exposure time. The film is then developed and fixed and can be examined macroscopically or through a microscope. With low or no magnification, usually called macroautoradiography, the presence of radioactive material is indicated by a blackening of the photographic emulsion; at higher power the blackening is seen to be produced by a larger or smaller number of black particles, the photographic grains. Under the electron microscope, the grains are seen to be filamentous.

The introduction of labelled material into an organism is usually in soluble form and the aim of an experiment can be to localize it in its original state (e.g. Na^+ ions) or after utilization as a precursor (e.g. labelled amino acids in protein synthesis or [H^3] thymidine in DNA synthesis). Many soluble compounds are lost or moved through diffusion during fixation, washing, dehydrating and embedding. Therefore autoradiography of a fixed and sectioned specimen can only be attempted when insoluble labelled substances are to be scored and the loss of labelled precursor is regarded as an advantage. When soluble material is to be retained, frozen sections are used. The techniques serving various combinations are shown in Table 6.

Certain practical points should be noted:

---

*We are grateful to the late Dr S. R. Pelc for his help with this section.

Table 6.

	Autoradiographic methods for:		
	Fixed compounds	Soluble compounds	Resolving power
Macro	Apposition	Ullberg (1962)	15–50 μm
Light microscopy	Stripping film method (Pelc, 1956)	Appleton (1964)	1–2 μm for ^{3}H
	Dipping method (Joftes and Warren, 1955)	Stumpf and Roth (1966)	2–5 μm for other isotopes
Electron microscopy	see Salpeter and Bachmann (1964) Budd and Pelc (1964) Caro (1962) Pelc et al. (1961)	No technique available	0.1–0.3 μm

Full details of these techniques will be found in Rogers (1967) and Gahan (1972).

1. The particles emitted by the isotope have different energies and therefore different ranges. The worst possible resolving power is determined by the range, which is of importance when ^{3}H is used whose β-particles are so weak, i.e. their range is so small, that resolving powers of 1 μm are easily achieved. The accompanying disadvantage is self-absorption within the specimen such that only β-particles emitted within a short distance from the emulsion are able to reach it (Figure 16).

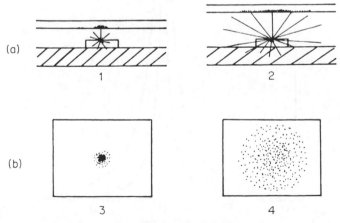

Figure 16. Diagrammatic representation of autoradiography (a) and (b) of the image of a point source of radiation in the specimen (x) as seen on the photographic film when there is little gap (1 and 3) or a large gap (2 and 4) between the emulsion and the specimen. The point source of radiation will be fairly well localized if the photographic emulsion is closely applied and poorly localized if the film and section are widely separated (also see Figure 17). The strength of the radiation will also affect the localization because stronger radiation penetrates further than does weaker radiation.

2. The geometry of emulsion and source of radiation are important for emitters of high-energy β-particles. As a general rule it can be assumed that the resolving power is one-half of the combined thickness of specimen and photographic emulsion if there is no gap between them. For example, a 3 μm section with a 1 μm emulsion gives 2 μm resolving power. The effect on resolution of the gap between the emulsion and the specimen is shown in Figure 17.

3. The size of the photographic grains limits the resolving power since obviously no detail smaller than the grains can be resolved. The usual grain sizes are between 0.1 and 0.25 μm and do not materially affect the resolving power for light microscopy but become important for electron microscopy work.

Autoradiographic methods and uses

The various methods of autoradiography are now so valuable and so widely used that a wide range of different techniques is available.

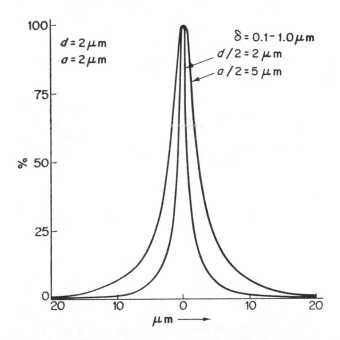

Figure 17. The calculated relative distribution of grains produced by a radioactive source at 0 in a section of thickness (*d*) of 2 μm and with an emulsion of thickness (*a*) of 2 μm. The size of the gap between emulsion and section (δ) is 0.1 μm in the inner curve and 1.0 μm in the outer curve. The resolution is defined as the width of the distribution curve at the half-way position (*d*/2). (From Doniach and Pelc, 1950. We are grateful to the *British Journal of Radiology* for permission to use this figure.)

For various methods, and some of the uses of autoradiography, reference should be made to Pelc (1958), Taylor (1956), Lima-de-Faria (1961), Fitzgerald (1959), Rogers (1967) and Gahan (1972). The basic stripping-film technique is as follows:

1. *Sections*. For most studies, which have been concerned with substances which can be fixed in cells, such as DNA or protein, fixed tissues have been embedded in paraffin and sectioned normally, but smears or squashes can be used. Certain metals, used in some fixatives, themselves react with photographic emulsions; consequently 1 : 3 acetic alcohol, or formaldehyde, is usually the recommended fixative. Cryostat sections of unfixed tissue can be used, but should be fixed in acetic alcohol or exposed at $-25\,^{\circ}$C.

   [*Note*: for water-soluble material it is essential to use unfixed cryostat sections.]
2. The *slides* on which the sections are to be picked up from the knife must be washed well in hot water (but not treated with acid) before being coated (or 'subbed') by being dipped briefly in a solution which contains 0.5 g gelatin and 0.05 g chrome alum in 100 ml distilled water. The slide must be dried for 2 days before it is used. Prolonged storage should be at $+4\,^{\circ}$C in a refrigerator.

   [*Note*: this coating has two advantages: it helps to hold the section onto the slide during all the subsequent processing and it stops the photographic emulsion from slipping once it is in position.]
3. If paraffin sections are used, they are caused to adhere to the coated slide and are then treated with xylol and taken through graded concentrations of alcohol to water.
4. The stripping film is prepared and handled in a dark room, with as little light as possible from a Wratten Series No. 1 red safelight. First, the edges of the photographic stripping film plate (AR 10 or AR 50) are trimmed by tracing a cut (with a sharp scalpel or razor blade) all around the plate and about ½ inch in from the edge. Other cuts are made as shown in Figure 18 to produce eight segments of film. Note that stripping film is a double layer of pure gelatine (nearest the glass) and emulsion. The emulsion must be nearest the specimen.

   With the scalpel, pick up the edge of each segment of film and very slowly strip that segment off the plate. Hold it with as little contact as possible and float it, face (emulsion) downward, onto the surface of water in a large bowl. The temperature of the water should be maintained at 21–23 °C to permit the maximum expansion of the film (2½ min). Six to eight pieces of film can be floated out together.
5. Place the slide, section uppermost, under the water and slowly raise it, under the film. Ideally the slide should be held slightly oblique to the film.

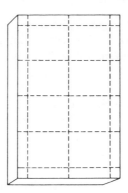

Figure 18. Diagrammatic view of a stripping film on its plate backing. Cuts (broken lines) are made in the photographic film to produce eight segments of film.

As the slide comes out of the water, it should be tilted so as to wrap the film closely about it (Figure 19). Care must be taken to ensure a tight fit of the film to the slide and to avoid pockets of air or of water above the section.

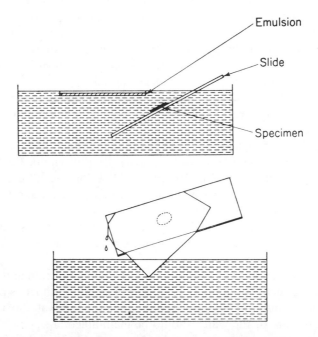

Figure 19. The photographic film is floated on the surface of the water, emulsion side downwards. When it has expanded, a section mounted on a slide is immersed obliquely into the water, beneath the film. The slide is lifted out of the water and is tilted to wrap the film around the slide.

6. Hang the slide (wrapped in film), pegged at one end, on a rack to drain and to dry.
7. When dry, store in the dark in a light-tight dry container at 4 °C or, for lowest background, − 25 °C. The film is now exposed to the radiation. The time of exposure can vary greatly, depending on the radioactive matter used and the concentration it has achieved in the tissue; it may be some days or some weeks.
8. The film is developed on the section, all the processing being done at 17–18 °C to avoid damaging the film. It is essential to use freshly filtered D 19 b developer; Johnson's 'Fix-Sol' (1 : 9) is satisfactory for photographic fixation.
9. The section can then be stained with haematoxylin and eosin, or with methyl green–pyronin. These processes too should be done at 17–18 °C.

Dipping film method

For many purposes, and for convenience, the use of the dipping film procedure may be recommended. As for the stripping film method, the slides should have been coated with chrome–gelatin (0.5% gelatin and 0.05% chromium potassium sulphate in distilled water) and dried overnight; the sections should be taken up onto these slides.

If cryostat sections are used, these may be fixed (e.g. in 1 : 3 acetic ethanol for 20 min), washed, and left to dry at room temperature. Overnight drying is often convenient.

Preparation of the film

Heat a water-bath to 46 °C. Stand a tall Coplin jar, with about 40 ml distilled water, in the bath to achieve 46 °C. Then, in the dark (with a suitable safelight), scrape about 15–20 ml of the Ilford K5 Nuclear Emulsion into the Coplin jar, stirring well to produce a fine, milky white suspension. Leave for a few minutes and then, with a Pasteur pipette, remove the bubbles that form on the surface.

Dip the slides, one at a time, by lowering smoothly, over a period of 3 s, into the emulsion; leave for 3 s, and then smoothly remove slowly taking another 3 s to do this. Clip the slide to a clothes peg held on a rail so that the emulsion drains off the slide. Leave overnight in total darkness to complete the drying. The covered sections are then placed in black, light-tight boxes and stored at either 4 °C or, for least background, − 25 °C. They are developed as for the stripping film procedure (above).

Result The photographic emulsion acts as a coverslip for the section but contains blackened regions or grains, caused by the radioactivity in the tissue section below it. Hence, at one focus, the tissue histology can be inspected, while at a slightly higher focus you should be able to observe how much radioactivity is contained in each tissue or cellular component. Quantitatively the number of photographic grains in the emulsion bears a definite (but not always exactly known) relationship to the number of radioactive atoms in the tissue below that point in the emulsion (for quantitation see Pelc, 1958; also Levi and Rogers, 1963). Care must be taken to assess the background count because some photographic grains may be found over those parts of the slide which lack tissue. This background depends on many factors, including the skill of the operator at many of the steps of the procedure; it may also be related to the glass used for making the microscope slide.

D 19 b developer (use Analar Reagents where possible)

Elon, 2.2 g; anhydrous sodium sulphite, 72 g; hydroquinone, 8 g; anhydrous sodium carbonate, 48 g; potassium bromide, 4 g; distilled water, 1000 ml.
    This should be filtered before use.

6

Enzyme Histochemistry

GENERAL INTRODUCTION

In studying the histochemical activity of an enzyme we have to base our studies on the properties of this enzyme that have been disclosed in detail by biochemical enzymology. But we also have to realize that the properties of an enzyme in its natural environment, perhaps attached to a solid matrix or bounded by a membrane (as in mitochondria or lysosomes) may be very different from that of the 'purified' enzyme, studied by conventional enzymology (e.g. Siekevitz, 1962). A detailed study on this subject is that of Quarles and Dawson (1969) who showed that the optimal pH of a phospholipase can be 2 pH units different when it is removed from its natural environment. Quantitative enzyme histochemistry can therefore play an important role in cellular enzymology. If done correctly, it can give information regarding the natural activity of a given enzyme acting within its normal environment in a particular cell; but, initially at least, it is necessary to try to reconcile the results obtained by histochemistry with those obtained by conventional biochemistry.

These considerations have practical implications. Conventional biochemistry extracts and purifies an enzyme and records the amount of this enzymatic activity per unit mass of tissue. However, it is becoming increasingly appreciated that, in life, the activity of many enzymes is subject to many diverse factors; purifying the enzyme and isolating it from these factors can give a totally misleading idea of the actual activity of the enzyme in the live tissue. This is one of the strengths of quantitative enzyme histochemistry, namely that it demonstrates how active the enzyme is when acting in its natural environment.

Another strength of quantitative enzyme histochemistry is that it can be done on very small samples. This is especially important when only small amounts of tissue are available, as in needle or punch biopsies of human tissue. For example, small biopsies taken from the ventricle during open-heart surgery could be used for assaying enzyme activities, and structural chemical changes, during the course of open-heart surgery. These have contributed to improvements in such surgery, which have decreased the mortality rate from 12 to 3%

(Braimbridge *et al.*, 1982). In one series of studies, such sequential biopsies indicated that there was some damaging agent acting during the course of surgery; moving the bright light, which had caused drying, stopped the deterioration of the cardiac tissue during the operation.

It should be noted that the application of rigorous quantitative enzyme histochemistry has been shown to yield results, on small biopsies or on samples of 20 cells, which were quantitatively equivalent to those that were obtained by conventional biochemical procedures (Chayen, 1978b; Olsen *et al.*, 1981, on isolated cells).

Possibly the clearest validation of the value of quantitative histochemistry (or quantitative cytochemistry) has been the development of the cytochemical bioassay of polypeptide hormones (Chayen, 1978a). These bioassays are now used fairly widely. The basic concept is that, when a polypeptide hormone acts on its target cells in an organized tissue, it induces a chain of biochemical events within the specific target cells. The great sensitivity of quantitative histochemistry is that it can measure biochemical changes induced specifically within the target cells. Changes induced by as little as 50 fg/ml (50×10^{-15} g/ml) of the hormone can be quantified and used for the bioassay of such hormones (Chayen, 1980). These bioassays combine (1) the great sensitivity of quantitative histochemistry; (2) the lack of distortion of the active moiety within the target cells so that small changes in such activity can be detected without the distortion induced by extraction and isolation of the active moiety from its natural environment; and (3) the ability of quantitative histochemistry to measure the activity of individual target cells within a mass of non-responsive cells that do not respond to the hormone. For example, parathyroid hormone acts on the renal distal convoluted tubules. The cells of these tubules may constitute only about 5% of the cells of the cortex of the kidney. Metabolic events influenced by this hormone, such as a rise in glucose 6-phosphate dehydrogenase activity, will not readily be measurable if the activity in the whole cortex is measured; they can be demonstrated quantitatively when the activity is measured by microdensitometry of the histochemical glucose 6-phosphate dehydrogenase reaction which allows measurement of the activity in histologically specified cells. This is the basis of the highly sensitive cytochemical bioassay of parathyroid hormone (as in Chayen and Bitensky, 1983).

The application of quantitative histochemistry to the measurement of a number of polypeptide hormones has been reviewed by various workers in Chayen and Bitensky (1983); its application in diverse fields has been reviewed in a book edited by Pattison *et al.* (1979).

Studies on 'soluble' enzymes

Many enzymes are 'soluble' enzymes: that is, they are not appreciably bound within cells so that, in biochemical studies, they occur in the supernatant fraction.

Their histochemical demonstration has cast opprobrium on histochemistry, because the enzyme that the histochemical procedures purport to demonstrate has been lost from the section within a few minutes of placing the section in the reaction medium. A typical example concerns the histochemical demonstration of glucose 6-phosphate dehydrogenase (G6PD) (Altman and Chayen, 1965; 1966): the dehydrogenase leaves the section within 1 or 2 min of the immersion of the section into the reaction medium. It reacts with substrate and coenzyme (NADP) in the reaction medium to generate NADPH, which then acts as the substrate for the tissue-bound diaphorase. Consequently the reaction localizes the NADPH-diaphorase, not the primary dehydrogenase; the amount of reaction product is limited by the activity of the diaphorase (Figure 7, p. 12).

It is now established that 'soluble' enzymes can be retained within the section if the reaction is done in the presence of a sufficiently high concentration of a colloid stabilizer (such as a suitable grade of polyvinyl alcohol or a commercially available partially degraded collagen such as Polypep 5115 from Sigma). Under these conditions there is no inhibitory effect of the hydrogen acceptor (e.g. neotetrazolium), as has bedevilled reactions done on non-stabilized sections (Figure 20).

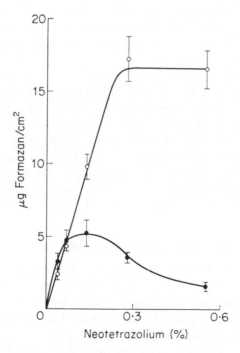

Figure 20. Activity of glucose 6-phosphate dehydrogenase (rat liver) with various concentrations of neotetrazolium: ●, in the absence of PVA; ○, in the presence of 20% PVA.

As discussed in detail on pages 13–14, a simple test (a transfer test) can be done to examine whether an enzyme is retained within the section. One section (or a set of sections) is reacted for sufficient time to yield a fairly strong reaction product. Suppose this requires 20 min. Another is reacted, in the presence of the colloid stabilizer, for 10 min; it is then removed from the reaction medium and transferred to fresh reaction medium, containing the stabilizer, for 10 min. If the enzyme has been solubilized, it will be left behind in the first reaction medium; consequently there will be virtually no increase in activity as a result of immersing the section in the second bath for 10 min.

The importance of the subcellular environment

In most biochemical enzymological studies the enzyme is first made fully available to its substrate. In contrast, one of the important potential contributions that can be made by histochemistry is in showing how available the enzyme is when in its natural environment, rather than how much of the enzymatic activity can be disclosed by 'suitable' (or 'unsuitable', physiologically) disruptive techniques. This opens the way to understanding the control of cellular metabolism, in that the enzymes that are present may not be working at full capacity all the time. The factors that control the degree of activity may be many and various and may be influenced by pharmacologically active factors; without studies on intact tissue, as can be done histochemically on sections or on isolated cells, they cannot be investigated.

Consequently, in histochemistry, it becomes pertinent to decide what is meant by 'a good reaction'. It could be that the 'best' reaction (in the sense of the reaction that is most informative of the state of the enzyme in life) might be the least stained preparation. For this reason it may be advisable to attempt to obtain two results for each enzyme reaction: the first, which shows how active the enzyme has been in life (the manifest activity) and the second, after 'suitable' disruption of the subcellular organization, which shows the total potential activity of the enzyme (the potential activity); the difference between the two will indicate the latent, or reserve, activity held by the cell for emergencies.

Characteristics of the histochemical reactions for enzymatic activity

Histochemistry differs from biochemistry in that it attempts to precipitate the reaction product at the site at which it is produced. Consequently the reaction product must be generated in sufficient concentration as to be insoluble in the aqueous environment of the incubation medium. This question of precipitation of the reaction product requires some explanation.

When you have a saturated solution of a sparingly soluble, or even of a virtually insoluble, salt such as calcium phosphate, some will be present in solution and, since it is a salt, what little of it that is dissolved will be completely ionized. Consequently an equilibrium is set up, as described by the following equation:

$$Ca_3(PO_4)_2 \rightleftharpoons Ca_3(PO_4)_2 \rightleftharpoons 3Ca^{2+} + 2PO_4^{3-}$$
$$\text{solid} \qquad \text{dissolved} \qquad \text{ionized}$$

This equilibrium is defined as follows:

$$K' = \frac{[Ca^{2+}]^3 \, [PO_4^{3-}]^2}{[Ca_3(PO_4)_2]}$$

where the square brackets indicate 'the concentration of' the material within the brackets, and the 'power to' is related to the relative concentrations of the ions within the square brackets. Thus, in the equation, for one molecule of calcium phosphate there will be three calcium ions (hence $[Ca^{2+}]^3$) and two phosphate ions.

Consider the alkaline phosphatase reaction, in which β-glycerophosphate is cleaved enzymatically to liberate phosphate ions. In order that K' be maintained constant, there must be a concomitant increase in the calcium phosphate, and this will proceed until the concentration of calcium phosphate exceeds its solubility constant, causing its precipitation.

Now consider the histochemical precipitation reactions. We have an enzyme that cleaves a substrate, AB, and we wish to precipitate B. [*Note*: B can be the phosphate derived from glycerophosphate, or the hydrogen from a dehydrogenase reaction, or a naphthol derivative in the many coloured or azo-dye reactions to be described later.] We can precipitate B by coupling it (or trapping it) to a 'coupler' C to yield the insoluble complex BC. A number of factors have to be considered in analysing this process:

(i) The solubility product of BC and the enzyme activity

As before, the solubility product K' is expressed by:

$$K' = \frac{[B] [C]}{[BC]}$$

In order that K' should be exceeded rapidly, so that BC should be precipitated immediately *in situ*, the concentration of B and C should be as high as possible. The concentration of B will depend on the activity of the enzyme, but the concentration of the coupler or trapping agent (C) is under our own control in that we put a determined quantity of it into the incubation medium. Consequently, it will be obvious that we require as high a concentration of C as is feasible. The only problem is whether it is stable at the temperature and

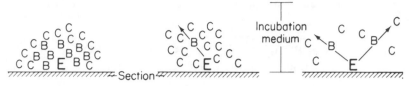

1. Active enzyme E liberates many molecules of B from AB and so exceeds the solubility product of BC. The high concentration of C in the incubation medium ensures that each molecule of B liberated is trapped by a molecule of C very close to the site of the enzyme.

2. This chemically fixed enzyme, E, shows only 10% of its unfixed activity. Hence in the same unit time, only one molecule of B is produced. Even though the concentration of C remains high, the solubility product of BC is not exceeded, and the molecule of B diffuses away before the next one is split off from AB by the enzyme.

3. In this case, the activity of E is weak but the concentration of C has been reduced by using an inadequate incubation medium. Consequently B does not encounter a molecule of C during its diffusion away from the site of its liberation (E).

Figure 21.

pH of the reaction. If it is relatively unstable, a 'pulsing' procedure, as is needed for the naphthylamidase reaction, should be used to maintain a high concentration of the active form of the trapping agent.

A less controllable factor is the concentration of B, since this is determined by the activity of the enzyme in the cells. If the tissue has been chemically fixed, the rate at which B is formed from AB can be greatly reduced (see Appendix 1). It is conceivable that, at any given moment, the concentration of B in a chemically fixed section or smear might never be sufficiently high, at the site of a weakly active enzyme, to cause the immediate precipitation of BC. Consequently B would diffuse away from its site of origin. It may finally become adsorbed onto or dissolved into some compatible material, as happens when lipophilic reaction products diffuse to fatty sites.

(ii) Diffusion gradients and 'substantivity'

When B is enzymatically liberated from AB, two antagonistic effects come into operation. The first is the tendency of B to become precipitated at the site of the enzyme. This is encouraged in histochemistry by having a high concentration of the trapping agent C in the reaction medium. The second is the tendency of B to diffuse away from the site at which it is liberated. This is a general phenomenon in physical chemistry: each molecule or ion tries to fill the whole available space; the rate at which it does this depends on the difference in the concentration of that molecule or ion at each point in that space. This can be understood more clearly by reference to simple examples. Suppose we have an enzyme acting on our substance AB in the absence of a trapping agent:

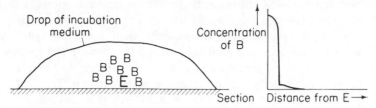

We begin with a high local concentration of B around the enzyme E, and some molecules of B will therefore begin to diffuse away from E to fill the incubation medium. Once this process has begun, a gradient of concentration of B will be established in the medium and this gradient will reduce the rate at which other molecules of B can move away from E.

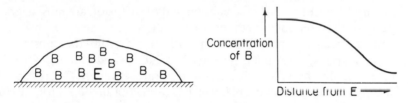

But suppose that, in the section and some distance from E, there is a site whose chemical composition or physical nature can bind B. Thus as each molecule of B reaches this site (S) it will be adsorbed on to it. In becoming adsorbed, B is no longer in solution at that point so that the gradient becomes broken at that point, i.e. the concentration of B *in solution* becomes zero. This will provide the equivalent of a 'vacuum' and so it will actively draw B away from E to S. Provided that the binding site (S) is large enough, much or even all of B can be drawn away from E to S.

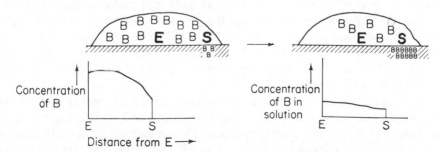

A similar situation will occur if insufficient trapping agent is added to the medium, so that B can diffuse without being captured adequately by the trapping agent. An analogous condition is one where weak enzyme activity (as occurs after certain types of fixation) produces too little of B to exceed the solubility product of BC at the site of the enzyme; the BC so formed can diffuse away to a site which adsorbs BC and so again breaks the concentration gradient which otherwise would restrain the free diffusion of the end-product (BC).

Many coloured end-products of histochemical reactions, such as some formazans and azo-dyes, are relatively insoluble in water but readily soluble in lipid. Where they are produced in low concentration, e.g. by the action of a weak (or of a chemically weakened) enzyme, they may diffuse into lipid and so set up a 'vacuum' in the diffusion gradient. For this reason many workers have been concerned over what they call the 'substantivity' of their end-products for lipids and for proteins. If the end-product has a high affinity ('substantivity') for lipid, you may expect this type of diffusion to cause a false localization of the dye at S instead of at E; if the end-product has a high affinity for protein, it will be expected to bind to the protein close to the site of its liberation and so yield a fair localization of E.

These problems of rate of precipitation and of diffusion are most aggravated in the azo-dye coupling methods to be discussed later in this section. In these the enzyme splits B (e.g. naphthol) off from the substrate (AB; e.g. naphthol phosphate) and B is relatively soluble—and so highly diffusible—until it is coupled by C (e.g. hexazonium pararosaniline). The rate at which B and C couple depends on the pH, which is rarely optimal both for the enzyme activity and for the coupling. Consequently a compromise pH has to be used, which reduces the activity of the enzyme as well as the efficiency of the coupling reaction. The actual solubility of the end-product is not usually known; as Burstone (1962) explained: 'although there is no actual knowledge about the submicroscopic nature of a "capture reaction" involving a stain precursor (e.g. liberated naphthol), we can assume that this reaction should take place as rapidly as possible. A rapid capture reaction, however, does not necessarily preclude poor localizations. For example, β-naphthol may couple very rapidly with a suitable diazo reagent but may produce diffusion artefacts because the dyes are known in many instances to remain in a colloidal state following coupling.' In our terms, the reaction B + C→BC *should* be rapid (but we have little evidence that it is) and the solubility product of BC *should* be exceeded as rapidly as possible (although this is difficult when the enzymes have been fixed and the reaction is done at a compromise pH) if diffusion to adsorbing sites (S) is to be avoided.

In practice, all these variables can be tested to some extent by the method of overlapping active sections over inactive sections and actually measuring the distance diffused by the reaction products. This technique, developed by Jacoby and Martin (1949) and particularly by Danielli (1953) has been used by various workers (e.g. McCabe and Chayen, 1965) in attempts to define the limits of these effects.

Validity of histochemical enzymatic reactions

In testing the activity of an enzyme it is always advantageous to test a comparable section or preparation with the medium lacking the specific substrate. It is also advisable to test whether the activity increases linearly with time of reaction. These precautions arise from the fact that you may get a good response with a short time of reaction but as good a reaction in the absence of substrate. However, the reaction continued for longer times will disclose the specific activity (Figure 22). The response at short reaction times may be due to endogenous substrates or other reactive moieties; this will be disclosed by the reaction done in the absence of exogenous substrate which, after achieving maximal activity at relatively short times, will remain constant, whereas that done in the presence of exogenously added substrate will continue to increase (as in Figure 22).

The specificity of the enzymatic reaction should be checked by testing the activity at different pH values, to determine the pH optimum of the enzymatic activity being disclosed, and also by the use of specific inhibitors.

Studies on bound or protected enzymes

The typical examples of 'bound' enzymes are the lysosomal enzymes (as reviewed by Bitensky and Chayen, 1977). These lie within small particles which may be as small as $0.3\ \mu m$ in diameter, and therefore adequately resolved only by objectives with high numerical aperture (NA). The lysosomal particle is bounded by a single membrane. Consequently, in unfixed preparations, the histochemical demonstration of the activity of the enzymes within the lysosomes may be restricted by the amount of the histochemical reagents that can penetrate through the lysosomal membrane. Some reagents readily penetrate these membranes so that the concentration of the histochemical reagents presented to the lysosomal

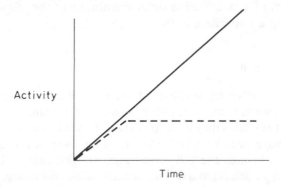

Figure 22. ----, activity due to endogenous substrate or reactive moieties. ——, activity due to exogenously added substrate. Note that initially the two activities may be identical.

enzymes will approximate closely to that of the reaction medium. Consequently the activity recorded for the enzymes will be the total activity of which the lysosomes are capable. In fact, this information, on its own, is not very valuable. In life, the activity of the lysosomes is regulated by the permeability of the bounding membrane. It is therefore frequently better to use a substrate that permeates only slowly into the lysosomes so that the activity recorded histochemically is that activity that can be manifested in spite of the bounding membrane. The total activity of which the lysosomal enzyme is capable can then be measured by pretreating serial specimens with a suitable buffer at a suitable pH (e.g. acetate at pH 5.0, as described by de Duve, 1969; Bitensky *et al.*, 1973; Bitensky and Chayen, 1977). Such measurements give the proportion of freely available (with intact membrane) to total activity. This proportion may vary considerably, depending on the state of the lysosomal membranes, which in many cases has been shown to be considerably modulated by disease, toxic material or hormones. For example, the highly sensitive cytochemical bioassay of thyroid stimulating hormone (TSH) and of the long-acting thyroid stimulating antibodies, depends on this quantitative comparison of freely available to total activity and how the ratio is altered by various concentrations of this hormone or of these antibodies (Bitensky *et al.*, 1974a).

The use of stabilizers for assaying bound enzymes

In most of the studies done up to now on lysosomal enzymes, it has been advisable not to use a tissue stabilizer, such as PVA or the collagen polypeptide. The aim of such studies has been to distinguish between the readily manifest and the total activity of these enzymes, as discussed above. For studies on mitochondrial enzymes some degree of stabilization has sometimes been found to help the demonstration of intramitochondrial enzymes. For other work it may be beneficial to destabilize the outer membrane of the mitochondria, for example by using a phosphate buffer.

Studies on intact cells

In studying whole cells, the possible role of the cell membrane may need to be considered. Some substrates are retarded by cell membranes so that increased concentrations of reactants may need to be used. In some cases five times normal concentrations have been required. On the other hand, with substrates that readily permeate through the cell membrane, or with cells with a weak cell membrane, it may be helpful to use some concentration of a collagen polypeptide stabilizer. The required concentration must be determined by trial and error. It may be noted in this context that PVA is a strong stabilizer of membranes and will retard entry of reagents into the cells.

Classification of enzymes

Since 1961, enzymes have been systematized into an internationally agreed classification with rules governing their systematic and trivial names. A numerical system has been developed by which each enzyme is characterized (Dixon and Webb, 1964). This system involves specific Enzyme Commission numbers to define each enzyme. For example, alkaline phosphatase is designated EC 3.1.3.1. This nomenclature has had little impact on histochemistry. Consequently, we retain the names of enzymes that are commonly used in histochemistry. Information regarding the biochemical properties of each enzyme has been derived from Dixon and Webb (1964), the *Biochemists' Handbook* (1961), Roodyn (1965) and from more recent publications dealing with specific enzymes, as listed in the text.

PHOSPHATASES

ALKALINE PHOSPHATASE

Biochemical background

A range of alkaline phosphatases (EC 3.1.3.1) is known, their action being related to the tissue or to the substrate which is preferentially (or more actively) used. They act on monoesters of orthophosphoric acid and have little effect on pyrophosphates, metaphosphates or phosphoric diesters. Intestinal alkaline phosphatase, which has a number of peculiarities, can hydrolyse amidophosphate bonds, as occur in creatine phosphate (Dixon and Webb, 1964). The general biochemistry of alkaline phosphatases has been reviewed by Butterworth (1983). All are glycoproteins which, in some cases, include sialic acid; the carbohydrate content may be as high as 20% of the dry weight of the enzyme. They are zinc metalloproteins, with four equivalents of Zn^{2+} per mole, which are essential for activity; in placental alkaline phosphatase the zinc may be replaced by cobalt. Most alkaline phosphatases are activated by magnesium ions, which attach at non-enzymatic sites. Some are entirely inactive in the absence of magnesium.

The active enzyme consists of two identical subunits which themselves are inactive. The active dimer can be dissociated by guanidine hydrochloride or by urea; acid pH causes dissociation and loss of activity which can be partially recovered by neutralization.

The reaction product, phosphate, is a good competitive inhibitor; zinc is a powerful inhibitor, apparently binding to the magnesium site. Some phosphonates are strong competitive inhibitors (Shirazi et al., 1981). Many non-competitive inhibitors are known. These include NADH, levamisole, theophylline, imidazole and anilinonaphthalene sulphonate. L-Phenylalanine is most potent against placental and intestinal alkaline phosphatases; levamisole and L-*p*-bromotetramisole are powerful inhibitors of the liver, kidney and bone enzymes.

The significance of alkaline phosphatase in clinical biochemistry has been reviewed by Moss (1983); methods used in routine haematology by Hayhoe (1983); and in electron microscopy by Borgers and Verheyen (1983).

One of the roles of the alkaline phosphatases may be to facilitate transport of active molecules into and out of cells. Many active molecules are transported as the soluble phosphates which, after permeating cell membranes, can be freed by the action of alkaline phosphatase. So, for example, a good substrate for this enzyme, in some cells at least, is pyridoxal phosphate.

Histochemical background

Alkaline phosphatase has been much studied by histochemists, and some of the basic concepts of enzyme histochemistry have been derived from such studies (e.g. Gomori, 1952; Danielli, 1953). Although many phosphatases may occur in tissues, the term 'alkaline phosphatase' refers to a characteristic enzymic activity which hydrolyses phosphomonoesters optimally at an alkaline pH so that greater activity is found at pH 9.4 than at pH 7.0. Consequently it is possible to show, in a single tissue, that glycerophosphate is hydrolysed optimally at pH 9.4, and also at pH 5.0, corresponding to the presence of acid phosphatases, with some extra optimal activity (against nucleotide phosphate or against glucose 6-phosphate) at around pH 6.5. Even though some overlap of acid phosphatase, 5′-nucleotidase, glucose 6-phosphatase and alkaline phosphatase activities can occur, the presence of each of these can be distinguished (Figure 23).

The histochemical demonstration of alkaline phosphatase activity may be of some importance in diagnostic haematology, as reviewed by Hayhoe (1983). It has also been used as an aid to investigations of human liver biopsies (Sherlock, 1958; Bitensky, 1967a). It is frequently found at cell membranes where active transport occurs (Danielli, 1953). It is found abundantly in the endothelium of blood vessels, in the brush border of renal convoluted tubules and at the

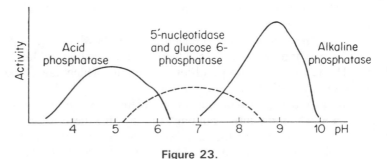

Figure 23.

lumenal margin of the intestinal epithelium. The activity in diverse tissues has been reported to be intestine > placenta > kidney > bone > liver (Fernley, 1971).

Inhibitors

According to Barka and Anderson (1963) inhibition can be achieved if sections are treated with Lugol's iodine (2–3 min); with 15% acetic acid (2–3 min); or with 5% trichloroacetic acid (1–2 min). Burstone (1962) listed beryllium, cysteine (2.5–5 mM), iodine (10^{-3} M) and sodium arsenate (10^{-2} M); polyanions such as polyoestradiol phosphate, polyphloroglucinol and polyphloretin were said to be strong inhibitors of the alkaline phosphatase of kidney and intestine. L-Phenylalanine is much used for inhibiting intestinal and placental alkaline phosphatase. Reduced glutathione is a strong inhibitor (0.9 mM) as demonstrated by Ahlers (1975; also Shedden *et al.*, 1976, on bone). Derivatives of xanthine and theophylline acting at 2 mM may also be inhibitory (Sugimura and Mizutani, 1979). However, the most useful general inhibitors (except for intestinal alkaline phosphatase activity) are either L-tetramisole (levamisole), acting at 0.1–0.5 mM, or L-*p*-bromotetramisole, acting at 0.1 mM (Borgers and Thone, 1975). The last is of particular interest in that D-*p*-bromotetramisole acts as an ideal control because it is totally inactive.

Rationale of the histochemical methods

The enzyme catalyses the reaction:

$$R-O-\overset{\displaystyle O}{\overset{\|}{\underset{\underset{\textstyle OH}{|}}{P}}}-OH \;+\; H_2O \;\longrightarrow\; ROH \;+\; HO-\overset{\displaystyle O}{\overset{\|}{\underset{\underset{\textstyle OH}{|}}{P}}}-OH$$

Consequently, the activity can be assessed by demonstrating the liberation of either phosphate or R, where ROH is coloured, as it is, if it is a derivative of naphthol.

THE GOMORI–TAKAMATSU PROCEDURE

Glycerophosphate, or any phosphomonoester (such as pyridoxal phosphate) that is soluble at very alkaline conditions and in the presence of a high concentration of calcium, is given to the section (or smear) at pH 9.4 in the presence of calcium and of the activating magnesium ions. The phosphate liberated by the enzymatic activity is trapped by the high concentration of calcium in the medium and precipitated as calcium phosphate which, at this alkaline pH, is virtually insoluble [but note that it is soluble at much lower pH values]. The colourless calcium phosphate is converted into the still insoluble cobalt phosphate, which is then transformed into insoluble black cobalt sulphide.

The main worry concerning this procedure was whether the enzyme, or the reaction products, could diffuse from their original site. Barter *et al.* (1955) used microscopic interferometry to show that, once the calcium phosphate had been precipitated, the subsequent procedures had no effect on its localization. However, there has been controversy as to whether the nuclear localization, reported by some workers, was artefactual or due to diffusion of the phosphate or of the enzyme itself. A high concentration of calcium ions in the reaction medium can protect against movement of the liberated phosphate. It was also shown (Butcher and Chayen, 1966a) that a sufficiently high concentration of calcium immobilized the enzyme, even in unfixed cryostat sections, without inhibiting the activity of the enzyme.

Method (Gomori, 1952; Butcher and Chayen, 1966a)

Unfixed cryostat sections should be used.

1. Incubate sections in freshly prepared medium at 37 °C for up to 60 min (e.g. 10 μm sections of rat kidney may require only 3 min, whereas those of rat liver may need 60 min).
2. Wash in running tap water or in several changes of alkaline water for 3–5 min. [*Note*: acidic pH will solubilize the calcium phosphate and cause it either to leave the section or to move to a spurious localization.]
3. Immerse in 2% solution of cobalt nitrate for 5 min, keeping the pH alkaline.
4. Rinse in distilled water.
5. Immerse in 0.5% (v/v) aqueous solution of ammonium sulphide for 1 min.
6. Rinse in distilled water.
7. Mount in an aqueous mountant (e.g. Farrants' medium or Aquamount).

Result

Enzyme activity is shown by a black precipitate of cobalt sulphide.

Measurement

By microdensitometry; light of 550 nm.

Solutions required for this method

2% solution of barbitone sodium (sodium 5,5-diethylbarbiturate)	10 ml
3% solution of sodium β-glycerophosphate	10 ml
2% solution of dried calcium chloride (or 2.7% $CaCl_2.2H_2O$)	20 ml

5% solution of magnesium sulphate (MgSO$_4$.7H$_2$O)   1 ml
Distilled water   5 ml
The final pH of this medium should be pH 9.4.

Control

Replace the 3% solution of sodium β-glycerophosphate with 10 ml distilled water.

Modified method

As above, but replace the calcium chloride by calcium glucoheptonate at 45 mg/ml in place of the calcium chloride. The advantage of this is that it gives a more even precipitate, which is preferable for microdensitometry (or photography).

THE NAPHTHOL PHOSPHATE METHOD (Gomori, 1952)

In this procedure it is the naphthol (or the R in the reaction equation above) of the phosphomonoester which is trapped by simultaneous coupling with a diazonium salt (to give the final colour). The reactions are as shown below. It may be noted that the validity of these naphthol-derived colour reactions has not been demonstrated as fully as has been the phosphate-precipitating procedure (and as discussed above). However, the method is particularly applicable to studies on bone (e.g. Shedden et al., 1976) where the calcium of the bone can interfere with the calcium glycerophosphate procedure.

Method

Fixation is not recommended (as discussed by Gomori, 1952). It is therefore advisable to use fresh unfixed cryostat sections, which give results with this technique comparable to those obtained by the less readily quantifiable calcium phosphate method.

1. Incubate sections in freshly prepared incubation medium in a Coplin jar (e.g. for up to 15 min).
2. Wash in distilled water.
3. Wash in 1% acetic acid for 1 min.
4. Rinse in distilled water.
5. Mount in an aqueous mountant (such as Farrants' medium).

Sodium α-naphthol acid phosphate + H_2O ⟶ α-naphthol + HO—P(=O)(OH)—ONa

α-naphthol + Fast Blue RR ⟶ azo dye + H_2O

Result

A purple-black precipitate denotes sites of alkaline phosphatase activity.

Solutions required for this method

Incubation medium

2% barbitone-sodium (sodium 5,5-diethylbarbiturate)	25 ml
α-naphthyl acid phosphate (sodium salt)	10 mg
10% magnesium chloride	0.2 ml
Fast Blue RR salt	25 mg

The final pH of the medium is 9.2.

Stir in the Fast Blue RR salt, filter into a Coplin jar and use **immediately**.

[*Note*: sections which were put into the incubation medium 20 min after it had been prepared were negative, assumably due to the decomposition of the Fast Blue RR salt.]

Control

As the test, but with no naphthyl acid phosphate in the incubation medium.

ACID PHOSPHATASE

Biochemical background

Acid phosphatase (EC 3.1.3.2) is typically a marker of lysosomes (as discussed by de Duve, 1959, 1969). The acid phosphatases from different animal tissues may behave differently (*Biochemists' Handbook*, 1961); their properties also seem to vary according to the substrate. Optimal activity is usually between pH 4.5 and 5.5, but that of the prostatic enzyme, acting on phosphorylcholine, is about pH 6.5; this enzyme catalyses the hydrolysis of most phosphomonoesters, including creatine phosphate and amido-phosphate; in contrast to the bone enzyme, prostatic acid phosphatase is inhibited by tartrate so that the source of the enzyme in serum can be identified. The acid phosphatase of erythrocytes is activated by magnesium and depends on free $-SH$ groups, but these properties are not shared by other acid phosphatases.

Sodium fluoride appears to be a potent inhibitor of this enzyme in many tissues (e.g. Gomori, 1952). The reaction sequence is the same as for alkaline phosphatase.

Histochemical background

The original method of Gomori (1952), produced before lysosomes had been discovered, was notoriously capricious. Holt (1959) made major changes in the technique; his procedure is the basis of the modern methods of Bitensky (1962; 1963a,c; Bitensky and Chayen, 1977). In most tissues this enzyme occurs inside lysosomes. These are cytoplasmic granules which, as primary lysosomes, may be as small as 0.3 μm in diameter. They retain their enzymes behind a relatively impermeable membrane so that, in normal cells, relatively little manifest (that is, immediately discernible) activity may be demonstrated with certain substrates. This is due to the fact that the substrate cannot readily permeate through the membrane. However, at the pH that is optimal for demonstrating acid phosphatase activity, the membrane becomes progressively more permeable, allowing increasing concentrations of substrate within the lysosomes. Fixation renders the membrane totally permeable.

Histochemistry has contributed considerably to the study of lysosomes. Firstly, it demonstrates which cells within a tissue were rich in lysosomal enzymes; secondly, it demonstrates whether these enzymes were freely available, or whether they were predominantly bound within an impermeable membrane. Consequently, histochemical procedures allow the demonstration of how much of the enzymes were active in life as against how much were retained in a latent state. This comparison has proved of considerable importance in assessing the state of cells and in studying the effects of disease (Bitensky, 1963a; Chayen et al., 1971; Chayen and Bitensky, 1971; Bitensky et al., 1974b); of immune reactions (Bitensky, 1963b, 1967b); and of cell damage (Chayen and Bitensky, 1968).

But it deserves to be emphasized that this advantage of the histochemical analysis of lysosomal activity depends on preparative procedures that retain the lysosomal membrane in its natural state; this excludes the use of freeze-drying or of chemical fixation.

The role of lysosomes, and of lysosomal enzymes, in plants has been reviewed by Gahan (1967a,b). In animal tissues the cells that are particularly rich in lysosomes are those that have a macrophagic function. Thus acid phosphatase activity is often used as an indicator of macrophages, giant cells and osteoclasts.

Procedures for measuring the degree of latency of lysosomal enzymes are discussed on pp. 190–193.

THE GOMORI LEAD PHOSPHATE METHOD (after Holt, 1959; Bitensky, 1962, 1963c)

Rationale

As for the alkaline phosphatase reaction, the substrate is β-glycerophosphate. However, the liberated phosphate cannot be trapped by calcium because calcium phosphate is not insoluble at the acidic pH at which this reaction must be done. Consequently, lead is substituted for calcium as the trapping agent. The lead phosphate is then converted directly into the brown-black lead sulphide by treatment with hydrogen sulphide (Bitensky, 1963c); this is preferable to ammonium sulphide in that it avoids the drastic effects on the cells or tissue caused by changing from an acidic to a very alkaline pH.

Much has been made (e.g. Barka and Anderson, 1963, p. 242) of spurious effects caused by 'metallophilia', namely the adsorption of lead itself onto sites in tissues or cells even in the absence of the substrate. In fact, such enzymatically 'non-specific' adsorption of lead may yield useful information related to the chemical nature of the adsorbing site (as in Ebel's test for polyphosphates, p. 98). It should not be a real worry provided the reaction is controlled by the use of fluoride or some equally effective inhibitor of acid phosphatase activity.

Acid phosphatase is a valuable marker of lysosomal activity. Short incubation often yields a discrete, granular localization, but prolonged incubation at the acidic pH may produce not only more intense activity but altered localization, even strong nuclear staining with clear cytoplasm. The importance of this will be discussed below.

For the histochemical reaction, β-glycerophosphate is used as the substrate. However, many commercially available samples of β-glycerophosphate may contain appreciable quantities of material, such as free phosphate, which will react with the lead nitrate. It is therefore essential to prepare the reaction medium well in advance, preferably overnight, to allow the precipitation of such material.

Method

Cryostat sections of unfixed tissue, freshly prepared air-dried smears or cytocentrifuge preparations of unfixed cells are required for this reaction. Holt (1959) showed that fixation in formol–calcium at 2 °C for 24 h caused up to 60% loss of activity even after extensive removal of the fixative. Paraffin-embedding can cause almost complete loss of activity.

1. Incubate serial sections, or preparations, for various periods of time at 37 °C in the reaction medium. The duration of the incubation will depend remarkably on the tissue and on its physiological state. As a rough guide, cryostat sections of rat liver may require 5–20 min; for sections of human endometrium 20 min may suffice for tissue taken at mid-cycle; 40 min for tissue from the proliferative phase; but less than 5 min is adequate for the secretory phase.
2. Wash thoroughly in distilled water to remove all the adsorbed lead. [*Note*: it is advisable not to use acetic acid for this wash since it can remove the specific precipitate as well as the 'background'.]
3. Place the sections (up to 1 min) in distilled water that has immediately previously been saturated with hydrogen sulphide.
4. Wash thoroughly in distilled water.
5. Mount in Farrants' or similar aqueous mountant.

Control

Incubate as for the test, but include 0.01 M sodium fluoride in the reaction medium.

Result

Black or brown-black deposits of lead sulphide that are found specifically in the test sections indicate acid phosphatase activity. With short incubation times, the deposit is likely to occur in small dots that are of the size of lysosomes.

Solutions required for this method

Reaction medium

To 400 ml 0.05 M acetate buffer, pH 5.0 (see Appendix 2) add 0.53 g lead nitrate and 40 ml of a 3% solution of sodium β-glycerophosphate. The response varies somewhat depending on the source of the β-glycerophosphate. The material from BDH gives clean results.

This reaction medium is left for 24 h at 37 °C, during which time some precipitate forms. The solution is allowed to cool to room temperature before it is filtered. This reaction medium should be used immediately after it has been filtered.

Solution of hydrogen sulphide

The hydrogen sulphide is made in a Kipp's apparatus by the action of concentrated hydrochloric acid on zinc (or ferrous) sulphide. It may also be obtained from a commercially available gas cylinder. The gas should be bubbled through a wash bottle containing very dilute hydrochloric acid (to wash the gas); it is then bubbled into a Coplin jar containing distilled water. If 50 ml distilled water are used, the gas should be passed into it for 2 min at a rate of about 120 bubbles per minute.

ACID PHOSPHATASE FRAGILITY TEST

One aspect of the power of histochemistry over conventional biochemistry is shown by the ability of histochemistry, if used correctly, to test the state of the lysosomal membranes. This has proved of very great value in many studies on the effect of disease or of experimental procedures (as listed in Bitensky and Chayen, 1977).

The histochemical analysis of the state of lysosomes (Bitensky, 1963a) is based on the fact that lysosomal membranes can exist in one of three states. The first is the stable condition, with a relatively impermeable membrane, so that the 'free' acid phosphatase activity is low relative to the activity that can be disclosed by suitable modification of the membrane bounding each lysosome. The third state is that in which the lysosomal membranes are so disturbed that they no longer function as a functional barrier. The second state lies in between the first and the third. This analysis is based on the 'fragility' test of Bitensky (1963a; also discussed by Allison, 1968).

The fragility test is performed as follows. Serial sections (or smears) of the test and control tissue are exposed to the reaction medium for increasing periods, for example 5–40 min. The shortest time required to give a clearly discernible reaction in the test specimen, as against that for the control, indicates the altered fragility in the test specimen. For example, with the synovial lining cells in human rheumatoid arthritis only 2–5 min was required, whereas in the equivalent normal synovial lining cells the required reaction time was generally 20–40 min (Bitensky and Chayen, 1977), indicating the greater stability of the lysosomal membranes in the normal synovial lining cells.

AZO-DYE ACID PHOSPHATASE METHOD (after Barka and Anderson, 1963)

Rationale

This method is comparable to the azo-dye technique for alkaline phosphatase. However, the acidic pH enforces certain changes. The efficient coupling of most diazonium salts requires an alkaline pH (Barka and Anderson, 1963). On the other hand, Barka (1960) showed that, if pararosaniline is diazotized just before it is used, the resultant hexazonium pararosaniline will couple with α-naphthol at pH 5.5 although its rate of hydrolysis (destruction) at this pH is greater than its rate of coupling. Consequently, a compromise is reached by doing the whole reaction at pH 6.5. There is remarkably little evidence concerning the speed of coupling at this pH; concerning how this pH affects the rate of enzymic liberation of the naphthol from the naphthol phosphate; or concerning the insolubility of the naphthol-diazotized pararosaniline complex (i.e. the rate at which it will precipitate at the site of the enzymatic reaction). The whole reaction is difficult to evaluate since Barka and Anderson (1963) state that at this pH the enzyme is only half as active as it is at pH 4.5–5.0, and at the pH of the histochemical reaction hexazonium pararosaniline causes 54% inhibition of the enzyme. Consequently, on their evidence, this reaction (at pH 6.5) will leave only a quarter of the activity of which the enzyme had been capable, even after the effects of chemical fixation (which they recommend). Moreover, as reported by Holt (1963) and Novikoff (1963), the acid phosphatase demonstrated by this technique is markedly different from that characterized biochemically. It is therefore difficult to assess its significance. However, because the method is widely used, it is included (below).

Method

Barka and Anderson (1963) recommended the use of frozen sections of tissue that had been fixed in cold formalin. We have obtained results with unfixed cryostat sections. The method given below is that of Barka and Anderson (1963).

1. Mount the frozen sections on slides. Dry at room temperature (1–2 h).
2. React at room temperature for 5–30 min, as required.
3. Rinse in water, dehydrate and mount in a synthetic resin.

Result

Red to brown coloration indicates a positive result. Fluoride does not seem to act significantly to control this reaction.

Solutions required for this method

Michaelis buffer

Sodium acetate $3H_2O$	9.714 g
Sodium barbiturate	14.714 g
Distilled water to make up to	500 ml

Solution A

4 mg/ml sodium α-naphthyl acid phosphate added to Michaelis buffer.

Solution B

Add 2 g pararosaniline hydrochloride to 50 ml 2 M hydrochloric acid. Heat gently. Cool and filter immediately.

Solution C

A 4% solution of sodium nitrite in distilled water.

Solution D

Diazotize the pararosaniline by adding 1 volume of solution C to 1 volume of solution B. Adjust the pH to 6.5 by adding sodium hydroxide. Filter.

Incubation medium

Add 5 ml solution A to 13 ml distilled water; to this solution add 1.6 ml solution D. [*Note*: Barka and Anderson warned that not all samples of pararosaniline are suitable; in some the impurities may be so excessive as to give 'non-specific background staining'.]

Barka and Anderson also recommended the use of substituted naphthols. For example, naphthol AS-TR can be used with hexazonium pararosaniline at pH 5.0. It is not clear how this overcomes the objections which Barka and Anderson (1963) themselves have expressed to the use of this coupling agent at such a low pH (as quoted above). However, the use of naphthol AS derivatives, including naphthol AS-TR and naphthol AS-BI phosphates, has been widely recommended. According to Lojda *et al.* (1964), the final concentration of the naphthol AS derivative should be 0.5 mg/ml; that of the hexazotized pararosaniline should be 0.15 mg/ml reaction medium (Lojda *et al.*, 1967). One word of caution may not be out of place: these naphthol AS derivatives are fairly large molecules and their ability to penetrate relatively intact lysosomal membranes does not seem to have been ascertained; this, of course, was of no consequence for studies on material that had been subjected to chemical fixation.

THE POST-COUPLING METHOD (Bitensky and Chayen, 1977)

Rationale

Many of the worries underlying the use of simultaneous coupling methods are overcome by this procedure, in which naphthol AS-BI phosphate is used as the substrate. The phosphatase, acting at its optimal pH, produces the insoluble naphthol AS-BI which, later, can be coupled to Fast Dark Blue R, or to Fast Garnet GBC, at a pH suited to such coupling, to yield a strong colour.

Method

Fresh cryostat sections of unfixed material, or unfixed smears of isolated cells, can be used.

1. React for various times in a Coplin jar at 37 °C.
2. Wash in cold distilled water.
3. Immerse in a cold, saturated solution (1 mg/ml) of Fast Dark Blue R or Fast Garnet GBC in 0.01 M phosphate buffer (pH 7.4). Leave in the refrigerator (e.g. at +4 °C) for up to 5 min, with occasional agitation.
4. Wash in distilled water at room temperature.
5. Mount in Farrants' or equivalent mounting medium.

Control

React in the same reaction medium to which 0.01 M sodium fluoride has been added to inactivate the acid phosphatase.

Result

Brownish stain often in small lysosome-like granules. There should be no colour in the preparations done in the presence of fluoride.

Solutions required for this method

Stock solution

11.4 mg naphthol AS-BI phosphate are added to 1 ml 5 mM solution of sodium bicarbonate (0.42 g sodium bicarbonate in 100 ml distilled water). To this are added 49 ml 0.2 M acetate buffer, pH 4.5. This solution may be stored at +4 °C.

Reaction medium

Take 2 ml of the stock solution and add 20 ml 0.1 M acetate buffer at pH 4.5 containing 0.1 g calcium chloride ($CaCl_2.6H_2O$).

5′-NUCLEOTIDASE

Biochemical background

This enzyme (EC 3.1.3.5) specifically catalyses the hydrolysis of phosphate esters on the carbon-5 of ribonucleotides and of deoxyribonucleotides; it is inactive against phosphate esters on the carbon-3 of these molecules, and against phosphates of sugars, glycerol or phenols (*Biochemists' Handbook*):

$$\text{adenosine 5-phosphate} + H_2O \rightarrow \text{adenosine} + \text{phosphate}$$

In fractionation studies it has been used extensively as a marker of the cell membrane fraction (Drummond and Yamamoto, 1971). However, it is now clear that two forms of this enzyme can occur (Zekri *et al.*, 1988): one which is membrane bound and one which is cytosolic. The former is a metalloprotein with optimal activity, which may vary between tissues, at between pH 6.5 and 7.5; adenosine monophosphate was the best substrate for this enzyme in some studies. It is possible that the main function of this enzyme may be related to the adenylate cyclase activity that is the response of many types of cell to many hormones:

$$\text{ATP} \xrightarrow[\substack{\text{adenylate} \\ \text{cyclase}}]{} \underset{\substack{\text{cyclic} \\ \text{AMP}}}{\text{cAMP}} + \underset{\substack{\text{pyro-} \\ \text{phosphate}}}{\text{PP}} \xrightarrow{\text{phosphodiesterase}} 5'\text{-AMP}$$

Adenylate cyclase occurs bound to the inside of the cell membrane; phosphodiesterase is a 'soluble' enzyme.

The cytosolic 5′-nucleotidase may occur in two forms: one showing maximal activity at around pH 6.5 and acting preferentially on purine nucleotides, especially inosine monophosphate, whereas the other acts preferentially on pyrimidine nucleotides and optimally at between pH 7.5 and 9.0. The cytosolic forms of this enzyme require exogenously added cations for maximal activity, magnesium being more effective than manganese, but calcium ions appear to be ineffective. Nickel, barium and zinc are strong inhibitors of both the cytosolic and the membrane-bound 5′-nucleotidase even though it has been claimed (Pilz *et al.*, 1982) that the enzyme is zinc-dependent.

The cytosolic enzyme is said to regulate the intracellular formation of adenine liberated during the catabolism of nucleotides. It has been shown to contribute to the formation of adenosine during ischaemia of the heart and in regenerating rat liver following partial hepatectomy (Zekri *et al.*, 1988).

The function of 5′-nucleotidase in the degradative metabolism of purines has been reviewed by Watts (1987). In this process it is involved in the conversion of nucleotides to nucleosides: e.g. xanthanylic acid to xanthosine, which can then be converted to xanthine; adenylic acid to adenosine; inosinic acid to inosine, which is then altered to hypoxanthine. Watts (1987) further reported that mature B-type lymphocytes may have about four times the activity of the corresponding T-lymphocytes.

Histochemical background

Barka and Anderson (1963) advised rat liver or adrenal as the test tissue for this reaction; in the latter it was said to occur mainly in the medulla. Burstone (1962) showed high activity in the posterior lobe of the hypophysis. Hardonk (1968) found strong activity in many tissues, with some tumours showing strong reactions (Hardonk and Koudstaal, 1968). According to Burstone (1962) manganese ions are stronger activators than magnesium ions (also Hardonk, 1968); nickel (Ni^{2+} at 10^{-3} M) and zinc almost completely inhibit 5'-nucleotidase activity, leaving the non-specific alkaline phosphatase relatively unaffected; 0.1 M sodium fluoride and 0.08 M borate are said to inhibit the 5'-nucleotidase reaction completely (Heppel and Hilmoe, 1951). Henderson et al. (1980) found markedly enhanced activity of this enzyme in the synovial lining cells of human joints afflicted with rheumatoid arthritis. These workers included L-p-bromotetramisole oxalate in their reaction medium to inhibit any alkaline phosphatase activity and used manganese as the activator of the 5'-nucleotidase activity; the addition of nickel (10^{-2} M) to the reaction medium virtually abolished the reaction.

Rationale of the histochemical procedures

Basically these procedures resemble the Gomori phosphate-precipitation methods for acid and alkaline phosphatases. There are two practical problems: (1) the reaction should be shown to be specific for the phosphate ester on the carbon-5 of the nucleotide (i.e. control the reaction with 3'-nucleotide as the substrate); (2) when tested at alkaline pH values, non-specific alkaline phosphatases will also hydrolyse the substrate so strongly that if both 5'-nucleotidase and alkaline phosphatase occur at the same locus in the tissue, an alkaline reaction medium will demonstrate both together unless an inhibitor of the latter is included in the reaction medium. To some extent the same applies when the reaction for 5'-nucleotidase is done at a more acidic pH; under these conditions it will be the acidic phosphatases which can confuse the results unless a specific inhibitor is included in the reaction medium. Since calcium phosphate is soluble at such pH values, the precipitating ion must be lead, but there is no evidence to show that lead ions may not be inhibitory as they are in the ATPase reaction. For this reason, the very recent lead method (see below) may be preferable, seeing that the lead is held within the lead ammonium citrate/acetate complex and therefore is not available for reacting with the enzyme: it is released from this complex by the phosphate liberated by the activity of the enzyme and immediately precipitated (as discussed in relation to the method for demonstrating sodium-potassium-ATPase activity).

Thus, depending on the pH at which the reaction is to be performed, there are two methods which can be used for the histochemical demonstration of 5'-nucleotidase activity, depending on whether the maximal enzymatic activity is believed to occur at an acidic or an alkaline pH.

CALCIUM METHOD FOR 5'-NUCLEOTIDASE ACTIVITY
(after Gomori, 1952; Henderson *et al.*, 1980)

In this procedure, calcium ions are used to precipitate the phosphate ions liberated by the action of the enzyme. The calcium phosphate is then converted to cobalt phosphate and the cobalt is visualized as cobalt sulphide.

Method

Use unfixed cryostat sections.

1. Incubate sections in freshly prepared medium at 37 °C for up to 30 min (or longer if necessary).
2. Wash in running tap water for 5 min. [*Note*: to avoid the possibility that the pH of the tap water may be slightly acidic, and so solubilize the precipitated calcium phosphate, it may be preferable to rinse for 5 min in several changes of distilled water, the pH of which has been made close to pH 8.0, e.g. by the addition of M/40 sodium bicarbonate.]
3. Immerse in a 0.1% solution of cobalt nitrate for 5 min.
4. Wash thoroughly in distilled water.
5. Immerse in a 0.1% solution of ammonium sulphide (BDH) in distilled water for 1 min. [Alternatively, a 2% solution of sodium sulphide, pH 8.0, may be used.]
6. Wash in distilled water.
7. Mount in an aqueous mountant such as Farrants' medium.

Result

Enzyme activity is shown by a black precipitate of cobalt sulphide.

Measurement

At 550 nm.

Solutions required for this method

Incubation medium

Adenosine 5'-monophosphate disodium salt	4 mM
Calcium chloride dihydrate	340 mM
Manganese chloride	0.06 mM
L-p-bromotetramisole oxalate	0.1 mM

These are dissolved in 0.1 M Tris-HCl buffer at pH 8.3. The use of the bromotetramisole is optional if it is clear that alkaline phosphatase activity in the tissue to be studied is negligible at this pH.

Control

The addition of 10^{-2} M nickel chloride to a separate aliquot of the full reaction medium should abolish the specific activity.

LEAD METHOD FOR 5'-NUCLEOTIDASE ACTIVITY (Chambers, 1991)

In this procedure the lead is sequestered within the lead ammonium citrate/acetate reagent and becomes manifest only when it is liberated by the phosphate ions released by the enzymatic reaction.

Method

Use fresh, unfixed cryostat sections.

1. Incubate sections at 37 °C in freshly prepared medium. For mouse cartilage, 5 min reaction time was sufficient, but longer times may be needed for sections of other tissues.
2. Wash well in distilled water.
3. Immerse for 2 min in distilled water that has been saturated with hydrogen sulphide.
4. Wash and allow to dry.
5. Mount in an aqueous mountant such as Farrants' medium.

Result

5'-Nucleotidase activity is shown by a brown-black precipitate of lead sulphide.

Measurement

At 550 nm.

Solutions required for this method

Incubation medium

Sodium acetate (anhydrous)	1 mM
Sodium chloride (24 mg/ml)	410 mM
Magnesium chloride (Analar: 4 mg/ml)	20 mM
Potassium chloride (Analar: 2.8 mg/ml)	37.5 mM
Adenosine 5'-monophosphate (0.16 mg/ml)	4 mM

These are dissolved sequentially in a 40% (w/v) solution of Polypep 5115 (Sigma) in 0.2 M Tris buffer at pH 7.4. The high concentration of Polypep 5115 is to retain even 'soluble' enzymes within the section. Finally the lead ammonium citrate/acetate (LACA, Sigma) is dissolved in a small volume of dilute ammonia (5 drops of 0.88 ammonia in a millilitre of distilled water) and added to the Polypep 5115 solution to give a final concentration of 32 mg/ml.

Control

The addition of 10^{-2} M nickel chloride to a separate aliquot of the full reaction medium should abolish the specific activity.

GLUCOSE 6-PHOSPHATASE

Biochemical background

This enzyme (EC 3.1.3.9) has maximal activity at pH 6; it is inactive against glucose 1-phosphate (*Biochemists' Handbook*). It is tightly bound to the microsomes of mammalian liver, kidney and intestinal mucosa (de Duve *et al.*, 1955; for many studies see Roodyn, 1965), from which it can be released as a lipoprotein by treatment with deoxycholate (Beaufay and de Duve, 1954). Traces have been found in the spleen, heart, skeletal muscle, brain, lung and gastric mucosa of rats (Dixon and Webb, 1964). Its activity in the liver has been reported to be increased in animals given large doses of glucocorticosteroids; it is elevated in diabetic rats but can be restored to normal levels by the injection of insulin; its activity is diminished in hypophysectomized rats. One of the hepatic glycogen-storage diseases is said to be due specifically to loss of this enzyme (*Biochemists' Handbook*).

Histochemical background

According to Chiquoine (1953; 1955) this enzyme activity is completely inhibited by fixation in 80% ethanol and embedding in paraffin wax. It is also completely inhibited by fixation in cold formalin. Fluoride, zinc, cyanide and alloxan also inhibit it.

Rationale of the histochemical method

This procedure depends on the ability of this enzyme to split phosphate from glucose 6-phosphate specifically at pH 6.5. At this pH, the phosphate must be trapped by lead, as discussed for the acid phosphatase or the 5'-nucleotidase methods. The precipitated lead phosphate is visualized by converting it into the coloured lead sulphide.

Method (after Wachstein and Meisel, 1956)

Fresh, unfixed cryostat sections should be used.

1. Incubate at 37 °C in freshly prepared incubation medium in a Coplin jar for up to 20 min.
2. Wash well in running water.
3. Rinse in distilled water.
4. Immerse in 0.5% (v/v) ammonium sulphide solution for 1 min.
5. Wash in distilled water.
6. Mount in an aqueous mountant such as Farrants' medium.

Results

Sites of this enzymatic activity are shown by a brown precipitate of lead sulphide.

Solutions required for this method

Incubation medium

0.1 M acetate buffer, pH 6.5	40 ml
Glucose 6-phosphate, sodium salt	26 mg
0.1 M lead nitrate	1 ml

Final concentration of glucose 6-phosphate is 2×10^{-3} M.

   Note: it may be preferable to use the hidden-lead reagent, LACA (as in the magnesium-dependent ATPase method, p. 167) instead of the lead nitrate.

Control

Replace the glucose 6-phosphate with 25 mg (2×10^{-3} M) sodium β-glycerophosphate to check the chemical specificity of the reaction. Alternatively use glucose 1-phosphate in place of the required substrate (glucose 6-phosphate, as discussed in the Biochemical background).

ADENOSINE TRIPHOSPHATASES

CALCIUM-ACTIVATED (MYOSIN-TYPE) ATPase

Biochemical background

Although these enzymes are often classed with the phosphatases, this association is not valid seeing that they are not esterases, as are the other phosphatases. They are better classed with the pyrophosphatases in that they act specifically on the phosphate-to-phosphate link, in contrast to the phosphate-to-carbon link as in the glycerophosphatases or nucleotidases:

$$A-P \sim P \sim P + H_2O \longrightarrow A-P \sim P + H_3PO_4 + energy$$

adenosine   adenosine

triphosphate   diphosphate

where $\sim P$ denotes the 'energy-rich' phosphate bond.

In the rat, myosin-ATPase (calcium-activated ATPase, EC 3.6.1.3) is found in very high concentration in the heart; it is highly active in skeletal muscle, kidney and lung; and occurs in the liver, pancreas and brain (Dixon and Webb, 1964). It has a molecular weight of 230 000 when isolated from the dog heart, but 540 000 when purified from rabbit muscle (Dixon and Webb, 1964). It is stimulated by dinitrophenol (DNP) apparently because DNP and actin change the configuration of the protein (Dixon and Webb, 1964). It is inhibited when more than half its −SH groups are combined with p-chloromercuribenzoate, although its activity may even be enhanced if fewer of the −SH groups are so combined. It is activated by calcium ions but may be inhibited by magnesium ions.

Histochemical background

There seem to be four classes of adenosine triphosphatases: myosin-like, calcium activated ATPase; magnesium-dependent ATPase; the sodium-potassium activated, ouabain-sensitive ATPase of the sodium–potassium exchange 'pump' of cell membranes; and calcium-ATPase. The histochemical demonstration of these enzymes depends on trapping the phosphate liberated by their activity.

According to Barka and Anderson (1963), formalin fixation causes 80–90% inhibition of these enzymes, and some subsequent histochemical procedures cause an additional 85% inhibition, leaving about 5% of the original activity. These strictures did not apply to the use of frozen sections and the calcium method developed by Padykula and Herman (1955a, b) and applied to their studies on myosin ATPase activity in muscle. In agreement with the biochemical background, these workers showed that the ATPase activity of the myosin was inhibited by p-chloromercuribenzoate (2.5×10^{-3} M) and could be reactivated by L-cysteine (2.5×10^{-3} M) or by BAL (2,3-dimercapto-1-propanol, 5×10^{-3} M). Moreover, using their type of histochemical reaction, Niles *et al.* (1964b) were able to show that the ATPase activity of myosin, in sections, was indeed activated by calcium ions and by dinitrophenol, but apparently inhibited by magnesium ions. They also showed that, given adequate morphological preservation, the histochemical reaction was precisely localized within the myosin of the cross-striations. This localization was dependent on sufficient calcium ions in the reaction medium to ensure rapid precipitation of the liberated phosphate; the rate of liberation was enhanced by the addition of dinitrophenol to the reaction medium.

The histochemical demonstration of myosin ATPase activity has proved of considerable practical benefit in assessing factors that influence the state of the myocardium during prolonged open-heart surgery (Niles *et al.*, 1964a; Darracott-Canković *et al.*, 1983). The orientation of the myosin, and its change under the influence of ATP, can be measured by quantitative polarized light microscopy and has been shown to be valuable in assessing the state of the myocardium, particularly of transplanted human hearts (Darracott-Canković *et al.*, 1987, 1989).

The method is basically the calcium-trapping technique, as discussed for the demonstration of alkaline phosphatase activity.

Method

Fresh cryostat sections, prepared as described previously in this book, must be used.

1. Incubate in the freshly prepared reaction medium at 37 °C (e.g. for 10–20 min).
2. Wash in three changes of 1% calcium chloride (2 min in each bath). [*Note*: the presence of calcium ions in the water used for washing the sections stops the precipitate from dissolving, as discussed in the general introduction to Chapter 6.]
3. Transfer to 1% cobalt nitrate solution for 2 min, and then sequentially into two baths of this solution, each for 2 min. [*Note*: this procedure changes the calcium phosphate into cobalt phosphate.]
4. Wash well in distilled water to remove adsorbed cobalt.
5. Immerse in dilute solution of ammonium sulphide (0.5% v/v) for 1 min.
6. Wash well in distilled water.
7. Mount in Farrants' medium or some similar aqueous mountant.

Result

ATPase activity is shown by a brown-black precipitate of cobalt sulphide. It should stain the cross-striations of cardiac and voluntary muscle.

[*Note*: the precipitation of the phosphate at its site of liberation depends on an adequate concentration of calcium. If the tissue is deficient in calcium, it may be necessary to incubate the section for 5 min at 37 °C in a 1% solution of calcium chloride before incubating the section with substrate (step 1 above). In extreme cases, it may be advisable to treat the intact tissue for a few minutes with a 1% or 5% solution of calcium chloride in 5% polyvinyl alcohol before it is chilled.]

Control

With glycerophosphate instead of adenosine triphosphate. Alternatively, when other phosphatases are present and active at the pH of this reaction, add 2.5×10^{-3} M *p*-chloromercuribenzoate to the reaction medium.

Solutions required for this method

Reaction medium

0.1 M sodium barbiturate solution (2.06 g/100 ml)	10 ml
ATP (trisodium salt, hydrated) (2.5×10^{-3} M)	75 mg
2,4-Dinitrophenol	30 mg
Calcium chloride solution (10.5 g $CaCl_2.2H_2O$/100 ml)	5 ml
Distilled water	35 ml

Prepare immediately before use.

MAGNESIUM-ACTIVATED ATPases (including Na⁺-K⁺-ATPase)

Biochemical background

There appear to be many kinds of ATPases that are activated by magnesium. That from skeletal muscle is inhibited by fluoride, by calcium, and by *p*-chloromercuribenzoate; its optimal pH is pH 6.8–7.0 (*Biochemists' Handbook*). An ATPase with similar properties has been found in mitochondria, except that its optimal pH is around pH 8.5. However, there are indications that there may be two or three mitochondrial ATPases with different pH optima. Intact mitochondria may show no ATPase activity until they are disrupted (Lehninger, 1965).

In biochemical studies the enzyme (EC 3.6.1.4) has been studied in homogenates by many workers (as listed by Roodyn, 1965). It has been found in highest concentration in the mitochondrial fraction, with some spread to the nuclear and microsomal fractions. The proportions found in the various fractions varied somewhat according to the different homogenization and fractionation procedures used. It must be remembered that biochemistry has diffusion and preparatory artefacts too; they are not confined to histochemistry.

Apart from the general magnesium-activated ATPases, there is a specific magnesium-activated ATPase that is activated by sodium and potassium ions (Na^+-K^+-activated ATPase) and characterized by the fact that it is inhibited by the cardiac glycoside, ouabain (at 10^{-4} M). This is the ATPase of the sodium pump, which appears to be vital for the transport of sodium and potassium ions, and of water, across cell membranes. It requires both Na^+ and K^+ ions, although the latter may be replaceable by other ions. It is inhibited by calcium, fluoride and agents that are known to reduce active transport in cells. This enzyme system has been widely reviewed (e.g. Skou, 1965; Charnock and Opit, 1968; Hokin and Dahl, 1972; Baker, 1972; Katz *et al.*, 1979; and Pérez-González de la Manna *et al.*, 1980).

Histochemical background

There has been considerable controversy over the histochemical identification of the magnesium-activated ATPases, especially of the Na^+-K^+-ATPase. The problems were as follows. Firstly, these enzymes require magnesium for their activity, and magnesium competes with calcium for the liberated phosphate and magnesium phosphate is relatively soluble. Secondly, the optimal activity of these enzymes is closer to neutrality than that of the myosin-like ATPase. At such pH values calcium phosphate is relatively soluble. Consequently lead has to be used to precipitate the phosphate liberated by the activity of the ATPase. But thirdly it was shown that lead ions caused the non-enzymatic hydrolysis of ATP, and strongly inhibited Na^+-K^+-ATPase activity. Some of the complex literature on this subject has been reviewed by Chayen *et al.* (1981).

The problem has been overcome by using a hidden-lead trapping agent, namely lead ammonium citrate/acetate (LACA), in which the lead is chelated to citrate. When ATP was added to a solution of this reagent the solution remained completely clear; the addition of free phosphate resulted in a precipitate. With this reagent, the characteristics of the Na^+-K^+-ATPase of the rat kidney were quantitatively very similar to those found in conventional biochemical studies.

LACA method (Chayen *et al.*, 1981)

Fresh cryostat sections must be used. The tissue should be sectioned and reacted on the day it is chilled or on the following day because Na^+-K^+-ATPase is not stable on storage for more than a few days (Pérez-González de la Manna *et al.*, 1980).

1. Immerse the section for 5 min in a 40% solution of Polypep 5115 in Tris buffer, pH 7.5, containing 0.1 M potassium acetate. The purpose of this is to remove free phosphate, which would confuse the final reaction.

2. Remove the potassium acetate solution (e.g. by sucking it off with a fine pipette). Add some of the full reaction medium. Remove this. [The object of this manoeuvre is to remove all free phosphates without disturbing the section.]
3. React some sections in the full reaction medium. For sections (10 μm) of rat kidney, a reaction time of 15 min at 37 °C was adequate. For measuring the Na$^+$-K$^+$-ATPase activity, other sections should be reacted in the same medium but containing 0.4 mM ouabain.
4. Transfer to a Coplin jar. Wash in several changes of 0.2 M Tris buffer, pH 7.4, at 37 °C until the Polypep 5115 has been removed entirely.
5. Immerse in a saturated solution of hydrogen sulphide (1–2 min).
6. Rinse in distilled water and allow to dry.
7. Mount in Z5 mounting medium (p. 44). The pH of this mounting agent should be pH 6.5.

Result

The sites of magnesium-activated ATPase activity are indicated by a brown precipitate of lead sulphide. The reaction done with the full reaction medium should indicate all the magnesium-activated ATPase activity. The result in sections reacted with the full medium but containing ouabain will show the ouabain-insensitive activity. Subtraction of the second value from the first gives a measure of the Na$^+$-K$^+$-ATPase activity.

Measurement

Measure with light of 585 nm wavelength.

Solutions required for this method

Incubation medium

Prepare a 40% (w/v) solution of Polypep 5115 (Sigma) in 0.2 M Tris buffer at pH 7.4 containing 1 mM sodium acetate. [For delicate tissue increase the concentration of sodium acetate to 2 mM.] Then add, sequentially:

Sodium chloride	24 mg/ml (410 mM)
Magnesium chloride	4 mg/ml (20 mM)
Potassium chloride	2.8 mg/ml (37.5 mM)
ATP (disodium adenosine 5'-triphosphate)	10 mg/ml (16.5 mM)
LACA (Sigma)	32 mg/ml

The LACA (lead ammonium citrate/acetate) is dissolved, with constant shaking, in the smallest possible volume of dilute ammonia (5 drops of 0.88 ammonia per millilitre) before adding it to the incubation medium.

The final pH of the incubation medium should be adjusted to pH 7.5 by adding either a mixture of sodium and potassium hydroxides (10:1 on a molar basis; 2.5 mM of each) or 0.1 M hydrochloric acid.

To economize on the reagents, it is best to do the reactions in Perspex rings (as described on p. 10).

Preincubation medium

Potassium acetate 0.1 M in 40% aqueous solution of Polypep 5115 in Tris buffer, pH 7.5.

Specificity

(i) Add 0.3 mg ouabain octahydrate (Sigma: 0.4 mM) to the full incubation medium to inhibit the activity of Na^+-K^+-ATPase. The activity of the Na^+-K^+-ATPase can be determined by subtracting the activity found in the presence of the ouabain from that found in its absence.

(ii) Where alkaline phosphatase is liable to interfere with measurement of these ATPases, add L-p-bromotetramisole (Aldrich: 0.4 mM) to both reaction media (i.e. with and without ouabain).

CALCIUM ADENOSINE TRIPHOSPHATASE

Biochemical background

Although movement of calcium ions plays a major role in cellular metabolism, relatively little is known about the specific calcium–magnesium ATPase (Ca^{2+}-Mg^{2+}-ATPase). Much of the movement of calcium may involve other enzymes. The whole subject is well reviewed by Carafoli (1982a, c). The specific Ca^{2+}-Mg^{2+}-ATPase has a molecular weight of about 105 000; 1 mol binds 2 nmol of calcium. During calcium transport the enzyme becomes phosphorylated by ATP, with an aspartyl phosphate residue forming at the active site of the enzyme (Carafoli, 1982b).

Most of the fundamental information concerning this enzyme has been derived from studies on the enzyme in erythrocytes. The first demonstration of the enzyme was by Dunham and Glynn (1961); its relation to a 'calcium pump' was shown by Schatzmann (1966). Its properties have been reviewed by Schatzmann (1982), the main features being the following: the enzymatic activity is markedly enhanced by calmodulin; it is very dependent on the presence of phospholipids. Lanthanides and ruthenium red inhibit the activity, possibly by linking with calcium-binding sites in the enzyme; vanadate inhibits the calcium pump (15 μM vanadate gave full inhibition but only in the presence of very low concentrations of ATP); quercetin (4–6 μM for half-maximal effect) inhibited both calcium transport and the Ca-Mg-ATPase activity.

Mircheff *et al.* (1977) showed a strong correlation between the calcium-activated ATPase activity and the absorption of calcium by rat ileum and intestine. This enzymatic activity was greater in the basolateral membrane than in the brush border (in contrast to alkaline phosphatase activity). However, the degree of stimulation of this calcium-activated ATPase activity, induced by 1,25-dihydroxyvitamin D_3 (1,25(OH)$_2$D$_3$) was more than 100 times the increased rate of calcium absorption, casting some doubt as to the significance of this enzymatic activity.

As Schatzmann (1982) pointed out, in intact nucleated cells it is more difficult to demonstrate, beyond reasonable doubt, the presence of a calcium-ATPase pump than in erythrocytes. However, he considered there was reasonable evidence for such a pump in 13 types of mammalian cells, including adipocytes, pancreatic islet cells, macrophages, Ehrlich tumour cells, hepatocytes, some cells of the kidney tubules, and cells of cardiac and smooth muscle, intestinal mucosa and various nerves.

Histochemical background

Basically this is a modification (Loveridge, unpublished data) of the new procedure for the Na$^+$-K$^+$-ATPase activity (described above). It has been used by Salmon *et al.* (1983) for studying the effect of calcitonin on bone and kidney; these workers demonstrated the specificity of the procedure.

Method

Fresh cryostat sections should be used. The tissue should be used shortly after it has been chilled.

1. Wash sections in 0.2 M Tris-HCl buffer, pH 7.4, containing the colloid stabilizer (40%; see below) for 5 min at 37 °C to remove endogenous free phosphates.
2. Incubate sections in the full reaction medium at 37 °C. As controls, react other sections in the same reaction medium but lacking calcium and containing 5 mM lanthanum chloride (to give a measure of the calcium-independent ATPase activity).
3. Wash in several changes of 0.2 M Tris-HCl buffer, pH 7.4, at room temperature.
4. Immerse in hydrogen sulphide–water for 1–2 min.
5. Dry.
6. Mount in Z5 medium (see p. 44).

Solutions required for this method

Colloid stabilizer

By absorption spectrophotometry, Polypep 5115 (Sigma) was found to contain at least 40 mM calcium. To remove the calcium, Salmon *et al.* (1983) stirred a 50% (w/v) solution of Polypep 5115 in double glass-distilled water with Amberlite MB-1 resin (Sigma: 1 g for each 10 g of the Polypep 5115) for 1 h. They then lyophylized the supernatant for storage before subsequent use. A Polypep low in calcium is now available from Molecular Design and Synthesis (University of Surrey, Guildford, UK).

Reaction medium

Magnesium chloride ($MgCl_2.6H_2O$)	5 mM
Disodium ATP	5 mM
Calcium chloride	50 μm to 5 mM, as required
Lead ammonium citrate/acetate reagent	32 mg/ml

To inhibit alkaline phosphatase activity, 0.4 mM L-*p*-bromotetramisole (Aldrich) may be added to this reaction medium; 0.4 mM ouabain octahydrate may be included to inhibit Na^+-K^+-ATPase activity.

Test for endogenous activity

Incubate in the full medium either lacking calcium chloride or including 5 mM lanthanum chloride. Any residual activity should represent the calcium-independent ATPase activity and should be subtracted from the activity recorded in the presence of the full reaction medium to give a measure of the true Ca^{2+}-Mg^{2+}-ATPase activity.

Measurement

The brown reaction product should be measured with an objective of at least ×40, at a wavelength of 585 nm (Salmon *et al.*, 1983).

PHOSPHAMIDASE

Biochemical background

There is relatively little known about enzymes that split phosphoamide bonds; often they are included in the acid phosphatases (as in the *Biochemists'*

$$\underset{\substack{\text{phosphocreatine} \\ \text{(the phosphoamide bond} \\ \text{is marked by the arrow)}}}{\text{HN=C}\underset{\text{N}-\overset{\text{O}}{\text{P}}-\text{OH}}{\overset{\text{NCH}_2\text{COOH}}{\big<}}} \xrightarrow{+\ H_2O} \underset{\text{creatine}}{\text{HN=C}\underset{\text{NH}_2}{\overset{\text{NCH}_2\text{COOH}}{\big<}}} + \underset{\text{phosphoric acid}}{\text{HO}-\overset{\text{OH}}{\underset{\text{O}}{\text{P}}}-\text{OH}}$$

Handbook). Dixon and Webb (1964, p. 620) point out that energy is stored in two forms, either as pyrophosphate bonds (as in ATP) or as phosphoamide bonds in creatine or arginine phosphates.

The enzyme that acts in this way on phosphocreatine (EC 3.9.1.1) is found apparently fairly widely in animal and plant tissues and in snake venom (see Dixon and Webb, 1964).

An additional point of interest is that it has been claimed by Holter and Li (1950, 1951) that crystalline pepsin, trypsin, chymotrypsin and rennin can cleave phosphoamide bonds; the optimal pH of phosphamidase activity in homogenates was around pH 4.6 (see Burstone, 1962). Such activity does not seem to fit in with the Enzyme Commission numbering, but would be of great value in histochemistry, because phosphamidase activity might be used to demonstrate proteolytic activity generally.

Histochemical background

The early demonstration of this enzyme by Gomori (1952) used *p*-chloroanilidophosphonic acid as the substrate. Only his results at pH 5–6 could be interpreted as being due to a specific phosphamidase enzyme. Others have used phosphocreatine at an alkaline pH as the substrate, but their results do not seem to be conclusive (Gomori, 1952, p. 195). The particular interest in phosphamidase activity, however it was demonstrated, has been its frequent occurrence in high concentrations in tumours, and there is evidence that the degree of malignancy may be related, to some degree, to the concentration of this enzyme (Gomori, 1952; Burstone, 1962).

Gomori (1952) did not find his own method 'entirely dependable'. Burstone (1962) pointed out that the phosphoamide bond is labile so that some decomposition occurs when sections are incubated with this substrate. Consequently, even if the section is held at an inclined angle during the reaction, 'indiscriminate precipitation of lead phosphate over the tissues' occurs. Other histochemical findings were reported by Meyer and Weinman (1955) and by Barka and Anderson (1963).

The indiscriminate precipitate, and the variability of the procedure, have largely been overcome by Butcher's modification (as below) of Gomori's method. In this method, polyvinyl alcohol is used to stabilize the substrate rather than the section. Unfixed sections are used; this is advantageous for the following reasons:

1. Chemical denaturation exposes more reactive, non-enzymatic, sites that can adsorb lead and its salts.
2. The enzyme is more active if the tissue is not chemically fixed. Consequently, the solubility product (see p. 139) of lead sulphide is exceeded quickly to yield a precipitate with less likelihood of diffusion to non-enzymatic sites.
3. The enzyme appears to occur within discrete cytoplasmic organelles so that chemical fixation will tend to confuse the histochemical result, very much as it does with acid phosphatase. However, phosphamidase activity is distinguishable from that of acid phosphatase in that it is not inhibited by fluoride.

Method (after Gomori, 1952; Meyer and Weinman, 1955;
Butcher, unpublished information)

Use fresh cryostat sections for this reaction.

1. Place the section in the incubation medium at 37 °C in a Coplin jar for up to 20 min. (Alternatively, react the section upside down in a staining trough.)
2. Wash the sections well in running tap water.
3. Rinse in distilled water.
4. Immerse in 0.5% (v/v) ammonium sulphide solution for 1 min.
5. Wash in distilled water.
6. Mount in an aqueous mountant such as Farrants' medium.

Result

Sites of phosphamidase activity are denoted by a brown precipitate of lead sulphide.

Solutions required for this method

Incubation medium

Solution A

0.05 M acetate buffer, pH 5.4	50 ml
0.1 M (3.3 g/100 ml) lead nitrate	1.9 ml
Sodium chloride	0.11 g
Polyvinyl alcohol (PVA; e.g. GO4/140 grade)	5 g

Add the lead nitrate and sodium chloride to the buffer. Then dissolve the PVA by heating (to not more than 60 °C), stirring continuously. Cool to 37 °C.

Solution B

p-Chloroanilidophosphonic acid	0.15 g
1 M sodium hydroxide	0.5 ml

(This must be prepared immediately before it is to be used.)
Add solution B to solution A in a Coplin jar.

Control

(i) Omit solution B. Any 'reaction' will be due to adsorption of lead.
(ii) Add 0.01 M sodium fluoride (21 mg/50 ml) to the incubation medium. True phosphamidase activity will not be abolished (in contrast, for example, to acid phosphatase activity).

ESTERASES

Biochemical background

It is necessary to consider something of the physical chemistry of the esterases if the behaviour and classification of these enzymes is to be understood even superficially. Careful work has been done on carboxylesterase by Webb (see Dixon and Webb, 1964, pp. 218–220) and reference will be made to this work in our review of the elements of the enzymology of esterases. Much of what will be described has application to other enzymes.

Elementary enzymology

An organic ester is the result of reaction between an organic acid and an alcohol (analogous, superficially, to the reaction of an acid and a base to yield an inorganic salt). A simple example is the reaction between acetic acid, $CH_3.COOH$, and ethanol, C_2H_5OH, to yield ethyl acetate ($CH_3.COOC_2H_5$):

$$CH_3.COOH + H.O.C_2H_5 \rightarrow CH_3.CO.O.C_2H_5 + H_2O$$

acetyl group   alkyl group

The ester bond is represented by:

$$-\overset{\displaystyle O}{\underset{\displaystyle |}{\overset{\displaystyle \|}{C}}}-O-\overset{\displaystyle |}{\underset{\displaystyle |}{C}}-$$

Similarly, esters can be formed between 'inorganic' acids and alcohols as, for example, between phosphoric acid and ethanol:

$$3H-\overset{H}{\underset{H}{C}}-\overset{H}{\underset{H}{C}}-O\lvert H + HO\rvert -\overset{O}{\underset{OH}{\overset{\|}{P}}}-OH \rightarrow H-\overset{H}{\underset{H}{C}}-\overset{H}{\underset{H}{C}}-O-\overset{O}{\underset{\underset{C_2H_5}{O}}{\overset{\|}{P}}}-O-C_2H_5 + 3H_2O$$

It will be seen that in this case the ester bond is denoted by:

$$-\overset{\displaystyle |}{\underset{\displaystyle |}{C}}-O-\overset{\displaystyle O}{\underset{\displaystyle |}{\overset{\displaystyle \|}{P}}}-$$

whereas in the ethyl acetate it was:

$$-\overset{\displaystyle |}{\underset{\displaystyle |}{C}}-O-\overset{\displaystyle O}{\overset{\displaystyle \|}{C}}-$$

Consequently, it will be seen that the phosphatases and esterases of histochemistry are both esterases which cleave somewhat similar ester bonds. The specificity depends on the substitution of phosphorus for carbon on one side of the bond. A phosphatase, therefore, is a specific type of esterase (e.g. a phosphomonoesterase).

The straight-chain esters hydrolysed by carboxylesterase can be written, in general formula, as:

$$acyl-\overset{O}{\overset{\|}{C}}-O-alkyl$$

(where acyl is the general form of acetyl in the specific acetic ester above). They become aligned at specific sites on the enzyme molecule (Figure 24).

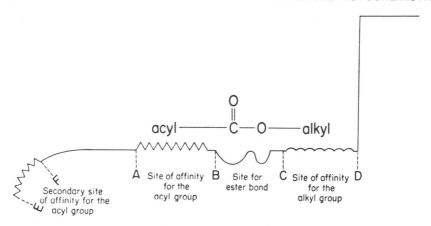

Figure 24. The orientation of the substrate on the active surface of the esterase.

Webb showed that the affinity and the reactivity of the ester were increased if the acyl group was extended up to a chain length of 4–6 carbon atoms; further increase in chain length gave increased affinity, by 'adhesion' at the secondary site (E–F) but this changed the orientation at B–C so that there was decreased reaction. A similar effect was found for the alkyl group: up to 4–6 carbon atoms gave greater affinity and reactivity, but increase in chain length above this reduced both affinity and reactivity; this indicates the size of C–D and the fact that there was no secondary site for the alkyl group as there was for the acyl group. The region B–C binds the ester bond and so locates the substrate molecule on the face of the enzyme molecule.

Since this enzyme acts on both aliphatic (straight-chain) and aromatic (benzene-ring) esters, this work of Webb (see Dixon and Webb, 1964) has relevance also to 'naphthol-acetate esterase' and to the lipases; in the latter the different chain lengths of the acyl group which can be bound on the enzyme depends assumably on the different lengths of A–B and of E–F in the different lipases.

Inhibition of esterase activity (see Dixon and Webb, 1964, pp. 346 *et seq.*)

Many enzymes which have esterase activity are inhibited in a highly specific manner by organophosphorus substances which have the general formula:

$$
\begin{array}{cc}
\begin{matrix} R{-}O & O \\ \diagdown & \| \\ & P \\ \diagup & \diagdown \\ R'{-}O & X \end{matrix}
& \text{and} &
\begin{matrix} R{-}O & O \\ \diagdown & \| \\ & P \\ \diagup & \diagdown \\ R' & X \end{matrix}
\end{array}
$$

where R and R′ are alkyl groups and X can be $-CN$, $-F$ or $-O.C_6H_4.NO_2$. These substances also inhibit trypsin, chymotrypsin, plasmin and thrombin. Their mode of action depends on the way esterases function. The substrate is first aligned on the esterase, as in Figure 24. The acyl group (Ac) is then transferred to the enzyme (Enz), so forming an acylated enzyme and liberating the alkyl group (Alk) as the free alcohol (1). The acylated enzyme then becomes hydrolysed (2):

$$
\underset{\text{Ac}-\overset{\displaystyle O}{\overset{\|}{C}}-\text{O}-\text{Alk} + \text{Enz.H}}{} \;\rightarrow\; \underset{\text{Ac}-\overset{\displaystyle O}{\overset{\|}{C}}-\text{Enz.} + \text{HO.Alk}}{} \qquad\qquad \mathbf{1}
$$

$$
\underset{\text{Ac}-\overset{\displaystyle O}{\overset{\|}{C}}-\text{Enz.} + \text{H}_2\text{O}}{} \;\rightarrow\; \underset{\text{Ac}-\overset{\displaystyle O}{\overset{\|}{C}}-\text{OH} + \text{Enz.H}}{} \qquad\qquad \mathbf{2}
$$

The organophosphorus substance mimics this process. Thus it too becomes aligned on the enzyme; the enzyme becomes phosphorylated and the X group is liberated (as was the Alk group of the normal ester).

$$
\begin{array}{c}
\text{R}-\text{O}\;\;\;\text{O} \\
\diagdown\!\diagup \\
\text{P} \\
\diagup\;\;\diagdown \\
\text{R}'\;\;\;\text{X}
\end{array}
+ \text{Enz.H} \longrightarrow
\begin{array}{c}
\text{R}-\text{O}\;\;\;\text{O} \\
\diagdown\!\diagup \\
\text{P} \\
\diagup\;\;\diagdown \\
\text{R}'\;\;\;\text{Enz.}
\end{array}
+ \text{HX} \qquad \mathbf{3}
$$

The difference between the behaviour of the normal ester and the organophosphorus substance now becomes apparent because the rate at which the phosphorylated enzyme can be hydrolysed is much slower than the rate at which the acylated enzyme is hydrolysed (2). In the weaker inhibitors this hydrolysis (4) does occur at a measurable rate and the enzyme can then become fully active again; in the more permanent inhibitors the hydrolysis occurs too slowly to have physiological value.

$$
\begin{array}{c}
\text{R}-\text{O}\;\;\;\text{O} \\
\diagdown\!\diagup \\
\text{P} \\
\diagup\;\;\diagdown \\
\text{R}'\;\;\;\text{Enz.}
\end{array}
+ \text{H}_2\text{O} \longrightarrow
\begin{array}{c}
\text{R}-\text{O}\;\;\;\text{O} \\
\diagdown\!\diagup \\
\text{P} \\
\diagup\;\;\diagdown \\
\text{R}'\;\;\;\text{OH}
\end{array}
+ \text{Enz.H} \qquad \mathbf{4}
$$

This very slow hydrolysis is the basis of the methods of Ostrowski and colleagues (Ostrowski and Barnard, 1961; Ostrowski et al., 1963) for the histochemical localization of acetylcholinesterase, in which the enzyme inhibitor is labelled radioactively, so that the enzyme is located, autoradiographically, by the site of the radioactive inhibitor.

The action of eserine and of prostigmine, which are powerful inhibitors of both cholinesterase and acetylcholinesterase, is due to their acting very much as do the organophosphorus substances (Dixon and Webb, 1964, p. 355, quoting work of Wilson et al., 1960). They become aligned on the enzyme at the active esterase site (B–C in Figure 24) and their carbamoyl group is transferred to the enzyme to give a carbamoyl-enzyme which is only very slowly hydrolysed (the reaction being that of 2 and 4). The enzyme finally does recover by its slow decarbamoylation by water (hydrolysis); the speed of this decarbamoylation can be considerably enhanced by hydroxylamine.

carbamoyl group
corresponding to Ac
in equations 1 and 2

Eserine
The heavier type
indicates the ester bond

Nomenclature and properties

The biochemistry of the esterases is very difficult and this is reflected in the very confused nomenclature and literature on the whole subject (see Dixon and Webb, 1964, p. 221). At one time the term 'ali-esterase' was used to denote an enzyme which attacked only aliphatic esters; later evidence (as discussed above) indicates that this is not a valid concept. Similarly, lipases are considered as different from other esterases only in the nature of regions E–F and A–B (Figure 24) and in that they act on substrates which generally are not in true solution.

The main classification of esterases depends on the distinction between those (B-type esterases) that are inhibited by a low concentration of organophosphorus compounds (i.e. those that perform reaction 4 at a very slow or negligible rate) and those that are not (A-type esterases which do perform reaction 4 reasonably rapidly). Cholinesterases are particular examples of B-type esterases in that they are also inhibited by low concentrations of eserine. The classification was reviewed by Aldridge (1956, 1961).

The cholinesterases are also of two types. The first, which used to be called 'true cholinesterase' but is better referred to as acetylcholinesterase (EC 3.1.1.7), is actually inhibited by high concentrations of acetyl choline. The second type, which used to be called 'pseudocholinesterase', is now referred to as cholinesterase (EC 3.1.1.8). Neither type of cholinesterase requires the charged nitrogen atom of choline for its activity, although cholinesterase is more specific for the choline structure than acetylcholinesterase (Dixon and Webb, 1964). Both types hydrolyse aromatic esters (like naphthol acetate used in histochemistry). The real difference between the two types seems to be that the rate of hydrolysis by acetylcholinesterase decreases sharply as the chain length of the acyl group is increased, whereas the reverse is true of cholinesterase (Dixon and Webb, 1964).

In tissues acetylcholinesterase (3.1.1.7) has been found in rat liver, occurring in greatest concentration in the microsomal fraction; similarly carboxylesterase (3.1.1.1) has been found in greatest concentration in the microsomal fraction of the liver of the mouse and rat (Roodyn, 1965).

Some of the chemical characteristics are listed in Table 7.

Table 7. Some properties of esterases

	Acetylcholin-esterase ('true cholin-esterase' 3.1.1.7)	Cholinester-ase ('pseudo-cholin-esterase' 3.1.1.8)	Lipases (glycerol ester hydrolase 3.1.1.3)	A-esterases	B-esterases
Effect of eserine $(1-5 \times 10^{-5}\,\text{M})$	Inhibited	Inhibited	Unaffected	Unaffected	Unaffected
Effect of organo-phosphorus compounds (usually $10^{-5}\,\text{M}$)	Inhibited	Inhibited	Some lipases are unaffected	Unaffected (these compounds are readily hydrolysed)	Inhibited
Occurrence	Nerve-tissue; erythro-cytes; liver	Serum; heart; intestinal mucosa; pancreas	Pancreas; oats		
pH optima	7.5–8.5	7.5–8.5	8.0 (but 6–7 in presence of tauro-cholate)	7.8	
Activators	NaCl; KCl; CaCl₂		Ca and Mg	Mn; glyoxaline; histidine; metal-chelating agents	

Histochemical background

Because the biochemistry of the esterases is rather complicated, it is of no surprise that the histochemistry of these substances has also been less than simple. The problem has been aggravated by the fact that some histochemists have not appreciated that phosphatases are also esterases. Consequently, in view of the lack of adequate controls in some histochemical studies, the histochemical evidence concerning the localization and activity of esterases is less than adequate. For all that, histochemistry proved of especial value in showing that esterases can occur within lysosomes. This was confirmed by the collaborative biochemical and histo-chemical study of Underhay et al. (1956) and by the studies of Tappel (1969).

Histochemically, esterase activity has been reported to be high in liver, pancreas, the 'chief cells' of the gastric mucosa, kidney tubules and bronchial mucosa (Burstone, 1962). There have been promising reports of significant esterase activity in tumours (e.g. Gomori, 1955). The very careful study by Willighagen *et al.* (1963) indicated that the histochemical reaction for this type of activity could be of prognostic value.

Cholinesterase, or acetylcholinesterase (by the Koelle method), has been successfully used as a histological stain for nervous tissue (e.g. Garrett, 1966a, b).

Various inhibitors of the different classes of esterase activity have been used to give some degree of specificity to the histochemical reaction. One of the most commonly used inhibitors is ISO-OMPA (tetraisopropylpyrophosphophoramide; Burstone, 1962); according to Barka and Anderson (1963), at 3×10^{-5} M this causes 90% inhibition of cholinesterase, whereas compound 62C–47, at concentrations of 3×10^{-5} to 1×10^{-4} M, completely inhibits acetylcholinesterase. Both groups of authors point out that the inhibitory effect of these, and of other inhibitors, varies in different organs, even in the same species; different species of animals respond to different concentrations. In a valuable series of studies on plant tissues, Gahan and co-workers (e.g. Gahan and Carmignac, 1989) showed that, when reacted with naphthol AS-D acetate (with Fast Blue BB as the chromogenic diazonium salt), carboxylesterase activity can be distinguished from other esterases by its inhibition by either diisofluoropropylphosphate (DFP) or diethylparanitrophenylphosphate (E 600), both at 10^{-4} M. It is resistant to 10^{-4} M eserine and to 10^{-4} M parachloromercuriphenylphosphate.

NAPHTHOL ACETATE METHOD FOR ESTERASE

Rationale

The ester used is that formed from acetic acid (acyl group) and α-naphthol (in place of ethanol on p. 174). As discussed above, esterases will act on aromatic or aliphatic esters; the advantage of having naphthol as the alkyl group is that as it is liberated, it can be coupled to form an insoluble azo-dye (as in the acid phosphatase method, see p. 155). The method is simple and much used. One note of warning must be added. In the discussion of the biochemistry of esterases we noted that these enzymes act on the specific ester bond. We also noted the similarity between the bonds:

$$\begin{array}{ccc} & O & & & O \\ & \| & & & \| \\ -C-O-C- & & \text{and} & & -C-O-P- \\ & & & & \end{array}$$

The basis of inhibition by organophosphorus compounds (see above) was their ability to locate themselves at the active site of the esterase molecule; they inhibited only because their rate of hydrolysis off the active site was so slow.

Thus it seems surprising that many histochemists apply naphthol acetate to one section and naphthol phosphate to another and claim, without further controls, that the former shows esterase activity while the latter shows phosphatase activity. With such staining procedures, some control and some caution in interpretation might be helpful.

Method (after Gomori, 1952)

Use fresh, unfixed cryostat sections.

1. Incubate sections at room temperature for up to 10 min in the reaction medium.
2. Wash in several changes of distilled water.
3. Mount in Farrants' medium.

Result

Enzyme activity is shown by a purple colour.

Reaction medium

Dissolve 10 mg α-naphthol acetate in 0.25 ml acetone. Add this to 20 ml 0.1 M phosphate buffer at pH 7.4. Shake well. Then dissolve 20 mg Fast Blue B salt in this solution. Filter the solution onto the section; use *immediately*.

Control

The control used will depend on what type of esterase activity is being investigated (see biochemical and histochemical data, above).

INDOXYL ACETATE METHOD FOR ESTERASE

Rationale

Barrnett and Seligman (1951) suggested that indoxyl acetate could be used as an ester substrate; esterase would liberate the indoxyl group, which would become oxidized to indigo. Holt (1958; Holt and O'Sullivan, 1958; Holt and Sadler, 1958; Cotson and Holt, 1958; Holt and Withers, 1958) developed this idea; there are few cytochemical methods which have been studied with such physicochemical precision. Basically the reaction sequence is as follows (Holt and Withers, 1958):

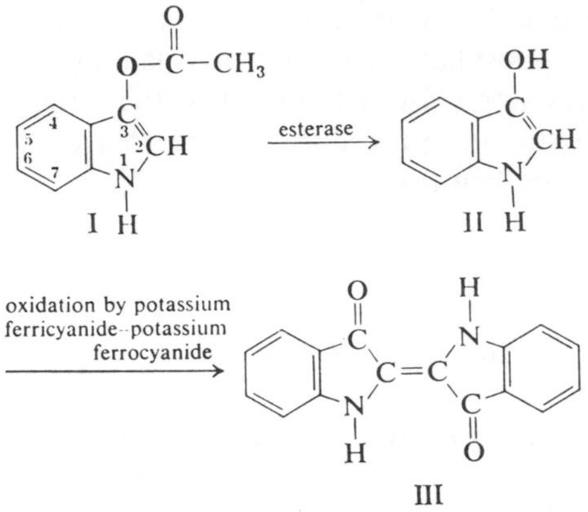

Of many indoxyl acetates tested, Holt and co-workers found that the 5-bromo-4-chloroindoxyl acetate gave the most precise staining. The concentration of the oxidant was stressed by Cotson and Holt (1958) because the step from II to III goes via a *leuco*-dye, and they suggested that if the rate of oxidation is too slow (e.g. if aerial oxidation is relied on) the *leuco*-indigo may diffuse to artefactual sites before it becomes coloured. The rate of oxidation is depressed at lower pH values (Cotson and Holt, 1958); the effect of the oxidant may be somewhat variable under a number of conditions.

There has been some criticism of the method as given originally by Holt, on the grounds that the formalin which he used as a fixative completely inhibits one form of esterase and that the ferri- ferro-cyanide oxidant is also inhibitory (see Shnitka and Seligman, 1960, 1961; Seligman, 1963).

Method

Unfixed cryostat sections should be used.

1. Incubate in the reaction medium at 37 °C for 15–30 min.
2. Wash in 30% alcohol containing 0.1% acetic acid.
3. Wash in distilled water.
4. Mount in Farrants' medium.

Result

Esterase activity is shown by a blue colour.

Solutions required for this method

Reaction medium

Tris buffer (0.2 M), pH 8.5   2 ml
Oxidant   1 ml

1 M calcium chloride	0.1 ml
2 M sodium chloride	5.0 ml
Distilled water	2.0 ml

Add this solution rapidly, with agitation, to 1 ml absolute ethanol containing 1.3 mg 5-bromoindoxyl acetate.

Oxidant

Potassium ferricyanide	210 mg
Potassium ferrocyanide	155 mg
Distilled water to make up to	100 ml

Control

As discussed for the naphthol acetate method.

CHOLINESTERASE

Rationale

Acetylthiocholine is hydrolysed by both acetylcholinesterase and cholinesterase at least as rapidly as acetylcholine (Gomori, 1952, p. 210). As discussed in relation to the biochemical data of esterases (above), longer-chain esters, such as butyrylthiocholine, are hydrolysed more rapidly by cholinesterase than by acetylcholinesterase; this distinction has been used in histochemistry (Gomori, 1952; Barka and Anderson, 1963). The thiocholine liberated by such enzymatic activity is precipitated by reaction with copper, present as copper glycinate in the reaction medium; the copper thiocholine is then transformed into brown copper sulphide by treatment with ammonium sulphide. Traces of copper thiocholine are added to the various fluids used in this reaction to ensure that any that is formed enzymatically will remain insoluble (i.e. enough copper thiocholine is added to ensure that the solubility equilibrium is kept from left to right in the equation on p. 139; in this way there is less danger of any enzymatically formed copper thiocholine being brought into solution to maintain the solubility constant K'). The chemistry of this reaction has been discussed by Malmgren and Sylvén (1955).

Useful variants of this technique are the gold–thiocholine and gold–thiolacetic acid methods for acetylcholinesterase and cholinesterase (Koelle and Gromadzki, 1966).

Method (after Koelle and Friedenwald, 1949)

Fresh cryostat sections should be used. Barka and Anderson (1963) emphasized that formalin fixation results 'in considerable loss of enzyme activity'.

1. Incubate in the reaction medium at 37 °C for 10–60 min, depending on the tissue. [It is best if the sections or smears are held horizontally so that the freshly prepared reaction medium can be filtered directly onto them.]
2. Wash in distilled water containing traces of copper thiocholine.
   [*Note*: this compound is almost insoluble; consequently only enough should be added to provide it in true solution. There must not be so much as to form a suspension which may become adsorbed onto the section or smear.]
3. Leave for 1 min in the ammonium sulphide solution (as for alkaline phosphatase).
4. Wash in distilled water.
5. Mount in Farrants' medium or an equivalent water-soluble mountant.

Result

Brown deposit, often crystalline, indicates cholinesterase or acetylcholinesterase activity.

Controls (see Garrett, 1966a)

(i) 3×10^{-5} M eserine should inhibit both cholinesterase and acetylcholinesterase activity. It can be added to the normal incubation medium.

(ii) ISO-OMPA at a concentration of 3×10^{-6} M should inhibit cholinesterase activity. It may be necessary to treat sections for 30 min with this, in the reaction medium which lacks acetylthiocholine, before incubating in the full reaction medium ($+$ ISO-OMPA $+$ acetylthiocholine). Any reaction after this procedure may be considered to be due to *acetyl*cholinesterase.

(iii) B.W.284C51 or 62C–47, at concentrations of 3×10^{-5} to 1×10^{-4} M are expected to inhibit acetylcholinesterase. As with ISO-OMPA, it may be necessary to preincubate for 30 min in the reaction medium containing the selected inhibitor but lacking the substrate, and then to incubate in the full reaction medium which contains both the substrate and the inhibitor. For this it is better to use butyrylthiocholine as the substrate. Any reaction after this procedure may be interpreted as being due to cholinesterase activity.

Solutions required for this method

Copper thiocholine

Dissolve acetylthiocholine iodide in a 1% solution of copper glycinate (to give an excess of copper glycinate). Leave this solution to stand overnight. Collect the precipitate (which should be copper thiocholine); wash it with water several times.

Solution A

Dissolve 3.75 g glycine in 18 ml 1 M potassium hydroxide. Make up to 100 ml with distilled water.

Solution B

0.1 M solution of cupric sulphate.

Solution C

Dissolve 14.5 mg acetylthiocholine iodide in 0.75 ml distilled water. Add 0.25 ml solution B. Cupric iodide is precipitated, leaving the acetylthiocholine in solution. Centrifuge and decant, retaining the clear solution.

Reaction medium

Distilled water	8.6 ml
Solution B	0.2 ml
Solution A	0.4 ml

To this add a trace of copper thiocholine (ideally this should be added to make the reaction medium 2×10^{-3} M with respect to copper thiocholine). Heat to 37 °C. Just before use, add 0.8 ml solution C. Filter onto the section (i.e. it must be used immediately).

LIPASES

Rationale

The substrate for lipases is a Tween; these compounds are esters of long-chain fatty acids and either sorbitan or mannitan (Gomori, 1952). The enzyme splits the fatty acid off and this is trapped as the insoluble calcium salt. The latter is then converted into the lead salt to produce a dark sulphide which stains the tissue. There is some doubt as to whether the rate of capture of the fatty acids by the calcium is sufficiently rapid, and whether the calcium soaps so produced are sufficiently insoluble (in both water and lipid) to provide rigorous intracellular localization of the site of the reaction. However, clear reactions have been obtained, which relate to specific parts of tissues known to contain these enzymes, and different responses have been observed to Tweens which have fatty acids of various chain length.

Method

Fresh cryostat sections should be used.

1. Incubate in the reaction medium at 37 °C for 2–3 h.
2. Wash well in distilled water, preferably containing 1% calcium chloride to ensure that the calcium soaps remain insoluble.
3. Place in a 1% solution of lead nitrate for 15 min.
4. Wash in running tap water for 5 min.
5. Immerse for 1 min in water saturated with hydrogen sulphide (as for the Gomori acid phosphatase method, p. 154).
6. Wash well in tap water.
7. Mount in glycerine jelly or in Farrants' medium.

Result

Brown-black precipitate of lead sulphide indicates lipase activity.

Control

As for the test, but leave the Tween out of the reaction medium (any colour is then due to 'metallophilia').

Solutions required for this method

Stock solutions (to be prepared fresh on the day of use)

A.   5% aqueous solution of the selected Tween (e.g. Tween 60).
B.   0.5 M Tris buffer, pH 7.2.
C.   10% solution of calcium chloride.

Reaction medium

Tris buffer (B)	5 ml
Tween solution (A)	2 ml
Calcium chloride solution (C)	2 ml
Distilled water	40 ml

PROTEASES

Biochemical background

Proteases, or peptidases, are of two types. The first requires a terminal carboxyl or amino group on one side of the peptide bond; these are the exopeptidases of which leucine aminopeptidase is a typical example. The second type, the endopeptidases, may even be inhibited by such groups but require the presence of a particular amino acid. For example, trypsin attacks the peptide bond produced by the carboxyl group of arginine (or lysine) but depends on the second amino group of these residues remaining free; its activity is inhibited by the close proximity of carboxyl radicals (as discussed by Bergmann and Fruton, 1941; Bergmann, 1942; Dixon and Webb, 1964, p. 225 *et seq.*). Enzymes such as trypsin, which act preferentially on bonds within the protein chain, are classified as endopeptidases; their presence may be inferred by detecting associated activity, such as phosphamidase or esterase activity, or demonstrated by allowing the enzymes to leak out of the section, or cells, into a film of a suitable protein and then demonstrating the areas of the film which have become affected. (See Daoust (1965) for a review of such methods as applied to the detection of deoxyribonuclease, ribonuclease, amylase and proteases.)

One other point deserves to be stressed. In contrast to pepsin, some proteases, particularly trypsin (Dixon and Webb, 1964, p. 243), split amides (cf. phosphamidase) and esters even more rapidly than they hydrolyse peptide bonds. Thus, even though trypsin is a powerful endopeptidase, it rapidly hydrolyses lysine ethyl ester. Chymotrypsin also hydrolyses amides and esters, with preference for aromatic esters (Dixon and Webb, 1964, pp. 243–245). Cathepsin-C is another protease which acts on amides and on esters (Dixon and Webb, 1964, p. 246). It could be that the histochemical demonstration of esterase activity encompasses the demonstration of protease activity.

Leucine aminopeptidase is a true exopeptidase (or aminopeptidase) in that it requires the presence of a free α-amino group. It is not specific for leucine. Unlike trypsin, it will not act on esters of amino acids (although it will hydrolyse amides) and, correspondingly, it is not inhibited by diisopropyl phosphofluoridate. [*Note*: organophosphorus compounds inhibit enzymes which have esteratic activity; see p. 177–179.]

It may be relevant to note that, apart from the proteolytic enzymes discussed above, plasmin and fibrinolysin act as endopeptidases (Burstone, 1962, p. 395).

Histochemical background

The histochemistry of endopeptidases has been fully reviewed by Smith and van Frank (1975) who have contributed considerably to this subject particularly by the synthesis of special substrates. These involve the amino acids and peptide bonds that are critical for the activity of a particular peptidase or dipeptidase, linked to β-naphthylamine. The activity is shown by the reaction of the liberated naphthylamine, together with its associated amino acids; this is reacted with Fast Blue B, which they particularly recommend.

AMINOPEPTIDASES AND NAPHTHYLAMIDASES

Biochemical background

Two quite dissimilar classes of enzyme are demonstrable by the ability to hydrolyse an amino acid naphthylamide, such as leucine 2-naphthylamide. The first class comprises the aminopeptidases; the second includes the lysosomal arylamidases (or amino acid naphthylamidases).

Aminopeptidases

These occur predominantly in the small particulate (non-lysosomal) fraction of homogenates; they are often used as 'marker' enzymes to characterize the cell membrane fraction of homogenates (e.g. De Pierre and Karnovsky, 1973; Peters, 1976). A good source of aminopeptidase is the kidney. Patterson *et al.* (1963, 1965) showed that hog kidney aminopeptidase hydrolysed leucine amide about 15 000 times more rapidly than conventional chromogenic substrates; this is a useful way of discriminating between such aminopeptidases and naphthylamidases (or arylamidases), which show little or no activity against leucine amide (Marks *et al.*, 1968; Behal *et al.*, 1969).

The nature and assay of aminopeptidases have been reviewed by Smith (1951, 1960; Smith and Spackman, 1955; Smith and van Frank, 1975). These enzymes are activated by manganese or cobalt and show optimal activity between pH 7 and 8.5, depending on the substrate (*Biochemists' Handbook*).

An amidase (EC 3.5.1.4) which can also hydrolyse leucine 2-naphthylamide occurs in the supernatant fraction of homogenates (Roodyn, 1965). Its activity is optimal at around pH 8.0 and, like the aminopeptidases, it acts preferentially on leucine amide.

Naphthylamidases (or arylamidases)

These enzymes have been found exclusively in the lysosomal fractions of homogenates (Mahadevan and Tappel, 1967; Tappel, 1969). They are not manganese dependent; they act preferentially on either leucine 2-naphthylamide or on arginine 2-naphthylamide; and they have little or no activity with leucine amide as substrate (Patterson *et al.*, 1963, 1965; Marks *et al.*, 1968; Behal *et al.*, 1969). The pH optimum of liver arylamidase and of pituitary arginyl arylamidase was 6.8 (Ellis and Perry, 1966). More recently it has been shown that the lysosomal enzyme, cathepsin H, acts strongly on amino acid naphthylamides with optimal activity at pH 6.0 (Kirschke *et al.* 1977).

The optimal activity of various arylamidases has been shown to be between pH 5.5 and 6.8 (Ellis and Perry, 1966). Tris buffer can cause inhibition (McDonald *et al.*, 1968).

Histochemical background

Histochemically the main interest in enzymes that hydrolyse leucine naphthylamide or arginine naphthylamide has been in those that act at acidic pH values. Such enzymatic activity has been shown to be associated with the same sites as are coloured by the acid phosphatase reaction (Chayen *et al.*, 1971) and with autophagic vacuoles which show acid phosphatase activity (Niemi and Sylvén, 1969). When the naphthylamidase reaction is done at an acidic pH, the reaction has been shown to occur in subcellular particles that resemble lysosomes in size (e.g. Sylvén and Lippi, 1965), in osmotic characteristics (McCabe and Chayen, 1965), and in their latency related to the state of the lysosomal membranes (Bitensky *et al.*, 1973).

The quantitative histochemical study of lysosomal naphthylamidase activity has been of value in the study of rheumatoid arthritis (Chayen and Bitensky, 1971; Bitensky *et al.*, 1973) and in the investigation of inflammation and of the potency of anti-inflammatory drugs (Chayen *et al.*, 1970). There is also evidence that this activity is of consequence in many pathological conditions. Fibrocytes are said to have considerable naphthylamidase activity. This subject was reviewed by Burstone (1962) particularly in relation to the study of malignant growths. Sylvén and Malmgren (1957) related the degree of naphthylamidase activity to the invasive properties of malignant growths. Willighagen and Planteijdt (1959) suggested that this activity could be used diagnostically; this suggestion was followed up by Millett *et al.* (1980).

Rationale of the histochemical procedure

In the cytochemical reaction, the substrate is either leucyl or arginyl naphthylamide:

leucyl   naphthylamide   + H₂O

The arrow marks the amide bond

leucine   β-naphthylamine

The enzyme releases naphthylamine, which has to be trapped by a diazonium salt, present in the reaction medium, to yield an insoluble azo dye. Studies on rates and distances of diffusion have shown that Fast Blue B is more effective than either Fast Garnet GBC or Corinth V (Nachlas *et al.*, 1957a; McCabe and

Chayen, 1965). Since such compounds decompose readily, it is necessary to replace the medium with fresh medium every 15 min.

Low concentrations of cyanide may activate the naphthylamidases, although high concentrations inhibit this activity. Inhibitors of this enzymatic activity include zinc (which is often present in commercially available diazonium salts), copper, mercury, lead and nickel (Burstone, 1962). The addition of cyanide to the reaction medium may be, in part at least, to remove some inhibitory ions.

Concept, and evaluation, of latency

Histochemistry has proved of especial value in the study of lysosomes. These organelles, which may be as small as $0.2 \mu m$ in diameter, are bounded by a semipermeable membrane which retards entry of certain substrates, such as glycerophosphate or leucine 2-naphthylamide. Other more lipophilic substrates, such as naphthyl phosphate or leucine-4-methoxy-2-naphthylamide, although excellent substrates, pass freely across the membrane and therefore are useless for studies involving changes in the state of the membrane. Yet it has been the investigation of changes in the state of the bounding membranes that has proved of value in many applied problems. These include the effect of anti-inflammatory drugs (Chayen *et al.*, 1972) and the effect and bioassay of thyroid stimulating hormone and the long-acting thyroid-stimulating γ-globulin (Bitensky *et al.*, 1974a; also Chayen *et al.*, 1976).

The concept of latency, as measured histochemically, may be explained as follows. Suppose we take sections, or smears, from a control and from a treated population. After a 'reasonable' time of reaction, the test may show greater activity of either naphthylamidase or acid phosphatase (as in Figure 25a); these remarks are applicable to both enzymes. Assuming that the activity was linear with time, this would imply that the test cells had more activity than those of the control. However, because lysosomal enzymes normally occur within a bounding membrane, this need not necessarily be true. The different activities could be due to the same concentration of enzyme but, in the control cells, with access of substrate to the enzyme restricted by the lysosomal membranes. Consequently to test for this possibility, the activities should be tested against different times of reaction. If the membranes of the test cells were totally permeable, the activity in these cells would increase linearly with time of reaction (as in Figure 25b). In contrast, in the control cells, where the lysosomal membranes would be intact, there would be no recordable activity for a short time. During this time, the relatively acidic medium would render the lysosomal membranes permeable, progressively allowing more substrate to reach the enzyme inside the lysosomal particle.

In practice, the lysosomal membranes of the test cells are rarely fully labilized so that what is recorded is a shorter latent time before the substrate reaches the enzyme (as in Figure 25c).

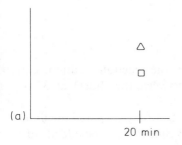

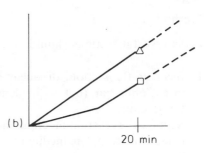

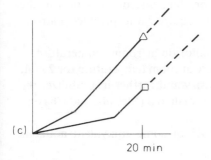

Figure 25. Diagrammatic representation of the effects of latency: △ , test; □ , control.

Measurement of latency

Serial sections (or smears) are reacted for a range of times to determine how long (t) is required to obtain sufficient coloration for assessment. For assessing latency, a second set of sections is then pretreated with the buffer alone (e.g. acetate buffer at pH 5.0) for various times and then reacted with the full chromogenic reaction medium for the determined time (t). In normal cells the amount of activity (stain) will increase as a consequence of the acidic pretreatment up to a maximum, and then decline. This decline is due to excessive labilization of the lysosomal membranes, allowing the enzyme to escape from the lysosomes.

The activity in the sections which had not been subjected to the acidic pretreatment is taken as the 'freely available' or 'manifest' activity (relative to the particular substrate used). The maximum recordable activity achieved by suitably prolonged acidic pretreatment is taken as a measure of 'total' activity. Then, if the activity is measured microdensitometrically, the 'bound' or 'latent' activity is recorded as:

$$\frac{(\text{Total activity} - \text{Manifest activity}) \times 100}{\text{Total activity}}$$

Method

Fresh cryostat sections should be used.

1. Incubate the section, or smear preparation, in the reaction medium, e.g. in a Coplin jar (but with plenty of space between the slides) at 37 °C for 15 min.
2. Remove the reaction medium and replace it with freshly prepared medium, namely the complete medium to which the Fast Blue B has been added, just before the section (or smear) is immersed in it.

   Repeat this process until sufficient reaction product is obtained. [*Note*: 8 μm thick sections of rat kidney produce a strong reaction in the first 15 min whereas those of rat liver may require 2 h to produce a strong reaction.]
3. Rinse well in 0.85% solution of sodium chloride at room temperature.
4. Transfer to 0.1 M solution of copper sulphate at room temperature for 2 min. [*Note*: this step is to chelate, and so intensify and further insolubilize, the dye that has been formed. The localization is unaltered (McCabe and Chayen, 1965).]
5. Mount directly in Farrants' medium, or Z5 or some equivalent aqueous mountant.

Specificity

Expose serial sections, or smears, to the full reaction medium but with the addition of leucine amide at four times the concentration of the chromogenic substrate (e.g. leucine naphthylamide). This should 'inhibit' aminopeptidase and amidase activity (as discussed above).

Control

There is no adequate control. Sections can be incubated in the full reaction medium lacking the substrate.

Measurement

Because of the size of the lysosomal particles that are liable to be the sites of this naphthylamidase activity, the activity must be measured with an oil immersion (NA 1.25) objective on a Vickers microdensitometer. Valid measurements can be made with this instrument and with an objective of this numerical aperture (NA), since it is essential to have the diameter of the scanning spot as small as 0.25 μm. Measurements with a larger scanning spot can be totally invalid, as discussed by Bitensky *et al.* (1973).

Solutions required for this method

Reaction medium

Prepare immediately before use:

0.1 M acetate buffer, pH 6.1	10 ml
0.85% solution of sodium chloride	8 ml
0.02 M potassium cyanide (32 mg in 25 ml of water)	1.0 ml
Leucyl β-naphthylamide hydrochloride (barium salt)	8 mg in 1.0 ml distilled water
Just before use add Fast Blue B	10 mg

Adjust the final pH to 6.5.
To check specificity, add 32 mg leucine amide in 1.0 ml distilled water.

Chelating solution

Copper sulphate ($CuSO_4.5H_2O$, Analar)	2.5 g
Distilled water	100 ml

β-GLUCURONIDASE

Biochemical background

This enzyme (EC 3.2.1.31) hydrolyses β-D-glucosiduronic and β-D-galacto-siduronic acids; it does not act on glucosides or on α-D-glucosiduronic acids (*Biochemists' Handbook*, p. 227).

One of its main physiological functions is the hydrolysis of steroid glucosiduronic acids and of acidic mucopolysaccharides (*Biochemists' Handbook*). On the other hand, Fishman (1961; also Fishman and Baker, 1956) showed that, at physiological pH and optimal concentrations of substrate and acceptor, its predominant function may be to act as a transferase:

$$R - O - C_6H_9O_6 \ + \ R'OH \longrightarrow R' - O - C_6H_9O_6 \ + \ R - OH$$

glucuronic acid	acceptor	new glucosiduronic acid	aglycone
(donor substrate)	(which can be a steroid or poly-saccharide, e.g. hyaluronic acid)	(e.g. steroid glucuronide)	of donor substance

This enzyme seems to occur in most tissues of mammals and has been reported in high concentrations in liver, kidney, spleen, lung, adrenal, thyroid and uterus. The richest source is the preputial gland of the female rat. In many tissues, particularly in the uterus, vagina and mammary gland, its activity has been shown to be markedly affected by gonadal hormones (as in Fishman, 1963). Within cells its location is predominantly lysosomal, with some activity present in the endoplasmic reticulum (Barrett and Heath, 1977).

The pH–activity curve shows a broad peak with the maximal activity between pH 4.3 and 5.0. Very low concentrations of glucarolactones give competitive inhibition. The enzyme may dissociate readily into inactive components; this dissociation may be reversed by deoxyribonucleic acid, protamine and diamines so that, to overcome dissociation, one or other is frequently added to the reaction medium as an activator (Bernfeld *et al.*, 1954; Fialkow and Fishman, 1961). Parenteral administration of testosterone-like steroids has produced greatly enhanced activity in the kidney of mice (Fishman and Lipkind, 1958; Fishman, 1961).

The specific inhibitor of this enzyme is D-glucosaccharo-1,4-lactone, which is present in solutions of saccharate. It is inhibited by copper, especially in the presence of ascorbate, but it is not inhibited by fluoride, cyanide or thiourea (*Biochemists' Handbook*).

Many studies show that β-glucuronidase activity is very much greater in neoplasms than in the adjacent normal tissue (e.g. Fishman and Anlyan, 1947; also Fishman and Bigelow, 1950, for gastrointestinal tumours). The work of Kasdon *et al.* (1950, 1953) drew attention to the elevated levels of this enzyme in human cervical cancer; this led to many attempts to use this activity as a diagnostic criterion for distinguishing early malignancy (e.g. Odell and Burt, 1949) but the results were not decisive (e.g. Fishman, 1951; Lawson and Watkins, 1965).

Histochemical background

This enzymatic activity is readily inhibited by fixatives (Appendix 1, p. 296; Barka and Anderson, 1963, p. 281; Nachlas *et al.*, 1956) and by some diazonium compounds (Fishman and Goldman, 1965). Its pH optimum is between 3.5 and 5.5 (Fishman and Goldman, 1965); diazonium compounds do not couple efficiently at these pH values (see Barka, 1960; also Burstone, 1962, pp. 95–108, for full discussion of diazonium coupling reactions). The histochemical demonstration of this activity involves the following steps:

1. The retention, at its natural site, of this readily inhibited labile enzyme.
2. The retention of this readily soluble enzyme (Barka and Anderson, 1963, p. 281) at its natural site during the histochemical incubation procedure.

Hayashi *et al.* (1964) used a simultaneous coupling method involving hexazonium pararosaniline. The reaction was done at a compromise pH of pH 5.2 even though, at this pH, the coupler is not very efficient (Barka, 1960); at higher pH values the enzyme is very much less active. Consequently the post-coupling method of Fishman and Goldman (1965) is preferred. In this, the naphthol AS-BI-glucosiduronic acid is used as the substrate in the absence of a diazonium compound. The reaction is done at pH 4.5, which is optimal for the enzyme. The enzymatic activity liberates the very insoluble naphthol AS-BI. The section is then transferred to a solution containing the most efficient coupling reagent, irrespective of whether it may be inhibitory to the enzyme, because the reaction has already been completed. Moreover the coupling can be done at a pH that facilitates the coupling of the reagent with the insoluble 'precipitate' of naphthol AS-BI.

The present authors found that the enzyme is readily liberated into the general cytoplasm when this becomes coagulated during fixation. When care was taken to avoid the coagulation of the cytoplasm the enzymatic activity was located in discrete granules and could be activated by as short as 1 min exposure to acidic conditions (in the absence of a stabilizer).

Fishman *et al.* (1964), who first synthesized naphthol AS-BI β-D-glucosiduronic acid, showed that the kinetics of its hydrolysis fulfilled the criteria required of a biochemical and histochemical substrate. Olsen *et al.* (1981) showed that results obtained by the histochemical method (below), done on samples of 20 cells (measured microdensitometrically) agreed well with those obtained by the conventional biochemical procedure in which each assay required 10^6 cells.

POST-COUPLING METHOD (Fishman and Goldman, 1965)

Method

Cryostat sections, preferably of unfixed tissue, should be used. Fishman and Goldman (1965) also used frozen sections of thin slices of tissue which had then been fixed for 24 h at 4 °C in a 4% w/v solution of formaldehyde containing 1% anhydrous calcium chloride and then treated with an 0.88 M solution of sucrose containing 1% gum acacia for 24h at 4 °C (to remove the fixative).

The unfixed cryostat sections should be used immediately once they are dry.

1. Immerse in the reaction medium at 37 °C. An incubation time of between 30 and 60 min should be sufficient for most tissues.
2. Wash in cold distilled water.
3. Immerse in the cold coupling solution which had been left in a Coplin jar in a refrigerator. Leave, in the refrigerator, for 5 min, with occasional gentle agitation.
4. Wash in distilled water at room temperature.
5. Mount in an aqueous mountant such as 1% aqueous polyvinylpyrrolidone or Farrants' medium.

Result

Mainly small dots or granules. Any contribution from a cytosolic enzyme will appear light, owing to the lower concentration of the dye.

Measurement

Because of the size of the granules, measurement must be done with on oil-immersion lens of high numerical aperture. With Fast Blue RR as the coupler, measurement should be done at 600 nm (Olsen *et al.*, 1981).

Solutions required for this method

Substrate stock solution

Dissolve 11 mg naphthol AS-BI-β-D-glucosiduronic acid in 1.0 ml 0.05 M sodium bicarbonate. Make this solution up to 100 ml with 0.1 M acetate buffer, pH 4.5. This stock solution (2×10^{-4} M) is stable for several months provided it is kept in a refrigerator. The optimal concentration of the substrate is said to be 1×10^{-4} M.

Reaction medium

Dilute the stock (2×10^{-4} M) solution with an equal volume of 0.1 M acetate buffer, pH 4.5 (to give a final concentration of 1×10^{-4} M).

Diazonium coupler

A saturated solution of any suitable diazonium salt, in 0.01 M phosphate buffer, pH 7.4, can be used. Among such salts are Fast Garnet GBC, Fast Dark Blue R, Fast Blue RR, Fast Blue 2B and Fast Red Violet 2B.

MODIFIED POST-COUPLING METHOD

The procedure we prefer is as follows:

Method

1. Use unfixed fresh cryostat sections. Leave at room temperature only so long as is required to ensure that the surface moisture has gone.
2. Immerse in the reaction medium at 37 °C in a Coplin jar in a water bath. Leave for 30 min to 3 h depending on the tissue. Rat liver required about 2 h incubation unless it had been activated (see below).

3. Wash in 0.1 M acetate buffer, pH 4.5, which contains 0.5% of calcium chloride ($CaCl_2.6H_2O$).

  [*Note*: calcium was found to be essential for retaining the integrity of the sections and of the reaction.]

4. Immerse for 5 min in the cold coupling solution in a refrigerator.
5. Rinse in distilled water at room temperature.
6. Mount in glycerine jelly which has been warmed to 40 °C to render it sufficiently fluid.

  [*Note*: in our studies the mountant seemed to be fairly critical. For this reaction, Farrants' medium was less satisfactory. It was possible to see changes in the distribution of the dye with prolonged immersion in some mountants.]

Result

Dark blue reaction, generally in discrete granules, indicated β-glucuroinidase activity; they were absent in controls (see below).

Solutions required for this method

Stock substrate solution

11.4 mg naphthol AS-BI glucuronide are dissolved in 1 ml 0.05 M solution of sodium bicarbonate (0.42 g sodium bicarbonate in 100 ml distilled water). To this is added 49 ml 0.2 M acetate buffer at pH 4.5. This stock solution was stored indefinitely at +4 °C.

PVA solution

Dissolve 0.5 g calcium chloride ($CaCl_2.6H_2O$) in 100 ml 0.1 M acetate buffer, pH 4.5. Add 10 g polyvinyl alcohol (PVA). It is not readily soluble and the solution has to be heated—but not boiled—with frequent mixing for about 5 min to dissolve the PVA. It is cooled before it is used.

Reaction medium

PVA solution   20 ml
Stock substrate solution   2 ml

Coupling solution

Dissolve 8 mg Fast Dark Blue R (Gurr) in 20 ml 0.01 M phosphate buffer (disodium hydrogen phosphate/potassium dihydrogen phosphate), pH 7.4 at +4 °C.

This solution, at $+4\,°C$, is filtered through a Whatman No. 1 filter paper and should be used immediately.

Glycerine jelly

Dissolve 15 g white gelatine in 100 ml distilled water (with gentle heating if necessary). Add 100 g glycerine and warm for 5 min in a water bath at 37 °C. Filter the warm solution through glass wool.

This jelly should be stored frozen in small quantities.

Control

0.05 g of D-saccharic acid 1,4-lactone (Sigma) is added to 22 ml of the reaction medium and the sections are incubated directly in this. Complete inhibition of β-glucuronidase activity has been observed with this lactone.

Activation

Immerse sections in 0.1 M acetate buffer, pH 4.5, which contains 0.5% calcium chloride $(CaCl_2.6H_2O)$. Leave for 2 min and then transfer to the normal reaction medium (step 2 above). Continue as for the normal reaction.

Measurement

As for the previous procedure.

N-ACETYL-β-GLUCOSAMINIDASE

Biochemical background

This is a lysosomal enzyme that acts against N-acetylglucosaminides and N-acetylgalactosaminides. It was first described as β-N-acetylglucosaminidase (EC 3.2.1.30; later as a β-N-acetylhexosaminidase, EC 3.2.1.52: Frohwein and Gatt, 1967; later it was called hexosaminidase). Two forms of this enzyme, A and B, are known (Dance *et al.*, 1970). In human tissues their molecular weights are similar (127 000–140 000); they have pH optima of 4.0–4.2 and 4.6–4.8. They are inhibited by acetate ions, acetamide, sulphated polysaccharides, silver and mercury, dithiothreitol and 4-hydroxymercuribenzoate, but not by other reagents that block thiol groups. Various forms of this enzyme have been found to be wanting in patients with Tay–Sachs and with Sandhoff disease. The Tay–Sachs disease is characterized by the storage of GM2 ganglioside and lack of hexosaminidase A. In the Sandhoff disease, both globoside and GM2 ganglioside are stored and both types A and B forms of this enzyme are lacking. This subject has been reviewed by Barrett and Heath (1977). The deficiency has been explained by a lack of a converter enzyme operating between forms A and B (Robinson and Stirling, 1968) and by other models (Neufeld *et al.*, 1975).

Histochemical background

Earlier methods involved both chemical fixation and simultaneous coupling with the coupling agent present in the enzymatic reaction. Shannon (1975) found 40–60% inhibition of this enzymatic activity after fixation in formol–calcium, even after post-fixation treatment with gum–sucrose; in liver the activity was inhibited by only 25%. The presence of the coupling reagent in the reaction medium added to the inhibition of the enzyme (Hayashi, 1965). These deleterious effects were overcome by Shannon's (1975) post-coupling method, which used fresh, unfixed cryostat sections protected by the presence of a high concentration of polyvinyl alcohol (PVA) in the reaction medium.

Both Hayashi (1965) and Shannon (1975) commented on the fact that, although maximal activity was produced at pH 4.0–4.5, the characteristic granular reaction was best seen when the reaction was done at pH 5.0. However, Robertson (1980) found maximal granular reaction at pH 4.5 when Fast Garnet GBC was used as the coupling reagent, at pH 7.4, after the enzymatic activity was terminated.

Method (after Shannon, 1975)

Fresh cryostat sections should be used as soon as they have come out of the cryostat and are dry. Because the substrate is rather expensive, less reaction medium can be used if the reaction is done in Perspex rings (as in Chapter 2 of this book).

1. The reaction medium should be at 37 °C.
2. Pour it into the ring, placed around the section and leave it (under a coverslip if necessary) for up to 2 h. (Shorter times may be satisfactory for some tissues.)
3. Shake off the medium and rinse briefly in two changes of 0.1 M citrate buffer containing 0.5% $CaCl_2.6H_2O$ (at the same pH as the reaction).
4. Transfer the sections for 5 min to the post-coupling solution which has been kept at 4 °C. The solution should be kept at 4 °C and should be agitated occasionally.
5. Wash in three changes of distilled water.
6. Mount in glycerine jelly (which has been warmed to 40 °C to render it sufficiently fluid) or in Farrants' medium or an equivalent aqueous mountant.

The mountant, especially if it is glycerine jelly, should be allowed to settle and harden, even for a day or so, before the section is inspected. This is to allow the mountant to fully impregnate the section and the lysosomes, and so to assert a suitable refractive index for microscopy.

Result

A predominantly granular stain is generally produced. With lower concentrations of PVA, or at pH 4.5 with some coupling reagents, the intensity of the reaction may be increased but a more diffuse stain has been reported (as described by Shannon, 1975).

Solutions required for this method

Reaction medium

Add 10 mg naphthol AS-BI-*N*-acetyl-β-glucosaminide to 1.0 ml 2-methoxyethanol (ethylene glycol monomethyl ether) and leave this at 37 °C until it is completely dissolved. This is then dispersed vigorously into the buffer (below).

Buffer

0.1 M citrate buffer (citric acid/trisodium citrate) pH 5.0 or 4.5)	19 ml
$CaCl_2.6H_2O$	0.5%
Polyvinyl alcohol (w/v)	20%

[*Note*: the use of citrate buffer is essential for the adequate demonstration of this activity.]

Post-coupling solution

0.01 M phosphate buffer (disodium hydrogen phosphate/potassium dihydrogen phosphate) containing 1 mg/ml of either Fast Garnet GBC (after the reaction done at pH 4.5) or Dark Blue R salt (when the reaction has been done at pH 5.0) at pH 7.4.
   This should be cooled to 4 °C before it is used.

Control

The specific inhibitor, *N*-acetylglucosaminolactone, at 5 mM, fully abolishes the activity (Shannon, 1975). It can be prepared by the method of Findley *et al.* (1958). If it cannot readily be prepared, control sections can be obtained by immersing the sections in water at 100 °C for 2 min. Alternatively, sections may be reacted with the full sequence but lacking the substrate.

Measurement

Because this reaction occurs predominantly in lysosomes, which may be as small as 0.3 μm in diameter, measurement must be done with a $\times 100$ oil-immersion objective. The wavelength of light used for these measurements will vary with the coupling agent used. For Fast Garnet GBC, the wavelength is 550 nm (Robertson, 1980).

PHOSPHORYLASES

Biochemical background

Phosphorylase activity is the culmination of a number of enzymic reactions. The biochemistry of each enzyme involved will be discussed separately before considering the histochemistry of 'phosphorylase activity'. It should also be noted that many enzymes show transglycosilase activity, i.e. they will transfer a sugar moiety. The true phosphorylases transfer glycosyl groups to and from phosphate itself; other transglycosylases (for example, that enzyme listed as EC 2.4.1.7) can act as phosphorylases but in fact can use a number of acceptors other than orthophosphate (e.g. some enzymes transfer parts of glycogen to parts of amylopectin to form a branched structure; others use uridine diphosphate as the acceptor; others use disaccharides, etc., as discussed by Dixon and Webb, 1964, p. 188).

Phosphorylase a

The best known phosphorylase enzyme is glycogen phosphorylase, which is also called α-glucan phosphorylase (EC 2.4.1.1); it catalyses the reversible reaction:

$$\text{Glycogen} + n\text{H}_3\text{PO}_4 \rightleftharpoons n(\text{glucose 1-phosphate})$$

The equilibrium is pH dependent and favours the synthesis of glycogen, i.e.

$$n(\text{glucose 1-phosphate}) + (\text{glycogen})_m \rightarrow (\text{glycogen})_{m+n} + n\text{H}_3\text{PO}_4$$

where $(\text{glycogen})_m$ represents the primer of chain length m which becomes extended, by the phosphorylase activity, to a chain length of $m + n$. Thus it will be noted that the effect of this enzyme, in this reaction, is to transfer glucose residues from glucose 1-phosphate to the primer carbohydrate chain.

For maximum activity adenosine 5-phosphate (AMP) is required as the main activator (in its absence the activity drops to 65% of maximal) but traces of cysteine or of EDTA are also necessary. The pH optimum is 6.1–6.9. Glucose is a competitive inhibitor; phlorhizin and p-chloromercuribenzoate are noncompetitive inhibitors (see *Biochemists' Handbook*, p. 479).

Phosphorylase activity is found predominantly in the supernatant fraction (Dixon and Webb, 1964).

It has been reported that it occurs in the microsomal fraction in well fed rats but is released into the supernatant fraction when the rats are starved (Tata, 1964, quoted by Roodyn, 1965).

Phosphorylase *a* has a molecular weight of 495 000 and contains four molecules of pyridoxal phosphate (which do not participate directly in the enzymic activity) per molecule of the enzyme. It has four active sites per molecule and can bind one molecule of AMP to a serine residue at each active site (hence the effect of AMP as an activator).

Phosphorylase *b*

Some, or much, of the potential phosphorylase activity of tissues can be present in an enzymatically inactive form, known as phosphorylase *b*. This has a molecular weight of 242 000; it contains two molecules of pyridoxal phosphate per molecule of the enzyme, and can bind two molecules of AMP per molecule of phosphorylase *b*. In the presence of magnesium ions it forms a dimer (i.e. two molecules of phosphorylase *b* linked together). However, its conversion to phosphorylase *a* (and hence to the enzymatically active form) requires enzymatic phosphorylation of the phosphorylase *b*; this is effected by a specific enzyme, phosphorylase kinase (EC 2.7.1.38), which adds two phosphate groups to each molecule of phosphorylase *b*. (See Figure 26.)

Phosphorylases appear to play a major role in controlling the metabolism of glycogen in animal cells generally; they are found in highest concentration in voluntary and heart muscle and in the liver (*Biochemists' Handbook*).

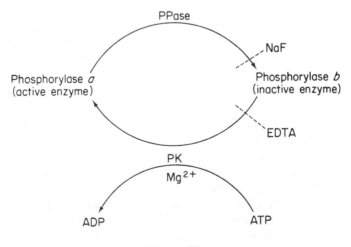

Figure 26.

Interconversion of phosphorylases *a* and *b*

$$\text{Phosphorylase } b \ + \ Mg^{2+} \ + \ 2ATP \ \xrightarrow[\text{(2.7.1.38)}]{\text{kinase}} \ \text{phosphorylase } a$$
$$\quad\text{inactive} \qquad\qquad\qquad\qquad\qquad\qquad\qquad\qquad \text{active}$$

The phosphorylase kinase that causes this reaction is stimulated by epinephrine and by glucagon (*Biochemists' Handbook*). In the presence of manganese or magnesium ions (the former seems to have been used in studies on muscle and the latter on liver) the kinase transfers two phosphate groups (from ATP) to each molecule of the inactive phosphorylase *b* which, in the presence of Mg^{2+} or Mn^{2+}, forms dimeric phosphorylated molecules of active phosphorylase *a*.

$$\text{Phosphorylase } a \ \xrightarrow[\substack{\text{phosphatase}\\ \text{(3.1.3.17)}}]{\text{phosphorylase}} \ 2(\text{phosphorylase } b) \ + \ 2H_3PO_4$$

The phosphorylase phosphatase is found in the soluble fraction of homogenates (*Biochemists' Handbook*).

Histochemical background

The particular interest in this enzyme has been its remarkably decreased activity in tumours, as shown in a long series of detailed studies by Godlewski (e.g. 1962, 1963). There has also been some discussion of whether activity of the 'branching enzyme' (mentioned in the biochemical background, above) can be seen in histochemical phosphorylase reactions by its products giving a violet-red or brown-purple colour with the iodine used in the reaction (see Burstone, 1962, p. 368, discussing work by Takeuchi).

In the histochemical method the section is given glucose 1-phosphate and a little glycogen; the tissue phosphorylase is encouraged to increase the size of the glycogen by accretion of glucose units from the glucose 1-phosphate (as in the equations above). This reaction depends on the presence of the active enzyme, phosphorylase *a*, and will be reduced if that enzyme is simultaneously being dephosphorylated, by the phosphorylase phosphatase, to the inactive form. Consequently, fluoride is added to the incubation medium to inhibit the action of this phosphatase. It is hoped that the extended molecules of glycogen will become bound to the section, and they are then demonstrated by their response to iodine. It is claimed that, with iodine, linear amylose stains blue but branched polysaccharide like glycogen stains red-brown. Consequently, if phosphorylase activity alone has occurred, the end-product should stain blue whereas, if the phosphorylase has worked in conjunction with the branching enzyme, a purple (blue + red-brown) colour should result. It has been found that the binding of the glycogen to the section is greatly enhanced by leaving the section at room temperature for 10 min before it is incubated. Insulin may be added to the incubation medium to enhance phosphorylase activity (Barka and Anderson, 1963, p. 291).

If the total potential phosphorylase activity is required, it becomes necessary to convert the inactive phosphorylase *b* into the active form (phosphorylase *a*). This conversion must be effected enzymatically (see above); it involves the addition of phosphate (from ATP) to the phosphorylase *b* by phosphorylase kinase, and the dimerization of phosphorylase *b* molecules by the action of magnesium ions. Hence, to demonstrate total potential activity, ATP and Mg^{2+} are added to aid the conversion of phosphorylase *b* to phosphorylase *a*, and fluoride is added to stop the degradation of the latter.

Method (after Takeuchi and Kuriaki, 1953; Godlewski, 1960, 1964)

Fresh cryostat sections, preferably 20 μm thick, should be used. They should be kept at + 4 °C for 1–3 h or left to dry at room temperature for 10 min before use.

1. Incubate in the selected test medium. For some tissues the incubation may be 20 min at 37 °C; others may require 60 min at 30 °C.
    [*Note*: the medium selected, I–V below, will depend on whether the activity to be studied is that which is already manifest in the specimen, or the total activity of which the specimen is capable.]
2. Wash briefly in distilled water.
3. Immerse in Gram's iodine, diluted 1 : 2 with distilled water, for 30 s or up to 5 min.
4. Mount in glycerine–iodine mixture.

Solutions required for this method

Medium: Enzymes demonstrated:	I P_a, P_b and BE	II P_a and P_b	III P_a and BE	IV P_a only	V Control
G1P (mg)	50	50	50	50	—
5′ AMP (mg)	10	10	10	10	10
Glycogen (mg)	5	5	5	5	5
EDTA (mg)	10	10	—	—	10
G6P (mg)	—	—	—	—	50
NaF (mg)	—	—	52.5	52.5	—
0.5 M acetate buffer, pH 6.0 (ml)	10	10	10	10	10
H₂O (ml)	15	2.5*	15	2.5*	15
40% ethanol (ml)	—	12.5*	—	12.5*	—

*Or replace the water and ethanol by 1 mg $HgCl_2$ in 15 ml water.

P_a, phosphorylase *a*; P_b, phosphorylase *b*; BE, branching enzyme; G1P, glucose 1-phosphate; G6P, glucose 6-phosphate.

Reaction medium for total potential phosphorylase activity

Use reaction medium III but add 10 mg ATP and 10 mg magnesium sulphate to activate phosphorylase kinase (PK). The presence of sodium fluoride should inhibit the breakdown of phosphorylase *a* to phosphorylase *b* by phosphorylase phosphatase (PPase).

Glycerine–iodine mixture

Add one volume of glycerine to one volume of Gram's iodine.

Results

Media I and III demonstrate phosphorylase activity and branching enzyme, the former producing a blue colour and the latter an orange-red colour.

Media II and IV show phosphorylase activity only, without branching enzyme. The colours produced fade in 12–24 h.

Other controls

For controls for branching enzyme see Burstone (1962) and Barka and Anderson (1963).

Other controls consist of pretreating sections with diastase (or saliva) to remove endogenous glycogen; any glycogen found at the end of the reaction is then solely that which has been formed, and bound to the section, during the course of the reaction. Diastase (see p. 111) can also be used at the end of the reaction to prove that the colour is indeed due to glycogen.

CARBONIC ANHYDRASE

Biochemical background

This enzyme catalyses the reaction $CO_2 + H_2O \rightleftharpoons H_2CO_3$. It is found very widely. In animals it has an important function in respiration (facilitating transport of carbon dioxide) and in the transfer of hydrogen and bicarbonate ions, and it is involved in the secretion of acid by the parietal cells of the stomach (Narumi and Kanno, 1973). In chloroplasts it may be involved in the fixation of carbon dioxide for photosynthesis. There are several forms of this enzyme, often occurring together, with high or low activities with very high turnover numbers. The molecular weight of all the forms of this enzyme is around 30 000, with one atom of zinc, which may be replaced by cobalt, in each molecule. Its activity is inhibited by many monovalent anions and by acetazolamide (Pocker and Watamori, 1973). Its activity is assayed by the rate at which it decreases the pH of a solution, saturated with carbon dioxide, from pH 8.3 to pH 6.3. The complex biochemistry of this enzyme has been reviewed by Maren (1967).

Histochemical background

The enzyme catalyses the reactions:

$$CO_2 + H_2O \rightleftharpoons H^+ + HCO_3^-$$

and

$$HCO_3^- \rightleftharpoons OH^- + CO_2$$

The histochemical procedure of Hansson (1967) depends on the trapping of the hydroxyl ions by cobalt, which is precipitated as cobalt hydroxide. Phosphate facilitates this precipitation. The cobalt hydroxide is then converted to the coloured cobalt sulphide. This method was investigated by Rosen (1972) and very critically by Muther (1972). The procedure was then modified and investigated by Loveridge (1978), who showed that the enzyme was bound and not lost into the incubation medium. The precision of the method (below) is shown in that it was used in the highly quantitative cytochemical section bioassay for measuring circulating levels of gastrin in the human (Loveridge *et al.*, 1980b) and was used for studying the effect of antibodies directed to the gastric parietal cells (Loveridge *et al.*, 1980a,b).

Method (Loveridge, 1978; Hoile, 1983)

Fresh cryostat sections (e.g. 18 μm) should be used, and left to dry for 10 min. To ensure that the reaction medium is fully active, it should be prepared during these 10 min so that it is fresh when applied to the sections. The sections should be placed horizontally in a suitable trough.

1. Incubate at room temperature, with the depth of the reaction medium about 1 mm. Agitate the trough gently throughout the reaction time, which should be as short as up to 3 min to avoid non-specific deposition of reaction product.
   Should the intensity of the reaction be too little, longer incubation periods can be used provided the reaction medium is replaced by fresh medium every 3 min.
2. Rinse in tap water (30 s).
3. Immerse in a saturated solution of hydrogen sulphide (1–2 min).
4. Rinse in distilled water.
5. Mount in Aquamount or some equivalent water-soluble mountant.

Result

Brown-black deposit.

Measurement

At 550 nm.

Solutions required for this method

Solution A

0.1 M cobalt sulphate ($CoSO_4.7H_2O$)	9 ml
0.5 M sulphuric acid	9 ml
0.067 M potassium dihydrogen orthophosphate	3 ml
Distilled water	4.5 ml

Solution B

Sodium hydrogen carbonate	1.125 g
0.1 M HEPES buffer, pH 7.4	60 ml

Reaction medium

Mix the two freshly prepared solutions immediately before use. Stir to remove most of the bubbles. After 3 min the pH should have risen to 6.8. Then pour onto the sections and leave for 3 min.

Inhibitor

Acetazolamide (10^{-5} M) is added to solution B to inhibit specific carbonic anhydrase activity.

ARYL SULPHATASE

Biochemical background

Depending on their relative substrate specificities, pH optima and response to inhibitors, three distinct aryl sulphatases have been distinguished in kidney and liver (Roy, 1960). Aryl sulphatase C was found to be microsomal, with a pH optimum of 8 (Dodgson *et al.*, 1955). Sulphatases A and B, with acidic pH optima, have characteristic lysosomal structural-linked latency (Viala and Gianetto, 1955).

Histochemical background

There appear to be several restrictions to the demonstration of these enzymes. They appear to be very substrate-specific (Roy, 1960); most histochemical procedures use *p*-nitrocatechol sulphate as the substrate in the presence of lead ions. The lead sulphate produced by the activity of the enzyme is then converted to lead sulphide by treatment with dilute ammonium sulphide. Goldfischer (1965) stressed that fixation is critical; for many tissues he advised fixation in cold calcium–formol for up to 1 week. The method appears to have been very capricious.

Naphthol AS substrates have been suggested for demonstrating this type of activity (e.g. Woohsmann and Hartrodt, 1965). Cautionary notes have been published by Roy (1962) and Gahan (1967).

DECARBOXYLASES: ORNITHINE DECARBOXYLASE

Biochemical background

Of all the decarboxylases, ornithine decarboxylase (ODC, EC 4.1.1.17) occurs most widely and is the most important metabolically, so that it has been said that it at least equals the significance of adenylate cyclase. It is the first step of the polyamine pathway:

$$[\text{arginine}] \longrightarrow \text{Ornithine} \xrightarrow[\substack{\text{ODC} \\ 1}]{\nearrow \text{CO}_2} \text{Putrescine} \xrightarrow{\quad 2 \quad} \text{Spermidine} \longrightarrow \text{Spermine}$$

Inhibitors: 1: α-difluoromethylornithine

2: S-adenosyl-1,8-diamino-octane

The *in vivo* half-life of ODC is only 8–30 min. The enzyme appears to be universally distributed and can be activated by many hormones, drugs and growth factors (Kaye, 1984; Bachrach, 1984). In mammals the molecular weight is generally of between 50 000 and 60 000; it is an acidic protein; in rat liver the pH optimum varies from pH 7.0 to 7.8 when assayed in the presence of sodium phosphate. Both the characteristics of this enzyme in plants and the therapeutic value of the main inhibitor of this enzyme have been reviewed by Bachrach *et al.* (1983). The clinical chemistry of the polyamines, including their increased excretion in the urine from patients with cancer, has been reviewed by R. Chayen (1984).

The enzymatic activity requires pyridoxal phosphate. An antizyme has been described which neutralizes ODC activity. ODC activity appears to be inactivated by phosphorylation (as reviewed by Bachrach, 1984).

Histochemical background

It was obviously difficult to demonstrate this enzymatic activity histochemically since the reaction from ornithine liberated a very similar molecule, putrescine; it also liberated carbon dioxide, which could be trapped only at very alkaline pH values, at which the enzyme could not operate. Consequently, earlier studies aimed at localizing the enzyme itself, either by 'labelled suicide enzyme inhibitors', labelled with a fluorescent or radioactive moiety to permit localization by fluorescence microscopy, or by autoradiography, or by the use of a fluorescent-labelled antibody. These procedures have been reviewed by Dodds and Chayen (1984).

The inherent difficulty in demonstrating the histochemical activity of this enzyme (or any other decarboxylase) was that of trapping the carbon dioxide liberated by this activity. This is because carbon dioxide is very weakly acidic, so that it will not displace most acidic ions, for example phosphate from lead phosphate. Dr G. T. B. Frost, together with ourselves, developed the use of lead hydroxyisobutyrate, with one atom of lead held between two molecules of hydroxyisobutyrate, so that the lead was not free to inhibit enzymatic activity. However, the hydroxyisobutyrate is so weakly acidic that it can be displaced by carbon dioxide liberated by that activity. The resulting lead carbonate could then be converted to the coloured lead sulphide (Dodds and Chayen, 1984). This method has now been used by Charlton and Daylis (1989) and fully developed by Dodds *et al.* (1990).

Method

Fresh cryostat sections must be used. All aqueous solutions must be prepared in carbon dioxide-free water and kept, when necessary, in a vessel sealed with a trap of soda-lime to exclude carbon dioxide.

1. Free phosphate is removed from the sections by immersing them for 5 min at 37 °C in a 40% solution of a carbon dioxide-free deionized polypeptide derived from collagen (Polypep 8350; obtainable from Molecular Design and Synthesis, Robens Institute, University of Surrey, Guildford, Surrey, UK). The pH of this solution should be brought to pH 7.0 with hydroxyisobutyric acid.
2. Suck off this Polypep 8350 with a pipette and replace it with the full reaction medium at 37 °C. A reaction time of up to 60 min may be required.
3. Remove the reaction medium; wash in several changes of distilled water at 19 °C.
4. Immerse in hydrogen sulphide–water (distilled water saturated with hydrogen sulphide gas) for 1–2 min at room temperature to convert the colourless lead carbonate to the coloured lead sulphide.

5. Rinse several times in distilled water. Dry. Mount in an aqueous mountant such as Farrants' medium.

Specificity

Sections are subjected to the same treatment but with a specific inhibitor of ODC activity, e.g. up to 40 mM α-difluoromethyl ornithine (Merrell Dow Research Institute, Strasbourg, France).

Measurement

By microdensitometry with light of 580 nm.

Solutions required for this method

Preparation of the reagent

Prepare a saturated solution of yellow lead monoxide (lead II oxide; Aldrich Chemical Co., Dorset, England) by strongly stirring 152 g into a solution of 126 g 2-hydroxyisobutyric acid dissolved in 500 ml deionized distilled water. Adjust the pH, if necessary, to between 7.2 and 7.4 with hydroxyisobutyric acid. Filter and store in a well stoppered bottle to avoid contact with atmospheric carbon dioxide. Before use, filter to remove any cloudiness.

Reaction medium

It is important to prepare all solutions in carbon dioxide-free water, which should be kept in vessels sealed with a trap containing soda lime to keep the solutions free of carbon dioxide.

Using carbon dioxide-free water, and working whenever possible with the solutions maintained under an atmosphere of nitrogen, prepare a 60% (w/v) solution of the deionized, carbon dioxide-free Polypeptide 8350 in carbon dioxide-free 0.2 M triethanolamine (Sigma) brought to pH 8.0 with a small volume of 5 M hydroxyisobutyric acid (Aldrich). The pH is then adjusted to pH 8.3 with the triethanolamine.

Then, under nitrogen, the following reagents are added sequentially:

L-p-bromotetramisole oxalate (Aldrich)	0.3 mM
Pyridoxal 5-phosphate (Sigma)	0.38 mM
L-ornithine (Sigma)	11.4 mM

DL-Dithiothreitol (Sigma) may also be added at 0.6 mM.

Just before use, add 0.3 ml freshly filtered lead hydroxyisobutyrate (as above) to each 1 ml of this reaction medium. Then adjust the concentration of the Polypeptide 8350 to 46% and correct the pH to 7.0 by adding more of either the triethanolamine or the hydroxyisobutyric acid.

[*Note*: the inclusion of bromotetramisole in the reaction medium is to inhibit any alkaline phosphatase activity.]

OXIDASES

Oxidases are enzymes that catalyse the oxidation of a substrate by removing hydrogen from it and passing it to an acceptor, which can be oxygen. This is in contrast with the action of dehydrogenases, which also remove hydrogen from their specific substrate but donate it to another acceptor which cannot be oxygen. Consequently both oxidases and dehydrogenases are classified as oxidoreductases in that they catalyse the same general reaction:

$$SH_2 + A \rightarrow S + AH_2$$

(where S is the substrate, SH_2 is the reduced substrate, A is the acceptor, and AH_2 is the reduced acceptor).

For oxidases, A is oxygen or an acceptor which has an electrode potential close to that of oxygen. For dehydrogenases, A is a hydrogen acceptor with an electrode potential closer to that of free hydrogen. This concept is so fundamental to the histochemistry of oxidative reactions, and indeed to cellular energetics, that more detailed consideration of it will be given later (under Dehydrogenases). However, in this section, we will consider those oxido-reductases that use either oxygen or substances which can replace oxygen as the hydrogen acceptor.

Histochemical studies on oxidases and peroxidases have been based largely on methods involving the use of diaminobenzidine, as developed by Graham and Karnovsky (1966). These investigations, and the development of the electron histochemical techniques for dehydrogenases, were reviewed by Shnitka and Seligman (1971). It may, however, be noted that proof of the true localization of the activity by the diaminobenzidine procedures and of other methods used in electron histochemistry is remarkably lacking.

CYTOCHROME OXIDASE

Biochemical background

Cytochromes were so named because they were 'cell pigments'. Originally three were described and these were named cytochromes *a*, *b* and *c*. As each new

cytochrome was discovered it was named according to which of these three it most closely resembled. Consequently, each cytochrome 'genus' now has several 'species' which are distinguished by a subscript. So, for example, for cytochrome a (genus) there are cytochromes a_1, a_2, a_3 and a_4 (species). One species or other of cytochrome has been found in all organisms; in mammals, in 1961, they were listed as a, a_3, c, c_1, b and b_5 (West and Todd, 1961). To this list have been added the various forms of cytochrome P-450 and P-448, which are of considerable importance in detoxification mechanisms.

The standard electrode potential, E_0', of the cytochrome c system is $+0.26$ V at pH 7 (West and Todd, 1961, p. 853). There has been some doubt as to whether cytochromes a and a_3 can be distinguished clearly (Dixon and Webb, 1964; Lehninger, 1965, p. 77). Cytochrome oxidase is cytochrome a, with or without cytochrome a_3 (Dixon and Webb, 1964, p. 391). Active cytochrome oxidase is a complex between cytochromes c and a (Dixon and Webb, 1964, p. 389). Reduced cytochrome c alone will not react with oxygen at neutral pH values; similarly, cytochrome oxidase does not oxidize p-phenylenediamine, cysteine or ascorbate in the absence of cytochrome c.

The cytochromes consist of a particular porphyrin linked to protein. Iron is present in the porphyrin ring. In some cytochromes, particularly cytochrome c, copper appears to be an integral part of the enzyme (see West and Todd, 1961, p. 854). The linkage between the porphyrin and the protein has been determined in the case of cytochrome c (Figure 27).

Cytochrome oxidase (also called cytochrome c oxidase) forms a structural component of mitochondria (Lehninger, 1965). It can be 'solubilized' from the

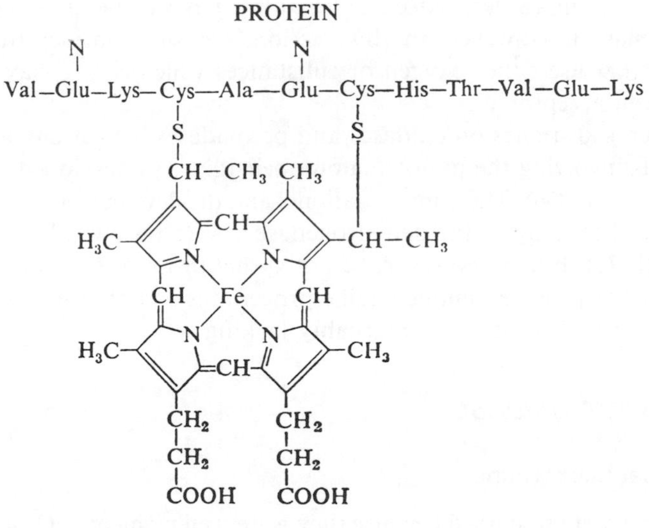

Figure 27. Linkage of porphyrin to protein in cytochrome c.

mitochondrial fraction of homogenates by treatment with deoxycholate and trypsin (*Biochemists' Handbook*, p. 382), when it is found to contain surprisingly large amounts of lipid. It is extremely sensitive to cyanide: 10^{-8} M cyanide gives 50% inhibition (in contrast to 80% inhibition given by 10^{-3} M azide); 10^{-4} M cyanide is usually sufficient to give total inhibition (Dixon and Webb, 1964; *Biochemists' Handbook*, p. 625). It is also very sensitive to hydrogen sulphide and to carbon monoxide.

Histochemical background

This subject has been so bedevilled by odd names and initials that some discussion of these must be given to clarify the nomenclature.

When a naphthol and an aromatic diamine are mixed in the presence of oxygen, a coloured indophenol is produced; this reaction is immediate if a strong oxidizer like hypochlorite is added (Gomori, 1952, p. 153 *et seq.*):

$$\underset{\substack{\text{aromatic}\\\text{diamine}}}{\underset{\text{NH}_2}{\overset{\text{H}_3\text{C}\diagdown\text{N}\diagup\text{CH}_3}{\text{(ring)}}}} \; + \; \underset{\text{naphthol}}{\overset{\text{OH}}{\text{(ring)}}} \quad \xrightarrow{\;+\;\text{O}_2\;} \quad \underset{\substack{\text{coloured}\\\text{indophenol}}}{\overset{\text{H}_3\text{C}\diagdown\text{N}\diagup\text{CH}_3}{\text{(ring)}=\text{N}-\text{(ring)}=\text{O}}}$$

Because this involves a *na*phthol with a *di*amine it was called the *nadi* reaction. When it was discovered that most tissues, if fresh, could take the place of the hypochlorite, that is to say they contained an oxidizing factor (or enzyme) which could produce this indophenol, the 'enzyme' was called indophenol oxidase. It was then discovered that tissues could form the indophenol in two different ways, one of which did not appear to be enzymatic. The reaction which was much investigated in *m*yeloid leucocytes (hence the M-nadi reaction) was insensitive to formalin (even after 6 years according to Gomori, 1952), to alcohol and to acetone; its maximal activity occurred at pH 12.5–13 and it occurred in the absence of oxygen. In contrast, the reaction found in fresh tissues ('tissue' is 'Gewebe' in German, hence this is the G-nadi or Gewebe-nadi reaction) was inhibited by acetone, alkali, acid, formalin and heat (55 °C), and it did not occur in the absence of oxygen. Finally it was shown that cytochrome oxidase could be responsible for the G-nadi reaction. It must be appreciated, however, that the ability to produce an indophenol through oxidation of a naphthol–aromatic

diamine mixture is not exclusively the prerogative of cytochrome oxidase; this is shown by the fact that it can occur non-enzymatically, as in the M-nadi reaction. Moreover, any oxidizing factor, enzymic or otherwise, which can operate at this electrode potential or replace hypochlorite in the original studies, will produce indophenol. In the histochemical method (which owes much to work by Burstone, e.g. 1959, 1960), the equivalent of oxygen is provided by oxidized cytochrome c, so giving some degree of specificity to the reaction; the function of the cytochrome oxidase (or, more correctly, of the cytochrome a part of the cytochrome c/cytochrome a complex) is to reoxidize the cytochrome c which becomes reduced in forming the indophenol.

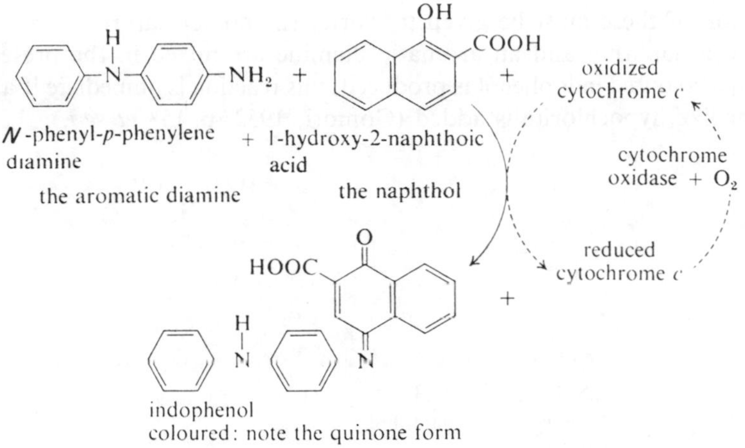

N-phenyl-p-phenylene diamine
the aromatic diamine

+ 1-hydroxy-2-naphthoic acid
the naphthol

oxidized cytochrome c

cytochrome oxidase + O_2

reduced cytochrome c

indophenol
coloured: note the quinone form

Cytochrome oxidase is of considerable significance in histochemistry. It has been much studied in malignant and benign growths (for a good review see Burstone, 1962). Many workers have found that the cytochrome oxidase activity of many cancers is markedly lower than that of the cells from which the growth originated (also see Butcher *et al.*, 1965).

Method (Butcher *et al.*, 1964)

Use fresh cryostat sections.

1. Incubate in the reaction medium in a Coplin jar at room temperature [to reduce the rate of spontaneous formation of indophenol]. Rat liver requires up to 60 min incubation; cardiac muscle or kidney is well stained after 20–30 min.
2. Transfer to Lugol's iodine (see p. 111) for 2 min. [This intensifies the colour and stabilizes it against the rapid fading which occurs if iodine is not used.]

3. Treat the section with a 5% aqueous solution of sodium thiosulphate for 4 min to reduce the non-specific iodine background coloration.
4. Place the section in a 1% aqueous solution of cobalt acetate for 60 min at room temperature. [This chelates the dye.]
5. Wash in distilled water.
6. Mount in Farrants' medium.

Result

Almost black reaction which should be restricted to mitochondria.

Control

Incubate as for the normal test but add 10^{-3} M potassium cyanide to the reaction medium (3.25 mg/50 ml).

Solutions required for this method

Reaction medium

Dissolve 10 mg N-phenyl-p-phenylenediamine and 10 mg 1-hydroxy-2-naphthoic acid in 0.5 ml absolute ethanol. Then add 35 ml distilled water containing 10^{-4} M cytochrome c. To this add 15 ml 0.2 M Tris buffer at pH 7.4. Filter into a Coplin jar or other suitable vessel for the reaction.

Tris buffer, 0.2 M, pH 7.4

Dissolve 2.42 g Tris in 17 ml of 1 M hydrochloric acid and make up to 100 ml with distilled water.

[*Note*: some tissues may have sufficient endogenous cytochrome c to make the addition of exogenous cytochrome c unnecessary. However, physiological or metabolic disorder may be related to the level of endogenous cytochrome c in the cells.]

DOPA-OXIDASE (PHENOLASE)

Biochemical background

One enzyme, which catalyses the oxidation of both monophenols (e.g. tyrosine) and orthodiphenols (such as adrenaline and catechol), is known by any of the following names: phenol oxidase, polyphenol oxidase, phenolase, DOPA oxidase, potato oxidase, catechol oxidase and tyrosinase. The enzyme (EC 1.10.3.1) has been purified and studied in detail: it has a molecular weight

of 100 000 and contains four atoms of copper in each molecule, the copper being at the active centre of the enzyme and essential for the oxidative catalysis. It is therefore inhibited by substances which complex with copper, including cyanide, diethyldithiocarbamate, glutathione, cysteine, BAL (British Anti-Lewisite) and potassium ethyl xanthate. 4-Nitrocatechol and 4-nitrophenol are competitive inhibitors. Although it is one enzyme, it behaves rather differently as a monophenolase than as a diphenolase. With monophenols there is a lag, or induction, period which can be eliminated or shortened if a trace of catechol or of ascorbic acid is added to the reaction medium (see *Biochemists' Handbook*, pp. 375–376).

This enzyme is very widely distributed throughout the plant and animal kingdoms. Its effect is prominent in that it is involved in the formation of the various melanin pigments. These are complex, high molecular weight polymers, generally combined with protein; all are formed from 3,4-dihydroxyphenylalanine (DOPA) as a result of a series of somewhat complicated reactions (e.g. see West and Todd, 1961, p. 1101). The oxidations in which phenolase is known to be especially significant are of two types:

1. In plants particularly, wounding or cutting often initiates the following reaction (Baldwin, 1959, p. 150):

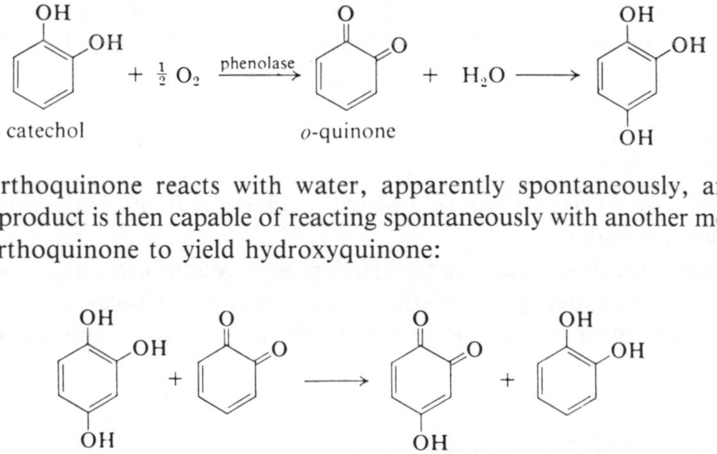

catechol   *o*-quinone

The orthoquinone reacts with water, apparently spontaneously, and the reaction product is then capable of reacting spontaneously with another molecule of the orthoquinone to yield hydroxyquinone:

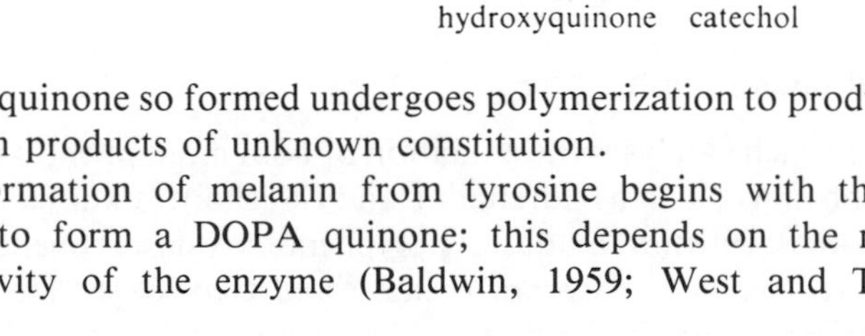

hydroxyquinone   catechol

The hydroxyquinone so formed undergoes polymerization to produce complex dark melanin products of unknown constitution.

2. The formation of melanin from tyrosine begins with the oxidation of tyrosine to form a DOPA quinone; this depends on the monophenol oxidase activity of the enzyme (Baldwin, 1959; West and Todd, 1961, p. 1101):

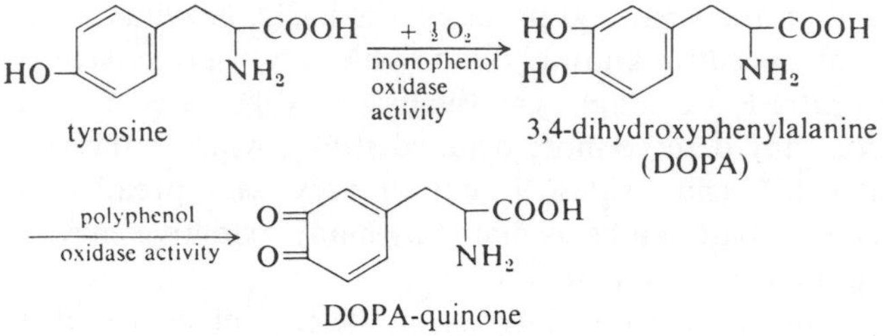

tyrosine

3,4-dihydroxyphenylalanine
(DOPA)

DOPA-quinone

The DOPA-quinone then undergoes spontaneous ring closure and reaction with water to go through intermediate stages to form first a red pigment and then the indole-5,6-quinone which polymerizes to form melanins (see Baldwin, 1959, p. 151; West and Todd, 1961, p. 1101):

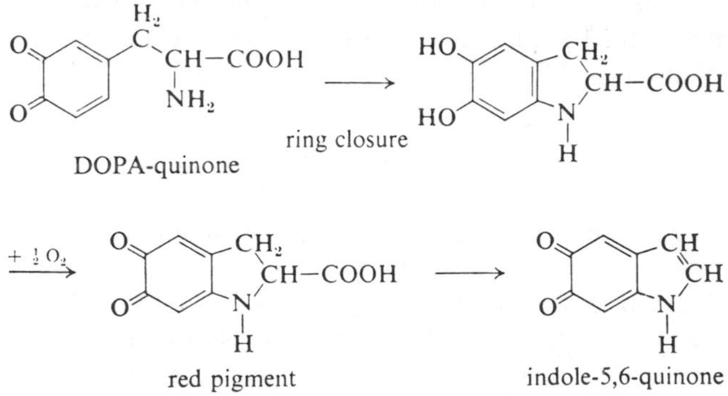

DOPA-quinone

ring closure

red pigment

indole-5,6-quinone

Similarly, phenol oxidase can oxidize adrenaline to produce the red adrenochrome which polymerizes to yield adrenaline–melanin.

Melanins are formed in pigment-forming cells (melanoblasts). The pigments occur in large quantities in tumours of these cells (melanosarcomata); melanin precursors are sometimes found in the urine of patients who have such cancers.

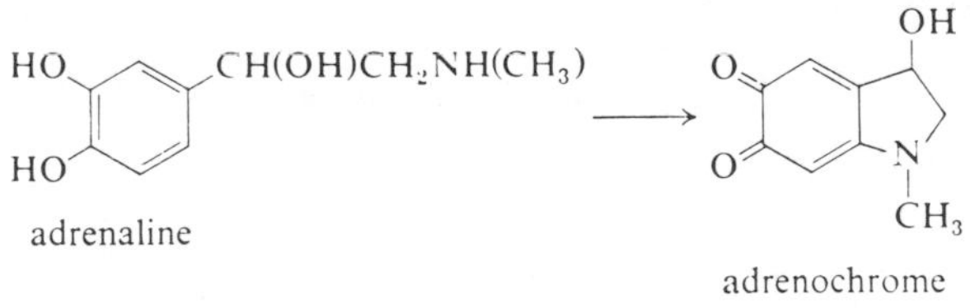

adrenaline

adrenochrome

Histochemical background

The histochemical demonstration of this enzyme may be of diagnostic significance in non-pigmented melanomata (see Burstone, 1962). A 'non-specific DOPA-oxidase' has been much reported in various blood cells and has been attributed to peroxidase activity (Barka and Anderson, 1963, p. 328).

In our opinion the first, and probably the basic, enzymatic question which the histochemist needs to know about a tissue is how effectively it metabolizes a given substrate. For example, can the tissue oxidize ascorbic acid? If it can, this oxidation may deserve more detailed study, which is necessary because, although ascorbic acid oxidase is not of very widespread occurrence, the oxidation of ascorbate can be mediated by many oxidative enzymes, including phenol oxidase (see Chayen, 1953).

So too with phenols. It is of interest, in the first place, to see if any phenol oxidation takes place; if it does, the phenomenon may be worthy of closer attention to decide which enzyme system is involved. The oxidation of phenols played a major part in the detection of a possible infective agent in human myocardium (Braimbridge *et al.*, 1967); the formation of adrenochrome-like pigment was significant in studies that re-evaluated the function of mast cells (Chayen *et al.*, 1966c); in each of these studies, the exact oxidative mechanisms were less important than the fact of oxidative activity with respect to phenols and catecholamines.

Rationale

Phenolic compounds, such as dihydroxyphenylalanine (DOPA), tyrosine, adrenaline and catechol or histamine, are given to sections of the tissue. If they are oxidized, a coloured quinone or an adrenochrome-like compound will form in the section and will be visible because of its inherent colour. Occasionally it may be helpful to add a hydrogen-acceptor like neotetrazolium which, acting at an electrode potential close to that of oxygen, may enhance the oxidative effect; the result will still be a melanin even though formazan will also be precipitated (by reduction of the neotetrazolium).

Method

Use fresh cryostat sections.

1. Incubate for at least 2 h in the reaction medium at 37 °C.
2. Wash in distilled water.
3. Mount in Farrants' medium.

Control

Before incubation, immerse for 30 min in 10^{-3} M potassium cyanide (3.25 mg/50 ml) in 0.1 M phosphate buffer at pH 7.4 at 37 °C. Then proceed to the normal incubation in a reaction medium to which 10^{-3} M potassium cyanide has been added.

Result

The test should yield a brown-black pigment, which should be lacking in the control.

Reaction medium

0.0056 M dihydroxyphenylalanine in 0.1 M phosphate buffer at pH 7.4.

[*Note*: at higher pH values, autoxidation of the DOPA occurs rapidly. The apparently specific reaction was found to be markedly depressed at pH 6.8.]

Variants

Similar molarities of catechol or other phenolic compounds may be used in place of DOPA. Also, neotetrazolium at a concentration of 0.25 mg/ml may be added together with the substrate to the reaction medium.

MONOAMINE OXIDASE

Biochemical background

This flavine-containing enzyme (EC 1.4.3.4), located in the outer membrane of mitochondria, catalyses the oxidative deamination of amines:

$$RCH_2NR_1R_2 + O_2 + H_2O \rightarrow RCHO + NHR_1R_2 + H_2O_2$$

where R_1 and R_2 are either hydrogen or methyl groups. The amine group must be attached to a methylene group, so α-methyl-substituted amines, including amphetamine (which may be used as a control), are not substrates. As described by Tipton (1986) its natural substrates are aromatic amine derivatives such as catecholamines and indolethylamines, although it is also active against aliphatic amines with a longer chain length. Although it is inactive against simple diamines, it can oxidize those in which the amine groups are separated by several methylene groups (e.g. long-chain ω-amino acids). It seems that polyamines do not act as substrates for this enzyme. It shows stereochemical specificity in that the l-($-$) forms of adrenaline and noradrenaline act as better substrates than do the d-($+$) enantiomers. More recently it was shown that this enzyme can also act on a different type of compound, MPTP (*N*-methyl-4-phenyl-1,2,3,6-tetrahydropyridine), indicating that it can catalyse the oxidation of a different class of compound. This neurotoxin, or its oxidation product, appears to be significant in Parkinson's disease. This subject, and the general clinical chemical

significance of monoamine oxidase, has been reviewed by Glover and Sandler (1986). The more commonly recognized substrates are:

adrenaline

tyramine

tryptamine
(5-hydroxytryptamine has an
− OH at the position marked 5)

histamine
(not oxidized by
monoamine oxidase)

It may be noted that it does not act on histamine, for which a pyridoxal phosphate-linked enzyme, histaminase, is required. The true monoamine oxidases appear to be glycoproteins, or to be closely associated with glycolipid. They contain covalently bound FAD (flavine adenine dinucleotide, a cofactor). There is no suggestion that metallic ions are required for their full activity, but there is some suggestion that sulphydryl groups may be essential.

There seem to be two main classes of monoamine oxidase (MAO) located predominantly on the outer mitochondrial membrane. These are designated MAO-A and MAO-B. These should be distinguished from benzylamine oxidase (or oxidases) situated in the cytosol and in blood vessels. They may be distinguished by the fact that MAO-A is often selectively and irreversibly inhibited by clorgyline, whereas MAO-B is often irreversibly inhibited by (−)-deprenyl. The complexities associated with these two forms of MAO, and of a 'mixed' form of the enzyme (which can be irreversibly inhibited by iproniazid), have been discussed carefully by Glover and Sandler (1986). It seems that there may be considerable differences in the response to the inhibitors in different species and even within the tissues of any one animal. The inhibition caused by clorgyline and by deprenyl may be through the formation of a covalent adduct with the coenzyme (FAD). The reversible inhibition by amphetamine involves only the MAO-A form of the enzyme. Normal MAO is not inhibited by carbonyl reagents, such as semicarbazide. (It does inhibit another type of amine oxidase, EC 1.4.3.6, the function of which is unknown; Tipton, 1986.)

The structures of ephedrine, which acts as a competitive inhibitor, and of two other inhibitors are:

$$\langle\!\!\!\bigcirc\!\!\!\rangle\!-\!CH(OH)\!-\!\underset{\underset{CH_3}{|}}{\underset{NH}{|}}{CH}\!-\!CH_3 \qquad\qquad \langle\!\!\!\bigcirc\!\!\!\rangle\!-\!CH_2\!-\!\underset{NH_2}{\underset{|}{CH}}\!-\!CH_3$$

ephedrine   amphetamine (benzedrine)

$$\underset{CH_3}{\underset{|}{CO}\!-\!NH\!-\!NH\!-\!\underset{\underset{CH_3}{|}}{CH}} \quad \text{iproniazid (marsilid)}$$

The antidepressant, clinical use of MAO-inhibitors has been discussed by Glover and Sandler (1986). Various aspects of the pharmacology of monoamine oxidase, and of its inhibitors, have been reviewed by Callingham (1986).

The enzyme has been found in most tissues. Its highest concentration has been found in liver, kidney, intestine and pancreas (Blaschko, 1961).

Histochemical background

The overall activity of this enzyme is summarized by the equation:

$$R-CH_2\overset{+}{N}HR_1 + H_2O + O_2 \rightarrow R.CHO + \overset{+}{N}H_2R_1 + H_2O_2 \qquad 1$$

However, Blaschko (1952) suggested that, in life, there could be an intermediate hydrogen acceptor, so that the reaction might be of the following type (Cohen *et al.*, 1965):

$$R-CH_2.\overset{+}{N}HR_1 + A \xrightarrow[MAO]{-2H} R-CH{:}\overset{+}{N}R_1 + AH_2 \qquad 2$$

$$R-CH{:}\overset{+}{N}R_1 + H_2O \longrightarrow R-CHO + \overset{+}{N}H_2R_1 \qquad 3$$

$$AH_2 + O_2 \longrightarrow A + H_2O_2 \qquad 4$$

It can be seen that the sum total of reactions 2, 3 and 4 is identical with 1. The difference is that this mechanism involves a hydrogen acceptor, A, which in the reduced form can reduce nitroblue tetrazolium (NBT) to yield a coloured formazan. That is, under anaerobic conditions, reaction 4 can be altered to reaction 5:

$$AH_2 + NBT \longrightarrow A + NBT \text{ formazan} \qquad 5$$

From these equations it follows that the activity of MAO can be measured either by its production of peroxide (equation 1) when the reaction is done

aerobically, or by the formation of the formazan if the reaction is done anaerobically. The subject has been reviewed by Darracott-Canković *et al.* (1986).

As a result of studies on monoamine oxidase activity in the human endometrium (Cohen *et al.*, 1965) it was postulated that this enzyme functions physiologically only inside the mitochondria. It was found that, at the onset of menstruation, the location of the enzyme was changed, so that it became predominantly cytoplasmic, outside the mito- chondria. It was suggested that this change in location could affect its physiological activity, for example because the enzyme had been separated from its physiological hydrogen acceptor (A, in the equations above).

It should be noted that different tissues metabolize the various substrates for monoamine oxidase at remarkably different rates. This may not be surprising considering the biochemical information that different tissues, of the same animal, will have different proportions of the forms of this enzyme. Consequently it is often advisable to do a preliminary study to decide which substrate should be used, and which inhibitor will be relevant to the tissue being studied.

PEROXIDASE METHOD (Ryder *et al.*, 1979,
after Graham and Karnovsky, 1965)

Method

Unfixed cryostat sections (18–20 μm thick) should be allowed to dry at room temperature for 15–60 min.

1. React at either 22 °C or 37 °C.
2. Wash in sodium chloride solution (0.9% w/v).
3. Post-fix in 10% formalin for 2 h.
4. Mount in glycerine jelly.

Solutions required for this method

Reaction medium

3-Amino-9-ethyl carbazole (Sigma)	2 mg
Dimethyl formamide (Analar)	0.5 ml
0.05 M phosphate buffer, pH 7.6	9.5 ml

Shake well and filter. Then add 10 mg peroxidase (Sigma type II) and 12 mg of the selected substrate (tyramine is recommended).

Inhibition studies

To test the specificity of the reaction, the sections should be preincubated for 15 min in the inhibitor in 0.05 M phosphate buffer, pH 7.6, and then transferred to the full medium, which also includes the inhibitor.

For MAO-A the activity elicited by this reaction medium should be sensitive to 10^{-7} M clorgyline but insensitive to 10^{-7} M deprenyl. For MAO-B, with tyramine as substrate, the sensitivity should be exactly the reverse.

With benzylamine (12 mg) as the substrate, MAO-B should be sensitive to deprenyl but benzylamine oxidase should be unaffected.

TETRAZOLIUM METHOD (modified from Glenner *et al.*, 1957; Cohen *et al.*, 1965)

Method

Unfixed cryostat sections must be used.

1. React sections at 37 °C. With tryptamine a reaction time of 2 h may be required; with adrenaline (e.g. in rat liver, kidney and heart muscle) the reaction time may be as short as 30 min. But the medium should be replaced with freshly prepared medium every 15 min because it may become discoloured.
2. Wash.
3. Mount in Farrants' medium or an equivalent water-miscible medium.

Solutions required for this method

Reaction medium

Dissolve 5 mg nitroblue tetrazolium (NBT) in 20 ml 0.05 M phosphate buffer at pH 7.5. Add 0.02 g (0.005 M) tryptamine hydrochloride.

When adrenaline (adrenaline hydrogen tartrate) is to be the substrate, first use alkali to adjust the pH of the 0.05 M phosphate buffer (pH 7.5) to pH 8.0. Then add the substrate. The substrate depresses the pH to 7.5. [*Note*: the pH cannot be adjusted with sodium hydroxide because this will precipitate the NBT.]

Controls

1. Include 0.01 M D-amphetamine in the normal reaction medium.
2. Include 0.01 M iproniazid in the normal reaction medium at pH 8.0. As with adrenaline hydrogen tartrate, iproniazid depresses the pH towards the

required pH of 7.5; the final pH can be adjusted to this pH by the addition of some dilute hydrochloric acid.

Iproniazid has been found to be an effective inhibitor of activity against adrenaline as the substrate, but not against activity against tryptamine as substrate.

PEROXIDASES

Wachstein and Meisel (1964) used a benzidine reaction to demonstrate endogenous peroxidase activity in a variety of tissues. Jensen's (1955) guiacol method seems to have been useful for showing such activity in plant tissues. The major advance was the diaminobenzidine (DAB) method of Graham and Karnovsky (1966). However, this procedure involved fixing the tissue and such fixation may decrease peroxidase activity (Rosene and Mesulam, 1978) or even abolish it (Hosoya et al., 1972).

The DAB method has been widely used, especially for electron microscopical studies; it has also proved valuable for immunochemical investigations (Nakane and Pierce, 1967). It is based on the fact that peroxidase oxidizes the substrate by transferring two electrons to hydrogen peroxide, so forming water. In the histochemical reaction DAB is oxidized and consequently polymerizes to form a highly coloured reaction product (as discussed by Ealey et al., 1984).

Previous cytochemical investigations were done on chemically fixed tissues. Ealey et al. (1984) established that the method can be used without fixation. They showed that the peroxidase of thyroid follicle cells can be retained in unprotected tissue sections (i.e. without the use of colloid stabilizers). Their modified DAB reaction yielded quantitative results which agreed with physiologically altered peroxidatic activity. [But not all peroxidases may be so well bound in the cells as was the thyroid peroxidase: it may be necessary to do the reaction in the presence of a colloid stabilizer.]

Method (Ealey et al., 1984; modified from Graham and Karnovsky, 1966)

Fresh cryostat sections, 16 µm thick, should be used. They should not be stored.

1. React at 37 °C: 25 min was required for sections of guinea-pig thyroid gland.
2. Stop the reaction by rinsing in distilled water.
3. Allow the sections to dry in the dark. Mount in an aqueous mountant such as neutralized Farrants' medium (at pH 6.5) or some equivalent water-soluble mountant adjusted to this pH.

Result

Diffuse brown reaction product.

Measurement

At 470 nm, with an oil-immersion objective.

Solutions required for this method

Reaction medium

Hydrogen peroxide	Either	15 mM
	or	1.5×10^{-4} M
3,3′-Diaminobenzidine tetrachloride (Sigma)		1.4 mM
in 0.1 M Tris-HCl buffer, pH 7.4		

Inhibitor

Either 1 mM potassium cyanide or 10 mM 3-amino-1,2,4-triazole (Sigma).

DEHYDROGENASES

Dehydrogenases are enzymes that oxidize their substrate by removing hydrogen (or 'reducing equivalents') and passing it to a suitable acceptor. They differ from oxidases in that the 'suitable acceptor' cannot be atmospheric oxygen; in almost all cases it is a coenzyme such as nicotinamide adenine dinucleotide (NAD), nicotinamide adenine dinucleotide phosphate (NADP) or a flavoprotein. The fate of the hydrogen is of particular interest in metabolic biochemistry. It may be passed through a hydrogen transport system (or 'electron transport' system), such as that of the mitochondria, to react ultimately with oxygen. On the other hand, it may be retained at a high negative electrode potential, the energy to be used for biosynthetic mechanisms. Typically the NADPH generated by the cytosolic pentose phosphate dehydrogenases is of the second type.

Simple biochemistry can determine very precisely the amount of reduced coenzyme produced, or capable of being produced, by an enzyme in a tissue. One of the special potentials of quantitative histochemistry is that it can determine how the 'hydrogen' may be utilized. Because of the practical advantages of such determinations (as discussed by Chayen *et al.*, 1986), some consideration of them will now be given.

Hydrogen transport and electrode potentials

As shown in Figure 28, hydrogen generated by mitochondrial dehydrogenase activity moves down the mitochondrial hydrogen transport system from a relatively high negative electrode potential to the relatively highly positive

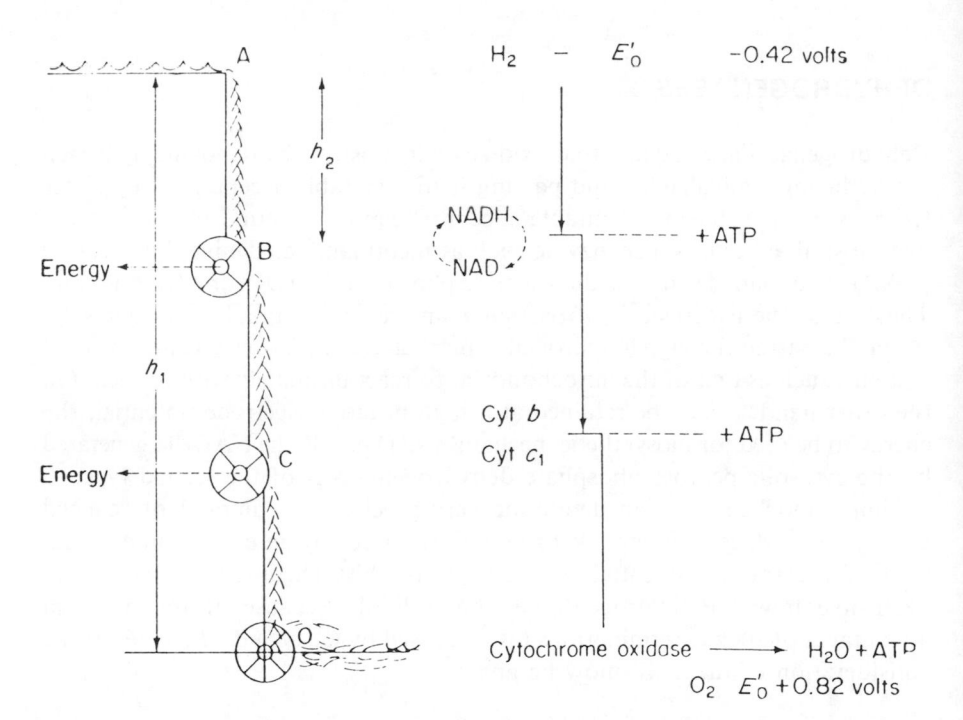

E'_0 { Histochemical hydrogen-acceptors

	NBT	MTT	NT
	+0.05	+0.12	+0.17

Respiratory chain: −0.32 −0.06 +0.04 +0.26 +0.29 +0.82

exogenous NAD

Pyruvate
α-keto-glutarate
S-S → FP → [Bound NAD] → FP → CoQ → cyt b → cyt c_1 → cyt c → cyt $a (a_3)$ → O_2
ATP   ATP   ATP
ADP + P   ADP + P   ADP + P
FP_2
succinate

Specific inhibitors: Arsenite   Amytal   Antimycin A   Cyanide

Figure 28. The hydrogen transport system of mitochondria.

H_2 − E'_0 −0.42 volts

NADH
NAD
- - - - - +ATP

Cyt b
Cyt c_1
- - - - - +ATP

Cytochrome oxidase ——→ H_2O + ATP
O_2 E'_0 +0.82 volts

Figure 29. Analogy between the energy liberated by a waterfall and by the hydrogen transport mechanism. Potential energy of a mass (m) of water at point A = mgh_1, where g is the acceleration due to gravity. Potential energy of the water at B = mg ($h_1 − h_2$), i.e. the energy which is liberated at B = mgh_2. At 0 all the potential energy has been finally liberated.

electrode potential of oxygen (as in Figure 29) to form water. So we can consider the activity of succinate dehydrogenase (SDH):

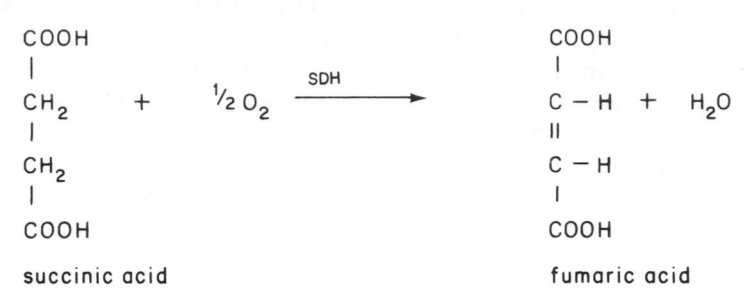

succinic acid   fumaric acid

This can be rewritten as follows:

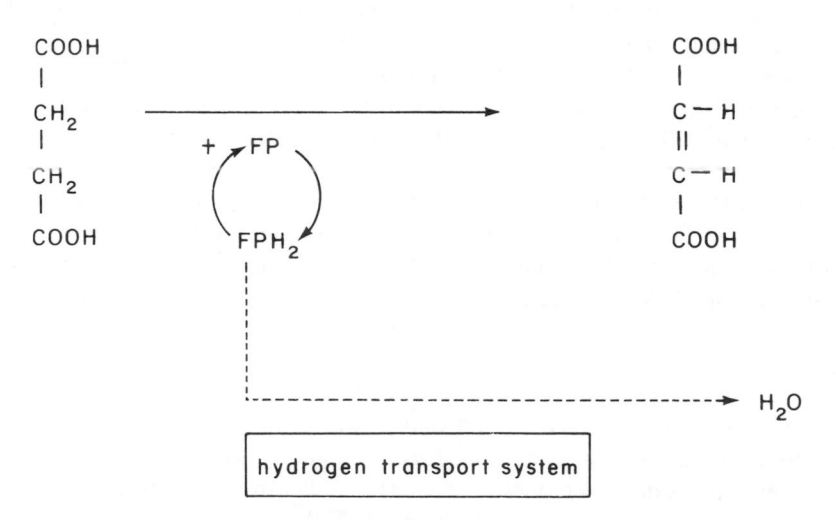

where FP represents the flavoprotein of SDH.

 The significance of this transport of hydrogen lies in the way energy is released from substrates by oxidation (or dehydrogenation). It could have been thought that the oxidation of a substrate, AH_2, itself yields energy. In common parlance, 'you get energy by burning sugar'. But, as was shown with great clarity by Krebs and Kornberg (1957), this first step, namely,

$$AH_2 + NAD^+ \longrightarrow A + NADH + H^+$$

or

$$BH_2 + NADP^+ \longrightarrow B + NADPH + H^+$$

actually sets free no appreciable amount of energy. (The exception is the oxidation of α-ketonic acids.) The hydrogen remains at the same electrode potential; energy is released only when the hydrogen (or electron) moves down

the hydrogen transport chain. Major changes in the free energy of the oxidation–reduction (redox) systems (corresponding to major alteration in the electrode potential of these systems) occur at only certain points in the hydrogen transport system. It is the energy liberated by these changes that is used for the production of ATP from ADP and inorganic phosphate. This linkage of energy from the oxidative system to provide the energy-rich phosphate bond of ATP is the process of oxidative phosphorylation.

So, to revert to common parlance, in a sense it is the hydrogen that carries the energy from the sugar. It begins at a high potential energy which it loses during its passage down the hydrogen transport chain in the mitochondria. In some ways it is comparable to the state of water in a waterfall which is being used to drive a turbine (Figure 29). In the case of the waterfall, the potential energy is converted into kinetic energy as it falls onto each successive ledge (A to B; B to C; C to O); in the case of hydrogen transport it is the change in electrode potential which is being converted into the chemical energy bond of ATP.

The reactions indicated by Krebs and Kornberg (1957) are as follows; NADP can be substituted for NAD. (The old form for the reduced coenzyme, i.e. $NADH_2$, was still in use in 1957; FP corresponds to flavoprotein.)

$NADH_2 + FP \longrightarrow FPH_2 + NAD$ (standard free energy change -11 kg cal)
$FPH_2 + $ oxidized cytochrome $\longrightarrow FP + $ reduced cytochrome (-16 kg cal)
Reduced cytochrome $+$ oxygen $\longrightarrow$ oxidized cytochrome $+ H_2O$ (-25 kg cal)
Total reaction: $NADH_2 + \frac{1}{2}O_2 \longrightarrow NAD + H_2O$ (total free energy change -52 kg cal)

One other concept requires to be discussed and this is the meaning of *electrode potential*, and how it affects the tetrazoles which will be used as quantitative cytochemical indicators of oxidation. It is most simply and pertinently explained by reference to oxidation–reduction dyes. (For fuller discussion reference should be made to Hewitt, 1950, 1961; Chayen, 1982.)

Very many dyes are known which change colour when they become oxidized or reduced; when the change is reversible they are called oxidation–reduction (or redox) indicators. Some which are in common use in biology and medicine are listed in Table 8. If an inert electrode is immersed in a solution of such an indicator which is in equilibrium between its oxidized and its reduced state, the electrode achieves a potential E. This potential (normally expressed relative to that of a normal hydrogen electrode) is directly related to an electrode potential which characterizes that dye (E_0) by a simple equation:

$$E = E_0 + 0.058 \log\frac{[\text{oxidized dye}]}{[\text{reduced dye}]}$$

where the brackets indicate 'concentration of' the component inside the bracket.

Table 8. Some useful redox indicators

Indicator	E_0' at pH 7·0
o-Bromophenol-indophenol	+ 0.230
o-Cresol-indophenol	+ 0.191
Guaiacol-indo-2:6-dibromophenol	+ 0.159
Janus green	− 0.255
Thionine (Lauth's violet)	+ 0.063
Methylene blue	+ 0.011
Methyl viologen	− 0.440
Neutral red	− 0.325
Phenol-indo-2:6-dichlorophenol (for vitamin C)	+ 0.217
Resazurin	Variable
Thymol-indophenol	+ 0.174

It is seen from this equation that the actual electrode potential measured (E) depends on the relative concentrations of the oxidized and reduced dye. When there is as much of the oxidized as of the reduced form, the equation becomes:

$$E = E_0 + 0.058 \log 1$$

or

$$E = E_0 \quad \text{(since } \log 1 = 0\text{)}$$

In practice the electrode potential which is characteristic for a given oxidation–reduction system is not expressed by E_0 because this is limited, on theoretical grounds, to a constant reaction of pH = 0; consequently it is expressed instead as E_0', which is the normal electrode potential of the system (when the concentration of the oxidized form and of the reduced form are equal, and at any pH other than pH 0).

The significance of these discussions is as follows. Suppose you have two oxidation reduction systems of different E_0', the one $X \rightleftharpoons XH_2$ having an E_0' of + 0.1 V and the other $Y \rightleftharpoons YH_2$ with an E_0' of − 0.1 V. Then it follows that the first system will be able to oxidize the second, i.e.

$$YH_2 + X \rightarrow Y + XH_2$$

The general rule is that oxidizing systems have the more positive potentials; reducing systems have the more negative potentials. In other words, the more negative the electrode potential, the greater is its reducing power. (Other conventions do exist but this seems to be the most generally used, e.g. see Burton, 1961; Hewitt, 1961; West and Todd, 1961, p. 821.)

The point that deserves to be stressed is that the reaction

$$YH_2 + X \longrightarrow Y + XH_2$$

does not require a catalyst (or an enzyme). Consequently the oxidation–reduction potential of a system can be estimated by adding to it various redox indicators of known E_0' and finding which are affected. This method has been used for investigating the potentials of living tissues. For example, if the tissue reduces an indicator of an E_0' of -0.03 V but does not reduce one of an E_0' of -0.09 V, the oxidation–reduction potential of the tissue must lie somewhere between these values.

Following this line of reasoning, it seems that the hydrogen passing down the hydrogen transport chain in, for example, mitochondria will reduce an oxidation–reduction (redox) indicator when its latest acceptor system has an electrode potential which approximates to (but is still more electronegative than) that of the indicator. The tetrazolium salts, which become reduced at particular electrode potentials to form insoluble and highly coloured formazans, come within this scheme (see Figure 30).

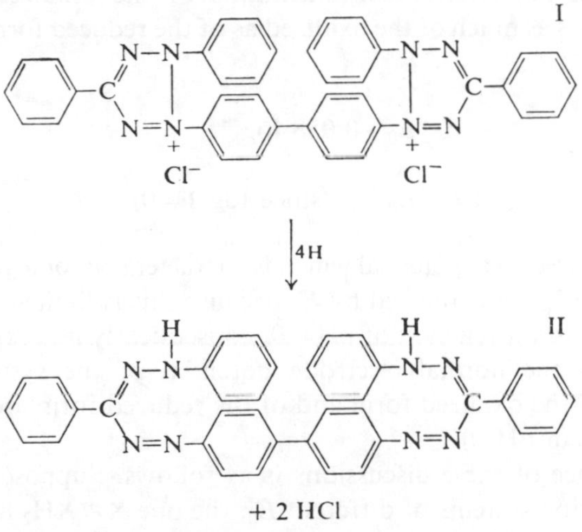

Figure 30. Formulae of neotetrazolium chloride (I) and its corresponding formazan (II).

General histochemistry of dehydrogenases

The activity of dehydrogenases is demonstrated, histochemically, by the use of one of the various tetrazolium salts that are commercially available. The requisite substrate is given, together with the cofactor, etc., that may be required by the particular dehydrogenase that is being studied. This is given at the pH that allows optimal activity, and in the presence of a tetrazolium salt. Hydrogen is removed by the enzyme and passed to one or more hydrogen carriers until it reaches a redox state that allows it to reduce the tetrazolium salts.

It is relatively easy to determine the redox potential of most water-soluble substances. It is more difficult with the tetrazolium salts because, while the tetrazole is readily water-soluble, its reduced form (the formazan) is very insoluble. As far as can be judged, the redox potentials of various tetrazolium salts/formazans is as follows:

triphenyltetrazolium/formazan	+ 490 mV
neotetrazolium/formazan	+ 170 mV
nitroblue tetrazolium/formazan	+ 50 mV

(Altman, 1972a, b). Against these values must be placed the electrode potential of the half oxidized : half reduced form of the NAD/NADH or NADP/NADPH couples of − 320 mV (Krebs and Kornberg, 1957). It follows, therefore, that a reaction for a dehydrogenase, done with a tetrazolium salt alone, will detect only that hydrogen which has passed through some hydrogen transport chain to such an electrode potential as will reduce the tetrazolium salt.

The measurement of the activity of which a dehydrogenase is capable is facilitated by the use of phenazine methosulphate (PMS). This takes the hydrogen from the reduced coenzyme, or other cofactor, and transmits it quantitatively to the tetrazolium salt to yield the insoluble, coloured formazan (Singer and Kearney, 1954; Farber and Bueding, 1956, for neotetrazolium). Evidence for this, and for the precision of modern quantitative histochemistry, was the demonstration of the same amount of activity (in μmol hydrogen/cm^3 of liver) of two dehydrogenases by conventional biochemistry and by quantitative histochemistry (Chayen, 1978b).

The total activity of a dehydrogenase can therefore be measured if the reaction is done in the presence of PMS, with neotetrazolium as the final hydrogen acceptor. (It is helpful to use neotetrazolium because it cannot react with the reduced cofactor of the dehydrogenase: the hydrogen is passed solely from the cofactor to the PMS.) This represents the total dehydrogenase activity (total generation of hydrogen in unit time) of which the enzyme is capable (T). It is often helpful to react serial samples (or sections) in the absence of PMS. The amount of hydrogen recorded will indicate the amount that has passed down some oxidation chain, such as occurs in mitochondria, or the microsomal respiratory pathway in the cytoplasm. This amount (in unit time) is called type I hydrogen. By subtracting this amount from the total of which the enzyme is capable (T) we get the amount of type II hydrogen ($T-$type I).

The ability to differentiate between the potential uses of the hydrogen, generated by a dehydrogenase, is one of the special advantages of histochemistry. Many consequences of such differentiation have been reviewed by Chayen et al. (1986; also Chayen and Bitensky, 1982). Two simple illustrations of the importance of making this distinction are the following: firstly the glucose 6-phosphate dehydrogenase (G6PD) activity (generating NADPH) of cells of rat

adipose tissue and of the zona reticularis of the guinea pig adrenal was almost identical. But in the former, where the NADPH might be expected to be required for biosynthesis, 96% was of type II; in the latter, where NADPH is required for hydroxylation of steroids, only 4% was of type II, the rest being of type I. The second sample concerns the G6PD activity of hepatocytes of rat liver, and of hepatocytes in experimentally induced hepatomas. The total activity was the same but, whereas in the normal hepatocytes it was divided equally between the type I and type II pathways, in the hepatocytes of the hepatoma only one-third was of type I and the rest was of type II. However, it should be noted (as emphasized by Chayen and Bitensky, 1982) that the histochemical demonstration of type I and type II hydrogen, from NADPH, indicates only the probability that any NADPH hydrogen that is generated will be used for one or other pathway. So, for example, in one tissue the total G6PD activity demonstrated by a histochemical reaction may be 200 units, of which 20 units go through the type I pathway (ratio of type II : type I of 9 : 1). Let us suppose that some condition causes a decrease in the total activity of this tissue, so that the total G6PD activity is now only 10 units. It may be that of these 10 units of activity nine pass through the type I pathway, giving a ratio of type II : type I of 1 : 9. These results indicate that initially there was a 9 : 1 probability that reducing equivalents from NADPH would be used for 'high electrode potential' functions, whereas now the conditions have not only diminished the activity of the enzyme but there is a high probability that what hydrogen is produced will be used for such functions as hydroxylation. An example of this is shown in the work of Smith and Wills (1981).

Factors in dehydrogenase histochemistry

As has been discussed (above; also Figure 28) the tetrazole traps the hydrogen when the latter reaches an oxidation–reduction system with an electrode potential that is compatible with the tetrazole being used. However, the localization of the formazan resulting from this reduction of the tetrazole depends on the following factors.

Rate of dehydrogenase activity

As has been discussed in relation to enzyme histochemistry generally (p. 139), the insoluble end-product—in this case the formazan—is truly insoluble only if its local concentration exceeds a certain value. Consequently, if the enzymic dehydrogenation is slow, or if the transport of the hydrogen is slow, the rate at which the formazan is produced may not exceed its solubility coefficient so that it will not be precipitated at the site at which it has been formed. It can then diffuse to other sites in the tissue which have an affinity for the formazan. Fat droplets are a well known example of this phenomenon.

This is an additional reason for avoiding chemical fixation, which is liable to diminish the activity of the dehydrogenase and so decrease the rate of formation, and hence the local concentration, of the formazan to such a level as may permit its diffusion from its site of formation.

Influence of oxygen

Neotetrazolium has such a positive electrode potential that oxygen will very successfully compete with it for the hydrogen (reducing equivalents) liberated by dehydrogenase activity. This forms the basis of a test for malignancy because malignant cells have some component which at least partially annuls this effect of oxygen (Altman *et al.*, 1970). Other tetrazolium salts may show less competition by oxygen. However, to avoid this effect, and to avoid the possible occurrence of deleterious free radicals during the reaction (e.g. Tappel *et al.*, 1963; Desai and Tappel, 1963; Slater, 1966), it is advisable to do the reactions in an atmosphere of nitrogen (see p. 11).

'Substantive' properties of the formazan

The formazans of some tetrazoles, such as that of neotetrazolium, tend to crystallize so that their ultimate localization may reflect the location of such sites; if fatty matter is in the vicinity, they may preferentially dissolve in this. On the other hand, the formazan of nitroblue tetrazolium does not crystallize but attaches to the nearest tissue element (see Nachlas *et al.*, 1957b). Consequently, it may be expected to give a fair indication of the site at which it accepted the hydrogen.

Site of hydrogen uptake

We have emphasized that tetrazolium salts accept hydrogen (or reducing equivalents) when it reaches a particular electrode potential; they cannot accept it directly from the dehydrogenase or its cofactor. This point has been stressed by Farber *et al.* (1956a, b; Sternberg *et al.*, 1956; also Nachlas *et al.*, 1958a, b). The implications of this deserve some consideration.

Enzymes bound in mitochondria An enzyme such as succinate dehydrogenase, which is tightly bound within the structure of the mitochondria, is in close proximity to a hydrogen transport system. Consequently, it is likely that the deposition of nitroblue formazan will indicate the true localization of such an enzyme, within the limits of resolution of the light microscope.

Cytosolic dehydrogenases If sections are reacted in the aqueous medium alone, those enzymes which are not tightly bound will be lost from the section.

Altman and Chayen (1965, 1966) showed that most of the glucose 6-phosphate dehydrogenase (G6PD) and of 6-phosphogluconate dehydrogenase was lost from unfixed sections within the first 2 min of immersion in a 'normal' reaction medium. In the reaction medium these enzymes reacted with their substrate and cofactor; the reduced cofactor then became the substrate for the tissue-bound diaphorase, generating formazan within the tissue. Consequently, the site at which formazan was deposited reflected that of the diaphorase, not of the dehydrogenase (as in Figure 7). So, in principle at least, with a reaction for G6PD, you could have two cells in a section, both with identical diaphorase activity, but one cell producing much more type II hydrogen; the 'G6PD' reaction would show them to be identical.

This had been a major stumbling block to the acceptance of histochemistry as a valid form of investigation. It was overcome when it was shown that certain colloid stabilizers, used at sufficient concentration, could retain all the measurable large molecules (particularly enzymes) inside the unfixed sections. The main stabilizers that have been investigated are polyvinyl alcohol (Altman and Chayen, 1965) and certain forms of degraded collagen (Butcher, 1971b) including Polypep 5115 (Sigma). The best proof of the retention of 'soluble' dehydrogenases is the 'transfer experiment'. In this, sections are reacted for one time, t_1, and then removed from the reaction medium and placed in a second bath of the medium for another time, t_2. If the dehydrogenase had left the section and gone into the reaction medium in time t_1, there would be no further reaction when it was immersed in the fresh bath for time t_2. If there was full retention, the results of $t_1 + t_2$ should be the same as when sections were reacted for time $t_1 + t_2$ but in the same reaction medium.

SUCCINATE DEHYDROGENASE

Biochemical background

This enzyme is part of the Krebs tricarboxylic acid cycle (Figures 31 and 32) and is therefore found in all aerobic cells. It is firmly bound to the inner membrane of mitochondria but it has been isolated, purified and investigated in considerable detail (as reviewed by Hatefi and Stiggall, 1976). It contains FAD (flavine adenine dinucleotide) covalently bound to the larger of its two subunits. It also contains $-SH$ groups that are essential for its catalytic activity. Reduced ubiquinone (e.g. ubiquinone-10) may activate the enzyme. The pH for its optimal activity is said to be about 7.6. It reacts with phenazine methosulphate (PMS) and with ferricyanide (but at concentrations not exceeding 3 mM for the isolated enzyme) so that these can be used as effective hydrogen-carriers from it to a tetrazolium salt. It has many competitive inhibitors: for histochemical studies the most useful of these is malonate ($COO^- -CH_2-COO^-$) which has two

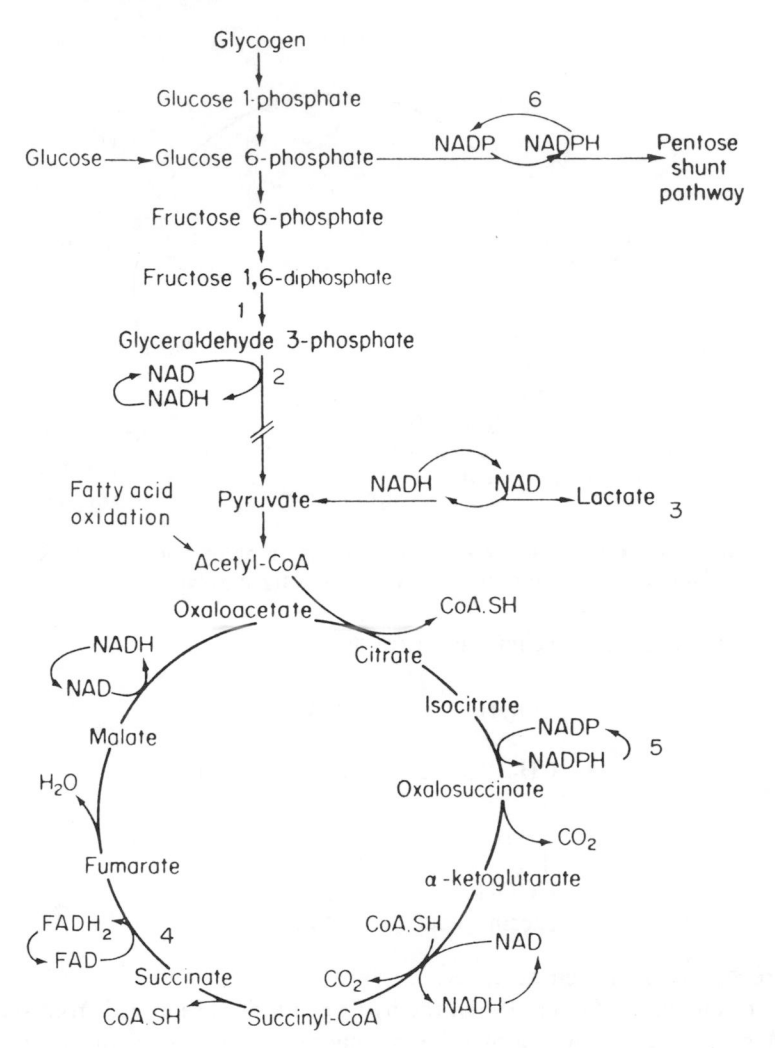

Figure 31. The Embden–Meyerhof pathway and the Krebs cycle: 1, aldolase; 2, glyceraldehyde 3-phosphate dehydrogenase; 3, lactate dehydrogenase; 4, succinate dehydrogenase; 5, isocitrate dehydrogenase; 6, glucose 6-phosphate dehydrogenase.

carboxyl groups with approximately the same spacing between them as in succinate but lacks the oxidizable $-CH_2-CH_2-$ lying between them (Dixon and Webb, 1964, p. 352). Other inhibitors include phosphate, fluoride, suramin and oxaloacetate (*Biochemists' Handbook*, p. 370). It is also inhibited by reagents that combine with $-SH$ groups.

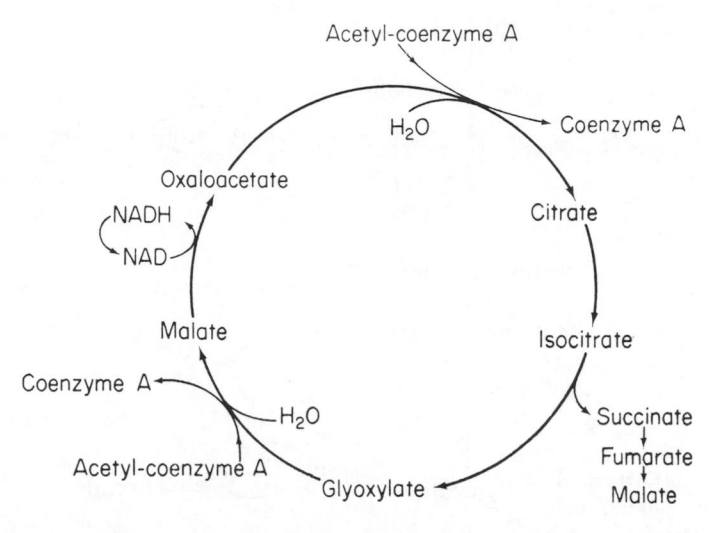

Figure 32. The glyoxylate cycle. Although this involves some of the Krebs cycle enzymes, succinate dehydrogenase is not directly involved.

The oxidation of succinate is as follows:

$$
\begin{array}{ccc}
\text{COOH} & & \text{COOH} \\
| & & | \\
\text{CH}_2 & +\text{A} \rightleftharpoons & \text{CH} \quad +\text{AH}_2 \\
| & & || \\
\text{CH}_2 & & \text{CH} \\
| & & | \\
\text{COOH} & & \text{COOH} \\
\text{succinic acid} & & \text{fumaric acid}
\end{array}
$$

where A is a hydrogen acceptor.

As discussed by Slater (1961), the hydrogen that is removed from succinate by this enzyme is moved along the mitochondrial hydrogen transport system, but this can be 'short-circuited' by the use of suitable hydrogen acceptors as shown below.

Succinate

$$2H\text{---}\text{---}\rightarrow \text{cyt.}\, b \text{---}\text{---} \text{cyt.}\, c_1 \text{---}\text{cyt.}\, a\,(a_3) \text{---}\text{---}\rightarrow O_2$$

Fumarate

Succinate
dehydrogenase

Phenazine   Methylene blue
methosulphate   2,6-dichlorophenolindophenol

This enzyme has been very widely studied, especially in muscle, particularly cardiac muscle, and in experimentally induced cancer.

Histochemical background

Succinate (or 'succinic') dehydrogenase is one of the most popular enzymes studied by histochemists, who use it as a marker for the Krebs cycle (Figure 32). It has been much investigated in studies on cancer (e.g. Burstone, 1962). Some of these studies have emphasized the value of histochemical investigation over reliance solely on conventional biochemical procedures, which can be confused by 'the tissue dilution artefact' (Chayen *et al.*, 1961). An example of this involved studies on experimentally induced liver cancers in rats. Conventional biochemical procedures showed conclusively that the succinate dehydrogenase activity of the affected liver dropped to about 50% of its normal value when the malignant change occurred, and that this coincided with a reduction in the number of mitochondria (Potter *et al.*, 1950; Fiala and Fiala, 1959). It should be noted that such results were 'per unit weight' of liver. Histochemical analysis showed that this decrease in activity per unit mass was, in fact, due to the considerable proliferation of cells from the bile duct, which have very few mitochondria and very low levels of succinate dehydrogenase activity even in normal, healthy liver. Histochemistry showed that the activity in the remaining hepatocytes was unchanged. However, with cells from the bile duct now constituting about half the volume, the activity 'per gram' of liver was decreased (Chayen *et al.*, 1961; Jones *et al.*, 1961; Jones, 1963).

Succinate dehydrogenase activity has been studied using quantitative histochemistry by Defendi and Pearson (1955), Jones *et al.* (1963) and Chayen *et al.* (1966a). It has also been used to study human myocardial dysfunction (Niles *et al.*, 1964a, 1966); these studies led to it being used to monitor open-heart surgery (Canković-Darracott, 1982; Braimbridge *et al.*, 1982).

Method

Fresh cryostat sections of unfixed tissue must be used. Because this enzyme is tightly bound inside mitochondria, no colloid stabilizer is required.

1. Incubate sections in the reaction medium in a Coplin jar, at 37 °C in an atmosphere of nitrogen. With neotetrazolium the reaction time may have to be as long as 2 h; with nitroblue tetrazolium it may be about 30 min.
2. Wash well in distilled water.
3. Mount in an aqueous mountant such as Farrants' medium.

Result

With neotetrazolium, red-blue granules and red colour, which are absent in the control, show the amount of succinate dehydrogenase activity.

With nitroblue tetrazolium, red to purple staining, which is not found in the control, indicates both the amount of succinate dehydrogenase activity and the site at which hydrogen, removed from the succinate, has been trapped by the tetrazolium salt.

Control

1. Incubate in a 0.05 M solution of sodium malonate in the buffer for 15 min at 37 °C.
2. Transfer to the complete medium to which 0.05 M malonate has been added (0.37 g sodium malonate in 50 ml of the reaction medium).
3. Wash in distilled water.
4. Mount in an aqueous mountant such as Farrants' medium.

This treatment should inhibit succinate dehydrogenase activity.

Measurement

Use light of 585 nm.

Solutions required for this method

Simple reaction medium

Add 0.1 g of either neotetrazolium chloride or nitroblue tetrazolium to 100 ml of 0.1 M phosphate buffer (or 0.05 M glycyl glycine buffer) at pH 7.8. Heat gently to dissolve. Cool. Filter. Adjust to pH 7.8 if necessary.

To 50 ml of this solution add 0.68 g sodium succinate ($6H_2O$). This gives a 0.05 M solution.

Bubble nitrogen through this solution at 37 °C before and during use.

PHENAZINE METHOSULPHATE METHOD

Method

React as for the simple method but add 2 mg fresh phenazine metho-sulphate (PMS) to 50 ml of the normal reaction medium. The reaction time will be considerably reduced. It is advisable to agitate the solution during the reaction.

Result

Phenazine methosulphate enhances the measurable activity because it accepts the hydrogen directly from the flavoprotein associated with the enzyme. When the hydrogen acceptor is nitroblue tetrazolium, the formazan should give fairly precise localization of the enzyme.

Note

Care must be taken to use fresh PMS because it may decompose on storage.

α-GLYCEROPHOSPHATE DEHYDROGENASE

Biochemical background

There are two α-glycerophosphate dehydrogenases which occur together in the same cell: that which acts within the cytoplasm requires NAD as cofactor, whereas that which acts within the mitochondrion does not. They are involved in the dihydroxyacetone phosphate shuttle. The significance of this shuttle is that NADH, generated in the cytoplasm, cannot penetrate the mitochondrial membrane. Consequently, cytoplasmic NADH is used to reduce dihydroxyacetone phosphate to form α-glycerophosphate, which apparently passes readily through the mitochondrial membrane and gives up its hydrogen, inside the mitochondrion, under the influence of the mitochondrial glycerophosphate dehydrogenase, which does not require NAD as cofactor:

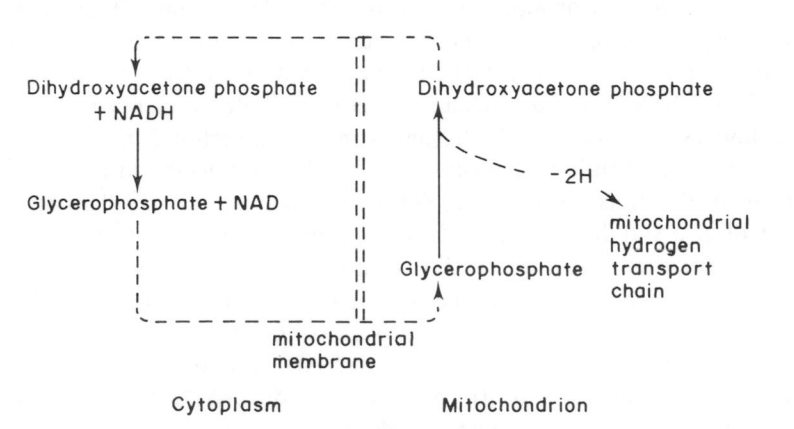

In this way (e.g. Lehninger, 1965) the reducing equivalents (i.e. the hydrogen of NADH) generated in the cytoplasm are transported by the reduced product (α-glycerophosphate) into the mitochondrion, where they are liberated to yield ATP energy via the mitochondrial hydrogen transport chain (by the normal process of oxidative phosphorylation).

The mitochondrial form of the enzyme, present in the outer phase of the inner membrane, contains FAD and non-haem iron. In biochemical studies phenazine methosulphate, ferricyanide, methylene blue and 2,6-dichloroindophenol have been used as hydrogen acceptors from this enzyme. The optimal pH appears to be around 7.6 (as is that for the NAD-linked enzyme). In some organs, but not in all, the mitochondrial form can be markedly increased by extracts of thyroid gland; thyroidectomy decreased particle-bound α-glycerophosphate dehydrogenase activity which was restored by triiodothyronine (as reviewed by Hatefi and Stiggall, 1976). The mitochondria of the flight muscle of the common house fly oxidize α-glycerophosphate at an exceptionally high rate.

The only certain way of distinguishing between the mitochondrial enzyme and the cytosolic (NAD-dependent) enzyme is by the need for NAD (*Biochemists' Handbook*). Histochemically they may be distinguished by the localization of the tetrazolium–formazan, provided a suitable tetrazolium salt is used (e.g. nitroblue tetrazolium).

Changes in the dihydroxyacetone phosphate shuttle have been implicated in the development of malignancy (e.g. Sacktor and Dick, 1960; Boxer, 1965; Stuart *et al.*, 1970; Chayen *et al.*, 1972a).

Histochemical background

The cytoplasmic enzyme normally reduces dihydroxyacetone phosphate to glycerophosphate. Fortunately it is readily reversible so that it can be assayed by its oxidation of α-glycerophosphate, yielding NADH. But, since it is not linked to a hydrogen transport chain, it is essential to include either phenazine methosulphate (as discussed for succinate dehydrogenase) or menadione sodium bisulphite salt $(0.6 \times 10^{-3}\,\text{M})$ in the reaction medium to trap the hydrogen of the NADH. The addition of phenazine methosulphate to the reaction medium also enhances the uptake of hydrogen from the mitochondrial enzyme.

To retain the soluble cytoplasmic enzyme it is necessary to add a colloid stabilizer to the reaction medium. Polyvinyl alcohol is an excellent stabilizer but it will also stabilize mitochondrial membranes and so not allow entry of substrate. For this reason it is better to use a collagen-derived stabilizer (e.g. Polypep 5115, Sigma). It may also be advisable to use a phosphate buffer because this assists penetration of substrate into mitochondria.

Two reactions should be done. One, with NAD, should disclose total activity (cytoplasmic and mitochondrial); the other, without NAD, should indicate the activity of the mitochondrial enzyme alone.

The suggested reaction medium is 0.3% neotetrazolium (or 0.4% nitroblue tetrazolium) in 0.1 M phosphate buffer (or 0.05 M glycyl glycine buffer), pH 8.0 in 40% (w/v) of Polypep 5115. To this solution add the substrate, α-glycerophosphate. (The concentration should be established by trial and error, e.g. try 10 or 50 mM.) For demonstrating NAD-dependent α-glycerophosphate

dehydrogenase activity, add NAD (e.g. at 5 mg/ml of final reaction medium). Then, just before use, add 0.2 mg/ml phenazine methosulphate (or 0.6×10^{-3} M menadione sodium bisulphite).

GLYCERALDEHYDE 3-PHOSPHATE DEHYDROGENASE

Biochemical background

This enzyme catalyses the reversible oxidative phosphorylation of glyceraldehyde 3-phosphate. This activity therefore involves the oxidation of an aldehyde and the synthesis of a high-energy phosphate anhydride, 1,3-diphosphoglycerate (see Figure 33). It occurs as a tetramer. Normally it contains three or four moles of tightly bound NAD to each mole of the tetramer. Thiol groups within the molecule are important; the essential cysteine residue is Cys-149, which is the most reactive of the thiol groups: it is the inhibition of this group that causes the marked inhibition induced by iodoacetate. Iodoacetamide causes considerably less inhibition (Racker, 1961). Optimal activity occurs around pH 8.6.

The activity requires exogenous phosphate (P_i):

$$\text{Glyceraldehyde 3-phosphate} + NAD^+ + P_i \rightleftharpoons$$
$$\text{1,3-diphosphoglycerate} + NADH$$

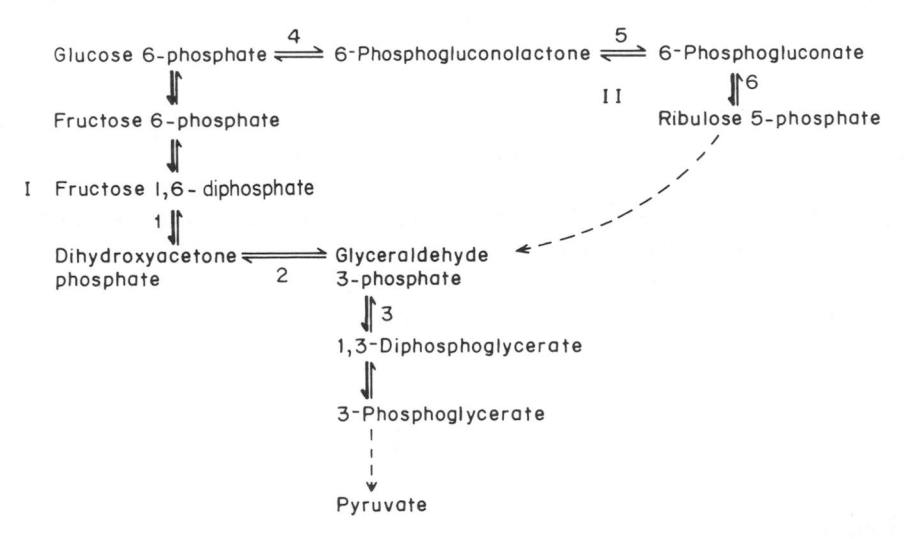

Figure 33. The relationship between the Embden–Meyerhof (I) and the pentose phosphate (II) pathways: 1, aldolase; 2, triose phosphate isomerase; 3, glyceraldehyde 3-phosphate dehydrogenase; 4, 5 and 6, enzymes of the pentose phosphate pathway (see later in this book).

This enzyme forms a link between the pentose phosphate and the normal glycolytic pathways.

Histochemical background

The reason this important enzyme has not been much investigated histochemically is that the substrate, glyceraldehyde 3-phosphate, is both expensive and very unstable. The simple way out of this problem was to use the cheap and stable fructose 1,6-diphosphate and add the relatively cheap enzyme aldolase (1 in Figure 33) to the reaction medium to convert it into the required substrate. This is the basis of the method of Henderson (1976) who found that, in the tissues studied, the endogenous aldolase activity was so strong that the conversion to the desired substrate was hardly affected by the addition of exogenous aldolase. However, this is not always the case; indeed, metabolic disturbances can be apparently due to deficient endogenous aldolase activity.

Consequently, it is advisable to add exogenous aldolase to the reaction medium to ensure that the activity measured is that of glyceraldehyde 3-phosphate dehydrogenase. The activity recorded with this medium minus that recorded when the reaction medium lacks exogenous aldolase will give an indication of the endogenous aldolase activity.

The quantitative histochemical method has been established by the detailed investigations of Henderson (1976). Because this enzyme is located in the cytosol, apparently unbound, a full concentration of a colloid stabilizer is essential.

Method

Fresh (unfixed) cryostat sections (e.g. 10 μm thick) must be used. It is convenient to do the reaction in rings placed around the section.

1. Incubate at 37 °C in an atmosphere of nitrogen. Often only a few minutes reaction time will be required.
2. Wash in distilled water.
3. Mount in a water-miscible mountant such as Farrants' medium.

Result

The activity is denoted by the deposition of formazan.

Control

React in the reaction medium containing 10^{-3} M iodoacetate. This should give complete inhibition. Otherwise subtract any value obtained with this medium from the value obtained without iodoacetate.

Measurement

Measure at 585 nm.

Solutions required for this method

Fructose 1,6-diphosphate	3 mg/ml
NAD$^+$	2.5 mg/ml
Potassium dihydrogen phosphate	0.5 mg/ml
Neotetrazolium chloride	3 mg/ml
Phenazine methosulphate	0.2 mg/ml

in 22% BO5/140 PVA or 40% GO5/140 polyvinyl alcohol (PVA) in 0.05 M glycyl glycine buffer, pH 8.5

The pH should be checked before the final solution is to be used and adjusted to 8.5 if necessary.

Before doing the reaction, add aldolase at a concentration of 25 units/ml reaction medium. Allow this to stand at 37 °C for 20 min before adding this reaction medium to the sections. (1 unit of aldolase converts 1 μM fructose 1,6-diphosphate per minute at pH 7.4 and 25 °C. It will be less effective at the pH of the reaction medium.)

LACTATE DEHYDROGENASE

Biochemical background

This enzyme reversibly oxidizes lactate to pyruvate:

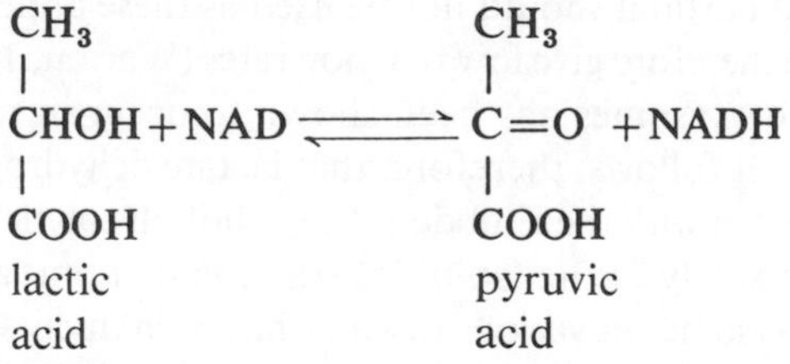

$$\begin{array}{ccc} \text{CH}_3 & & \text{CH}_3 \\ | & & | \\ \text{CHOH} + \text{NAD} \rightleftharpoons & \text{C}=\text{O} & + \text{NADH} \\ | & & | \\ \text{COOH} & & \text{COOH} \\ \text{lactic} & & \text{pyruvic} \\ \text{acid} & & \text{acid} \end{array}$$

It occurs very widely because of its crucial role in glycolysis. The formation of lactic acid from pyruvate allows cells to overcome a temporary lack of oxygen, the lactate being reoxidized to pyruvate when there is sufficient oxygen. It occurs in several isoenzyme forms, each of which is a tetramer made up of at least two different subunits, A or B. (A third subunit, C, is also recorded.) The association of these two subunits generates five tetrameric isoenzymes: A_4, A_1B_3, A_2B_2, A_3B_1 and B_4 (Markert, 1984). The genetical aspects of these isoenzymes have been reviewed by Whitt (1984). The measurement of lactate dehydrogenase activity has been used diagnostically for over 30 years (Skillen, 1984). For example, LD-1 and LD-2 predominate in the serum of patients with myocardial disturbances; LD-4 and LD-5 in hepatitis and disturbances of skeletal muscle; and LD-2/LD-3/LD-4 predominate in cancerous conditions. (The LD terminology for the different isoenzymes is that used in clinical biochemistry, LD-1 being that which moves closest to the anode on electrophoresis.) Such analyses are taken as indicative of cell damage in different organs. Indeed, leakage of lactate dehydrogenase (LDH) out of cells has been much used as a marker of cell damage in more general biochemical studies, as discussed by Danpure (1984), who also gave a valuable warning that apparently viable cells often release low levels of this enzyme without obvious deleterious effect.

The pH for optimal activity of the conversion of pyruvate to lactate is 7.4. It has been said that pH 10 is optimal for the oxidation of lactate to pyruvate (*Biochemists' Handbook*).

The LDH of animal cells acts on many α-hydroxyacids, but is specific for the L($+$) configuration. Originally it was thought to occur almost exclusively in the soluble fraction of tissue homogenates (Dixon and Webb, 1964, p. 627). It contains $-SH$ groups and can be inhibited by *p*-chloromercuribenzoate.

For over 20 years it has been known that plant cells contain cytoplasmic particles, glyoxysomes, which contain a series of enzymes, the glyoxylate cycle, that allows the conversion of fat to sucrose (Beevers, 1969). The typical or key enzymes of this cycle are malate synthetase and isocitrate lyase, the activity of which increases during germination. It was then shown that this cycle also operates in animal cells and involves the activity of lactate dehydrogenase (Warren, 1970) acting on glyoxylate; the optimal pH for the reduction with NADH was 6.9, whereas that for the oxidation of glyoxylate, with NAD, was 9.3. However, both reactions occur readily at pH 8.6. Oxamate and malonate inhibit these activities. Tris, borate, pyrophosphate and barbital should not be used as these appear to form adducts with glyoxylate and therefore give low reaction rates (Warren, 1970). The oxidation of glyoxylate by glyoxysomes has been shown to occur in rat liver and kidney (Gibbs *et al.*, 1977). It follows, therefore, that lactate dehydrogenase activity may not be solely associated with the Embden–Meyerhof glycolytic pathway and is not necessarily present solely in the 'soluble' fraction of homogenates of cells. Its importance in glyoxysomes in various diseases has been discussed by Watts (1973).

It appears that glyoxysomes can be distinguished from peroxisomes (as discussed by Gibbs *et al.*, 1977).

Histochemical background

The enzymatic activity of lactate dehydrogenase is very readily reversible:

$$CH_3.CHOH.COOH + NAD \leftrightharpoons CH_3.CO.COOH + NADH$$

lactic acid   pyruvic acid

Histochemically it is necessary to assay the dehydrogenase activity since this yields NADH which reacts with a tetrazolium salt to yield a coloured, insoluble formazan. Consequently, to upset the equilibrium (so that the reaction goes solely from left to right in this equation), it is advisable either to remove the NADH very rapidly, or to remove the pyruvate. The first is done by using phenazine methosulphate (PMS), which rapidly takes the hydrogen from the NADH. The second is achieved by including potassium cyanide in the reaction medium: this complexes with the pyruvate (with the ketone group).

This enzyme is very much studied histochemically. It is very active in most tissues. However, because it is generally considered to be a soluble enzyme, it is liable to be lost from the section, or cells, into the reaction medium. There the NADH generated by its activity will act as the substrate for the NADH-diaphorase which remains bound within the section. For this reason, many workers have used a reaction medium containing a colloid stabilizer such as polyvinyl alcohol (PVA).

The different isoenzymes have aroused some interest. McMillan (1967) inhibited the M-isoenzyme (muscle-type) by adding 4 M urea to the reaction medium; he inhibited the H-isoenzyme (heart-type isoenzyme; isoenzymes 4 and 5) by including pyruvate to give a concentration of lactate : pyruvate of 10 : 1. Alternatively, if the normal reaction is done in the presence or absence of potassium cyanide, the difference in activity recorded may reflect the activity of isoenzymes 4 and 5, which, unlike the other isoenzymes, are activated by potassium cyanide.

The association of lactate dehydrogenase activity with the glyoxylate cycle has received little attention. Cytochemical procedures have been described for studying two enzymes that are unique to this cycle, namely isocitrate lyase and malate synthase (Davis *et al.*, 1989a, b). These enzymes have been studied in epiphyseal cartilage and growth plate.

Method 1 (earlier procedure)

Use fresh cryostat sections of unfixed tissue.

1. Incubate in the reaction medium at 37 °C, in rings (or in a humidity chamber or a Coplin jar). The duration of incubation can be decided by inspection of the intensity of colour produced by the formazan. In the absence of a colloid stabilizer, the time required may be 30 min or longer, depending on the tetrazolium salt used.
2. Wash well in distilled water.
3. Dry.
4. Mount in an aqueous mountant such as Farrants' medium.

Result

Red, blue or purple formazan demonstrates lactate dehydrogenase activity.

Control

As the test, but excluding either the substrate or the coenzyme, or both, from the reaction medium.

Solutions required for this method

Reaction medium

Dissolve 1 mg/ml neotetrazolium or nitroblue tetrazolium in 0.05 M phosphate buffer at pH 8.0. Add 0.2 g sodium lactate to 25 ml of this solution. Check that the pH is 8.0; correct the pH, if necessary, to this pH. Then add 5 mg NAD to each millilitre of this solution.

Note: this method is not recommended, because lactate dehydrogenase is a soluble enzyme so that it will be lost rapidly from the sections. The final reaction will reflect the activity and localization of the NADH-diaphorase.

Method 2 (recommended method)

Fresh cryostat sections (or smears of cells) must be used.

1. Remove sections from the cryostat and allow them to dry for a few minutes.
2. Place a ring around each section.
3. Add PMS to the reaction medium, mix well, and then add this reaction medium to the sections at 37 °C in an atmosphere of nitrogen. For many tissues, a reaction time of 5 min has been sufficient.
4. Remove the ring and wash in warm running water to remove the 'sticky' reaction medium.
5. Rinse in distilled water.
6. Allow the section to dry and then mount in an aqueous mountant such as Farrants' medium.

Result

Red, blue or purple formazan demonstrates the activity of this enzyme.

Control

Expose serial sections (or cell preparations) to the identical medium lacking either the lactate or the NAD, or both.

Measurement

At 585 nm.

Solutions required for this method

Reaction medium

Neotetrazolium or nitroblue tetrazolium	4.5 mM
Sodium lactate (a 70% w/w solution,	
specific gravity 1.38; BDH)	50.0 mM
NAD$^+$	3.6 mM
PMS	0.65 mM
Potassium cyanide	1.0 mM

in 50 mM glycyl glycine buffer, pH 8.0,
   containing 30% (w/v) PVA (GO4/140; Wacker Chemicals).

Note: the inclusion of potassium cyanide is to remove pyruvate as it is formed during the reaction and so allow the full activity of those isoenzymes that are sensitive to pyruvate. An increase in activity (with or without potassium cyanide) will indicate the amount of these isoenzymes (numbers 4 and 5); in some studies the difference has been as much as 30%. However, if the full activity of this enzyme is not required, cyanide can be left out of the reaction medium.

GLUTAMATE DEHYDROGENASE

Biochemical background

This dehydrogenase, which is present in mitochondria (see Roodyn, 1965; Bendall and de Duve, 1960; Dixon and Webb, 1964, p. 627), has a key role in cellular metabolism and consequently it is found throughout the animal, plant and bacterial kingdoms (see *Biochemists' Handbook*, p. 333; also Dixon and Webb, 1964). In some tissues it depends on NAD, in others on NADP, causing the following oxidative deamination (West and Todd, 1961, p. 1060):

1. Oxidation:

$$CH_2-CH_2-COOH \qquad\qquad\qquad CH_2-CH_2-COOH$$
$$\big| \qquad\qquad\qquad\qquad\qquad\qquad\qquad \big|$$
$$H_2N-CH-COOH \quad +NAD \rightleftharpoons HN=C-COOH + NADH_2$$

L-glutamic acid $\qquad\qquad\qquad\qquad$ α-iminoglutaric acid

NADP can replace NAD in this reaction.

2. Deamination:

$$
\begin{array}{ll}
\underset{\text{α-iminoglutaric acid}}{\overset{\displaystyle\text{CH}_2-\text{CH}_2-\text{COOH}}{\underset{|}{\text{HN}=\text{C}-\text{COOH}}}} +\text{H}_2\text{O}\xrightarrow[\text{reaction}]{\text{spontaneous}} \underset{\text{α-ketoglutaric acid}}{\overset{\displaystyle\text{CH}_2-\text{CH}_2-\text{COOH}}{\underset{|}{\text{O}=\text{C}-\text{COOH}}}} + \underset{\text{+ ammonia}}{\text{NH}_3}
\end{array}
$$

In intact mitochondria, this enzyme is relatively inactive, becoming more active as the mitochondria are deliberately disrupted (see Bendall and de Duve, 1960). It is also involved in various transaminase reactions which have become of significance because the levels of serum transaminases are used in clinical biochemistry for the assessment of tissue death in the liver and in the heart (see e.g. Maclagan, 1964; King, 1965). Bitensky (1967b) has suggested that glutamic dehydrogenase activity may be closely linked with transaminase activity in damaged cells by the following type of scheme shown below.

In this scheme, increased activity of the glutamate dehydrogenase will be expected to shift the transamination equilibrium from left to right in the normal transamination process:

$$\text{Protein} \rightleftharpoons \text{amino acid} + \alpha\text{-ketoglutarate} \underset{\text{transaminase}}{\rightleftharpoons} \text{glutamate} + \text{keto-acid}$$

and so enhance the catabolism of protein.

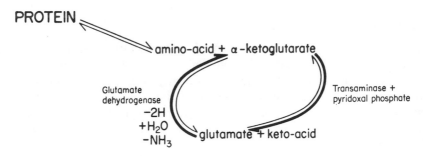

The pH optimum of glutamate dehydrogenase, when oxidizing glutamate, is about 8–8.5, and depends on the buffer and the ionic concentrations. The dehydrogenase is inhibited by p-chloromercuribenzoate, by Ag^+, Hg^{2+} and by Fe^{3+}.

Histochemical background

This enzyme reaction has been found to be a very sensitive indicator of cellular damage (see Kirkby, 1965; Bitensky, 1967b; Chayen and Bitensky, 1968).

The cause for the increased activity appears to be an increased permeability of mitochondria, in damaged cells, to the substrate and coenzyme (Chayen *et al.*, 1966d). The enzyme activity can be measured by quantitative histochemistry.

Method

Fresh cryostat sections of unfixed tissue must be used. Care must be taken not to damage the tissue sample mechanically: even pressure by forceps can be detected in the final reaction if it occurs sufficiently long before the chilling is done.

1. Incubate in the reaction medium in an atmosphere of nitrogen either in a ring or in a Coplin jar. Incubate for 2 h at 37 °C if neotetrazolium is used; with nitroblue tetrazolium, 30 min should be sufficient.
2. Wash in distilled water.
3. Mount in Farrants' medium.

Result

Blue, red or purple formazan (depending on the tetrazole used) indicates glutamate dehydrogenase activity if it is not present in the control.

Solutions required for this method

Reaction medium

Dissolve 0.1 g neotetrazolium chloride or nitroblue tetrazolium in 100 ml 0.05 M phosphate buffer (or 0.05 M glycyl glycine) at pH 7.8. Warm to dissolve. Cool. Filter. Then prepare the reaction medium as follows:

NAD	50 mg
Monosodium glutamate	170 mg
Tetrazole in buffer	10 ml

Control

1. Incubate as for the test but in a reaction medium lacking the glutamate.
2. Incubate as for the test but in a reaction medium lacking NAD.

β-HYDROXYACYL CoA DEHYDROGENASE

Biochemical background

Many tissues may derive half their energy from the oxidation of fatty acids. Neely and co-workers (Neely *et al.*, 1972; Neely and Morgan, 1974) showed that nearly 70% of the oxidative activity of well oxygenated cardiac muscle is due to the oxidation of fatty acids and this can increase in certain conditions. However, fatty acid oxidation is suppressed, with a concomitant stimulation of glycolysis, in hypoxic or anoxic hearts (Neely and Morgan, 1974). It was therefore of some moment to be able to measure fatty acid oxidation in the myocardium during prolonged open-heart surgery; this was done cytochemically on small biopsies taken during surgery (Canković-Darracott *et al.*, 1977).

The utilization of fatty acids (*Biochemists' Handbook*) is a complex process. They are first converted to acyl-coenzyme A:

$$R.CH_2.CH_2.CH_2.CH_2.CO.S.CoA$$

The next step involves a complex dehydrogenation process:

$$R.CH_2.CH_2.CH_2.CH_2.CO.S.CoA \longrightarrow R.CH_2.CH_2.CH{=}CH.CO.S.CoA$$

This reaction is mediated by acyl dehydrogenase. This product is then hydrated (by the action of enoyl hydrase; also called crotonase) to yield β-hydroxyacyl-coenzyme A, which is then oxidized by the enzyme β-hydroxyacyl dehydrogenase, with NAD as coenzyme:

$$R.CH_2.CH_2.CH(OH).CH_2.CO.S.CoA + NAD \rightleftharpoons$$
$$R.CH_2.CH_2.CO.CH_2.CO.S.CoA + NADH + H^+$$

The resulting β-ketoacyl-coenzyme A then undergoes thiolysis to yield acetyl-coenzyme A and acyl-coenzyme A, which is the original fatty acid minus two carbon atoms.

$$R.CH_2.CH_2.CO.|CH_2.CO.S.CoA$$
$$+ \qquad |$$
$$CoA.S|H \rightleftharpoons R.CH_2.CH_2.CO.S.CoA + CH_3.CO.S.CoA$$
$$\text{acyl-coenzyme A} \qquad\qquad \text{acetyl-coenzyme A}$$

The optimal pH for β-hydroxyacyl dehydrogenase is said to be between 9.6 and 10 (*Biochemists' Handbook*).

Histochemical background

The cost of the substrate, acetoacetyl-coenzyme A is too great for routine use. On the other hand, acetoacetylcysteamine is very much less expensive and is only about five times less effective as a substrate. However, it involves the use of NADH, and the resulting NAD^+ cannot be detected histochemically.

Consequently, the reverse reaction is utilized in histochemistry, with a new substrate (hydroxybutyryl cysteamine) which was especially produced for this purpose:

$$\text{hydroxybutyrylcysteamine} + NAD^+ \rightleftharpoons \text{acetoacetylcysteamine} + NADH$$

To overcome autolysis of the substrate at high pH values, nitroprusside is incorporated into the reaction medium to complex with any liberated cysteamine which otherwise might react with the tetrazolium salt and yield spurious amounts of the formazan. Furthermore, menadione was used instead of PMS; not only has it been shown to be as effective as PMS in a number of studies, but it also forms an adduct with thiols and therefore could remove any free cysteamine.

The stoichiometry of the histochemical reaction has been demonstrated by Chambers *et al.* (1982).

Method (Chambers *et al.*, 1982)

Fresh cryostat sections or fresh smears of whole cells should be used.

1. React at 37 °C in an atmosphere of nitrogen. With sections of rat and human heart a reaction time of 10–15 min was sufficient.
2. Wash well with distilled water.
3. Mount in an aqueous mountant such as Farrants' medium.

Result

Coloured precipitate of the neotetrazolium formazan.

Control

React a serial section under the same conditions but with cysteamine (5.0 mmol/l) instead of the hydroxybutyryl cysteamine. For quantification, the activity found with this control medium is subtracted from that obtained with the standard medium.

Measurement

Measure with light of 585 nm wavelength.

Solutions required for this method

Reaction medium

Dissolve 1% neotetrazolium chloride in a 30% (w/v) solution of polyvinyl alcohol (GO4/140 grade; Wacker Chemical Co.) in 0.05 M glycylglycine buffer, pH 8.0. Then add NAD at a concentration of 1.0 mmol/l (0.66 mg/ml) and sodium nitroprusside (4.0 mmol/l; 1.2 mg/ml). Menadione is dissolved in ethanol (17 mg in 1.0 ml absolute ethanol) and added to the medium to give a final concentration of 0.1 ml to 10.0 ml of the medium (i.e. a 1.0 mmol/l solution). Adjust the pH to 8.8. Then just before use 0.054 ml of the solution of β-hydroxybutyryl cysteamine (dissolved 1 : 1 in ethanol) is pipetted into a small tube and 3.0 ml of the medium is added to it. The whole is well shaken to ensure good mixing. This should achieve a final concentration of 50 mmol/l (0.009 ml/ml) of this substrate.

The pH should be adjusted to pH 8.5, if necessary. The medium is saturated with oxygen-free nitrogen (because oxygen competes with neotetrazolium chloride for reducing equivalents); its temperature is equilibrated to 37 °C and it is poured into Perspex rings positioned around each section. The reaction is then left at 37 °C in an atmosphere of moist nitrogen.

URIDINE DIPHOSPHOGLUCOSE DEHYDROGENASE

Biochemical background

This enzyme, uridine diphosphoglucose dehydrogenase (UDPGD) converts uridine diphosphoglucose (UDPG) to uridine diphosphoglucuronic acid (UDPGln) which is required for the biosynthesis of chondroitin sulphate and of hyaluronic acid. The UDPGln is also convertible to UDP-xylose, which is the first sugar nucleotide in the biosynthesis of the linkage region of glycosaminoglycans (as discussed by Balduini et al., 1973; De Luca et al., 1976). The UDP-xylose may also be converted into UDP-arabinose which is required in the biosynthesis of hemicellulose.

The enzyme has been studied in the 'soluble' fractions of liver and pea seedlings (*Biochemists' Handbook*), in the cornea and in epiphyseal plate cartilage (Balduini et al., 1973). It catalyses the irreversible four-electron oxidation of UDPG:

$$UDPG + 2NAD^+ \longrightarrow UDPGln + 2NADH + 2H^+$$

In plants, yeasts and bacteria as well as in epiphyseal plate cartilage it has been shown to be strongly and specifically inhibited by UDP-xylose (Balduini et al., 1973). Its activity is irreversible and depends on the presence of thiol groups; it is totally inhibited by 17 mM formaldehyde (Ordman and Kirkwood, 1977); it is also inhibited by NADH (De Luca et al., 1976). The pH optimum of this enzyme appears to be about 7.8–9.0.

A peculiarity of this enzymatic activity is that it appears to be inhibited by aspirin given *in vivo* (Palmoski and Brandt, 1983).

Histochemical background

The first histochemical study on this enzyme was by Balogh and Cohen (1961), who used nitroblue tetrazolium for the detection of its activity. They recorded that, at pH 8.4, there was too much 'non-specific' deposition of the formazan for their procedure to be of use on sections of liver. Because they did not use an intermediate hydrogen acceptor (such as phenazine methosulphate) they suggested that what they detected was a combination of the activities of the UDPGD activity plus (or modulated by) the activity of an unknown diaphorase.

McGarry and Gahan (1985) modified the method of Balogh and Cohen (1961) to produce a quantitative cytochemical study in plant cells. Zemel and Nahir (1989) slightly modified the procedure of McGarry and Gahan (1985) for application to human chondrocytes. They used a 20% solution of a grade of polyvinyl alcohol which did not give full stabilization but which gave the best demonstration of this enzymatic activity.

At first sight it may seem strange that, although both groups of recent workers used polyvinyl alcohol to stabilize their sections, both used less than the full concentration (30%) that would be required to achieve complete stabilization. The possible explanation is that, in our hands, the fully stabilized control sections, reacted with 2.4 mM NAD but no substrate, have yielded values comparable to those obtained with the full medium (with substrate). This was perplexing until it was realized that NAD itself may act as a substrate for some other enzyme (possibly xanthine dehydrogenase). Consequently, the studies were extended (Dr S. Mehdizadeh, personal communication) to determine the lowest concentration of NAD that was consistent with it acting as a coenzyme for UDPGD while yet being sufficient to yield good UDPGD activity.

One peculiarity of the reaction is that it cannot be enhanced by including phenazine methosulphate in the reaction medium.

The reaction is done at a pH lower than optimal because the tetrazolium salt tends to become reduced at high pH values.

Method (after McGarry and Gahan, 1985; Zemel and Nahir, 1989)

Use cryostat sections of fresh tissue.

1. React at 37 °C in an atmosphere of nitrogen. For human cartilage a reaction time of 90 min was required; 10 min was sufficient for mouse articular cartilage by method II.
2. Wash in distilled water.
3. Mount in an aqueous mountant such as Farrants' medium.

Result

Purple or blue deposit of nitroblue formazan.

Control

1. Because of the 'non-specific' deposition of the formazan it is essential to test another section with the full reaction medium lacking the substrate.
2. Use the inhibitor UDP-xylose, at 0.2 mM added to the full reaction medium.

Measurement

Microdensitometrically at 550 nm.

Solutions required for this method

Reaction medium I (after McGarry and Gahan, 1985; Zemel and Nahir, 1989)

Nitroblue tetrazolium	3.67 mM (5 mg/ml)
UDPG	2.28 mM (1.28 mg/ml)
NAD$^+$	1.5 mM (1 mg/ml)

dissolved in 0.05 M glycylglycine buffer in 22% (w/v) polyvinyl alcohol (GO4/140 grade)

The final pH should be 7.9.
  Bubble nitrogen through the medium. The reaction should be done in an atmosphere of nitrogen.
  The reaction medium recommended by McGarry and Gahan included 30 mM potassium cyanide.

Reaction medium II (Dr S. Mehdizadeh, personal communication)

Nitroblue tetrazolium	2.20 mM (3 mg/ml)
UDPG	5.34 mM (3 mg/ml)
NAD$^+$	1.05 mM (0.7 mg/ml)

dissolved in 0.05 M glycylglycine buffer, pH 8.0, in 30% (w/v) of polyvinyl alcohol (GO4/140 grade)

Check that the final pH is 7.8.

STEROID DEHYDROGENASES

These dehydrogenases are essential in the normal conversion of inert steroid precursors to physiologically active hormones. For example, a hydroxysteroid dehydrogenase is required for the conversion of pregnenolone to progesterone in the pathway that leads to the production of cortisol; it is used in the pathway to the formation of testosterone and oestrogens. Their biochemistry and histochemistry have been discussed by Baillie *et al.* (1966). In general they are more active with NAD than with NADP as cofactor.

The methods of Baillie *et al.* (1966) used the following reaction medium:

NAD	2 mg/ml
Nitroblue tetrazolium	0.5 mg/ml
The steroid, dissolved in dimethyl formamide	0.25 mg/ml
dissolved in 0.1 M Tris (or phosphate) buffer, pH 7.5	

The reaction time may be as long as 2 h.

The method of Robertson (1979) is based on that of Quattropani and Weisz (1973; Balogh, 1966). The reaction medium contains the following:

NAD	2.5 mg/ml
Nitroblue tetrazolium	1 mg/ml
dissolved in 30% polyvinyl alcohol in 0.05 M glycylglycine buffer, pH 8.0	

The steroid is dissolved in dimethyl formamide at a concentration of 1 mg/0.1 ml. It is then dispersed into the reaction medium to give a final concentration of 0.4 mg for each millilitre of reaction medium. The final concentration of the dimethyl formamide should not exceed 5%. The final pH of this reaction medium should be adjusted to 7.6.

Robertson (1979) showed that the reaction was linear with time for up to at least 25 min; he recommended reacting for 10 or 20 min.

The control should lack the steroid (substrate) but must include the same concentration of dimethyl formamide.

NADP-DEPENDENT DEHYDROGENASES: GLUCOSE 6-PHOSPHATE AND 6-PHOSPHOGLUCONATE DEHYDROGENASES

The main dehydrogenases that use NADP as coenzyme are the cytosolic NADP-dependent isocitrate dehydrogenase, the malic enzyme, an NADPH-glyceraldehyde 3-phosphate oxidoreductase (Wood, 1973), and the two dehydrogenases of the pentose phosphate pathway, glucose 6-phosphate and 6-phosphogluconate dehydrogenase. The isocitrate dehydrogenase has not been much studied; it can be demonstrated by the same procedures as are used for the two dehydrogenases of the pentose phosphate pathway. Because these have been extensively investigated, attention will be focused on these.

Biochemical background

The oxidation of glucose can go along one of two main paths. In normal glycolysis (the Embden–Meyerhof pathway) the phosphorylated six-carbon glucose is split into two three-carbon sugars, one form of which is oxidized by the NAD-dependent glyceraldehyde 3-phosphate dehydrogenase and converted ultimately to pyruvate (as in Figure 31). This can be fed into the Krebs cycle, or converted to lactate by the NAD-dependent lactate dehydrogenase.

In the alternative pathway, the pentose phosphate pathway (as fully reviewed by Wood, 1985, 1986a, b), the phosphorylated glucose (glucose 6-phosphate, G6P) is oxidized by an NADP-dependent enzyme, glucose 6-phosphate dehydrogenase (G6PD). The product of this oxidation, 6-phosphogluconolactone (6PGL) can be in equilibrium with the substrate (G6P), that is, the enzymatic activity is reversible. The 6PGL is then irreversibly converted to 6-phosphogluconate, which is subjected to oxidative decarboxylation by the enzyme 6-phosphogluconate dehydrogenase. This NADP-dependent decarboxylation converts the six-membered phosphorylated sugar to a five-membered phosphorylated pentose sugar, ribulose 5-phosphate, which yields ribose 5-phosphate (Figure 34). Consequently, in contrast to the Embden–Meyerhof pathway, this pathway depends on NADP as the coenzyme for the oxidative steps. It is often called the hexose monophosphate or pentose phosphate pathway (or shunt).

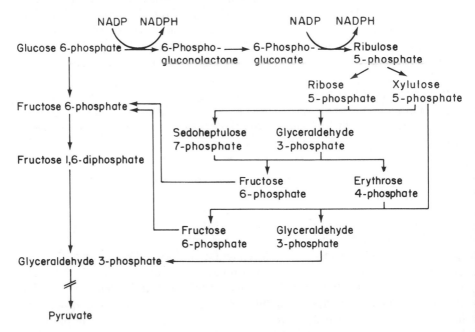

Figure 34. The pentose shunt and its various pathways in relation to the Embden–Meyerhof glycolytic pathway (on the left).

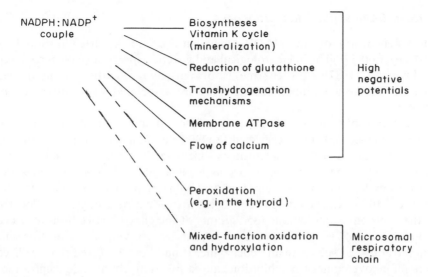

Figure 35. Some of the various uses of reducing equivalents from NADPH. (Those involving the type II pathway are shown with solid lines; broken lines indicate those involving type I hydrogen.)

The significance of this 'shunt' in cellular metabolism is considerable. Firstly, it produces pentose sugars which can be used in the synthesis of nucleic acids; secondly, the dehydrogenases are a major source of NADPH, as indicated in Figure 35. The 'biosyntheses' shown in the figure include the biosynthesis of steroids and their interconversion (Popjak, 1961; Savard, 1961), and the synthesis of fatty acids (Kornberg, 1961), glycogen (Chain, 1959), and various folates. The microsomal respiratory chain is involved in the many hydroxylation and detoxification mechanisms involving cytoplasmic cytochrome P-450 and P-448.

The pentose shunt dehydrogenases occur in high concentration in liver, adrenal, kidney, the lactating mammary gland and erythrocytes, as reviewed by Glock and McLean (1953, 1958), Horecker and Mehler (1955), Dickens (1956), and Kornberg (1957). They are widely distributed in other animal tissues, in plants and in bacteria. Because this pathway produces ribose sugars, as required for both nucleic acids which are involved in protein synthesis and cell division, these dehydrogenases have been studied in malignant growths (e.g. Glock and McLean, 1954; McLean and Brown, 1966). Based initially on histochemical findings (Chayen *et al.*, 1962), elevated activity of 6-phosphogluconate dehydrogenase has been used in the detection of human uterine cancers (Bonham and Gibbs, 1962; Bonham, 1964; Cameron and Husain, 1965; Husain and Cameron, 1966).

Glucose 6-phosphate dehydrogenase

The biochemistry of this NADP-dependent enzyme has been reviewed by Rosemeyer (1987). The haematological and genetic aspects have been reviewed by Luzzatto (1987); this enzyme is of particular significance in human erythrocytes, because inherited anaemias are associated with mutant forms of G6PD.

It is a cytosolic enzyme. It dissociates into subunits of molecular mass of about 50 000–60 000. The dimer seems to be the main native form (Rosemeyer, 1987), but it can associate into tetramers which may be several times more active than the dimers (Yoshida, 1966). It may be assayed at pH 8 with magnesium, and with maleimide to block any 6-phosphogluconate dehydrogenase activity (Deutsch, 1983). In the human, G6PD is the product of a single gene located on the X chromosome; about 185 variants of the enzyme have been recorded (Rosemeyer, 1987), which can be detected by assaying at non-saturating concentrations of substrate and coenzyme. It had been reported that oxidized glutathione overcomes the inhibition caused by relatively high concentrations of NADPH (Krebs and Eggleston, 1974), but this has been contested (Rosemeyer, 1987). Fatty acid acyl esters of coenzyme A (e.g. palmitoyl-CoA) may inhibit this enzymatic activity, but the inhibition may be reversed by the addition of spermine. It has been suggested that aspirin may inhibit G6PD (as reviewed by Rosemeyer, 1987). Cyanide, magnesium and calcium activate the enzyme; mercury and *p*-chloromercuribenzoate inhibit it.

6-Phosphogluconate dehydrogenase

The biochemistry of this NADP-dependent enzyme has been reviewed by Rosemeyer (1987). It is a cytosolic enzyme and catalyses both the oxidation and the decarboxylation of the substrate:

$$6\text{-phosphogluconate} + NADP^+ \rightleftharpoons \text{ribulose 5-phosphate} + CO_2 + NADPH + H^+$$

It dissociates into subunits of 50 000–60 000 molecular weight. The native dimers, as found in mammals, have a molecular weight of about 102 000, but considerable variation in different tissues has been reported. The presence of carbon dioxide in reactions can act as a substrate for the reversal of this enzymatic activity. The enzyme has been reported to be inhibited by fructose 1,6-biphosphate and by nucleotides; citrate has been reported to inhibit the activity of this enzyme. In human erythrocytes ATP and 2,3-diphosphoglycerate are inhibitors. As with G6PD, palmitoyl-coenzyme A is inhibitory and the inhibition can be reversed by spermine (as reviewed by Rosemeyer, 1987). Cyanide may activate; copper ions, which may occur as impurities in some materials that may be used in assaying this enzyme, are inhibitory (*Biochemists'*

Handbook). The pH optimum of 6PGD activity may vary from 8.0 to 9.0 (*Biochemists' Handbook*).

6-Phosphogluconolactonase

The activity of G6PD generates the phosphogluconolactone (6PGL):

$$\text{Glucose 6-phosphate} + \text{NADP}^+ \leftrightharpoons 6\text{PGL} + \text{NADPH} + \text{H}^+$$

Initially it was thought that 6PGL was so rapidly naturally hydrolysed to 6-phosphogluconate that the activity of the enzyme was fairly immaterial. In part, this attitude may have been stimulated by the difficulty of obtaining the substrate. Hofer and Bauer (1987) showed that the relatively stable γ-lactone gave very similar results to those of the natural δ-lactone. At the concentrations of the lactone likely to occur at any given moment in cells, they pointed out that the enzymatic hydrolysis was at least 100 times faster than the spontaneous hydrolysis. Consequently, and with the availability of the γ-lactone as substrate, it is possible that this enzyme may play a very significant function in regulating the activity of the pentose phosphate shunt.

It has optimal activity at pH 7.4; the rate of spontaneous hydrolysis increases rapidly at values above pH 8.0. It is not activated by divalent ions, nor is its activity impaired by chelating agents. The enzyme is specific in requiring the phosphorylated lactone and does not hydrolyse gluconolactone.

Histochemical background

It was the histochemical demonstration of exceptionally high activity of 6PGD in malignant cells (Chayen *et al.*, 1962) that led to the biochemical test used in the diagnosis of cervical cancer, and to a histochemical 'stain' for helping the cytological screening of human cervical cancer (Cohen and Way, 1966).

These dehydrogenases have been a critical test for histochemistry. They are 'soluble' enzymes, present in the cytosol, so that they are lost into the reaction medium within the first minute or two when fresh sections are immersed in a normal reaction medium (Altman and Chayen, 1965). The enzymes then act on their substrate and coenzyme in the solution above the section. The NADPH that is generated then acts as the substrate for the NADPH-diaphorase that remains bound inside the section. Consequently the activity recorded is that of the diaphorase, not the dehydrogenase (as in Figure 7). However, the enzymes are quantitatively retained within the fresh sections provided the reaction is done in the presence of adequate concentrations of either a suitable grade of polyvinyl alcohol or a commercially available degraded collagen (such as Polypep 5115, Sigma). Under such conditions, the amount of enzymatic activity agreed quantitatively with corresponding biochemical measurements on large samples of the same tissue (Chayen, 1978b).

As would be expected, there have been several reports showing that the activities of the two dehydrogenases were high in malignant cells. However, some of the earlier reports may have been confused by a potentially more valuable phenomenon. Normally it is recommended that these activities should be demonstrated with neotetrazolium chloride as the final hydrogen acceptor. Such reactions should be done in an atmosphere of nitrogen because the oxygen in air can compete very successfully with neotetrazolium chloride for the reducing equivalents (hydrogen), so eliminating the observable reaction. However, when the reaction is done in oxygen, many malignant cells still retain considerable G6PD activity (e.g. 20–50% of that found when the reaction is done in an atmosphere of nitrogen). This may constitute a useful test of malignancy.

Since these dehydrogenases are a major source of NADPH, it is not surprising that they have been extensively investigated. Studies on malignant cells include those of Heyden (1979), Butcher (1979), Ibrahim *et al.* (1983) and Bitensky *et al.* (1984). The involvement of the pentose shunt dehydrogenases in rheumatoid arthritis and the effect of glucocorticoids have been studied by Henderson and Glynn (1981) and Bitensky *et al.* (1977). G6PD activity in cartilage and growth plate of bone was measured by Dunham *et al.* (1983, 1986). The effect of oestradiol on G6PD activity in human breast cancer cells was measured by Monet *et al.* (1987).

Methods

These enzymatic activities can be measured with either nitroblue tetrazolium or neotetrazolium chloride as the final hydrogen acceptor. The use of neotetrazolium chloride is advocated (where possible) if full assays of type I and type II hydrogen are required (as in Altman, 1972b; Bachelet *et al.*, 1985; Chayen *et al.*, 1986).

Most assays have been done with the sections stabilized by the presence of 30% (w/v) of a suitable grade of polyvinyl alcohol (GO4/140; Wacker Chemicals Ltd.). The use of 40% (w/v) of Polypep 5115 (Sigma) is equally effective.

Fresh cryostat sections must be used. To achieve full activity, the sections, taken off the knife, should be stored in a desiccator until they are used. But they should not be kept for longer than is necessary to cut all the sections.

GLUCOSE 6-PHOSPHATE DEHYDROGENASE

Method

1. Incubate in the reaction medium in rings in an atmosphere of nitrogen. Between 10 and 20 min at 37 °C should be sufficient for most tissues (but it may be as long as 45 min for bone and cartilage), reacted in the presence of phenazine methosulphate (PMS). Longer times will be required for

reactions done in the absence of PMS. The activity should then be recorded as the activity for a given time, for example for 10 min incubation (i.e. divide the measured activity by the time of incubation, and multiply the result by 10; ideally it should be established that the activity is indeed linear with time, as has been found in many studies).

2. Wash in distilled water.
3. Mount in an aqueous mountant, such as Farrants' medium.

Result

A red-purple stain is obtained when nitroblue tetrazolium is used; blue grains and a red colour are obtained with neotetrazolium chloride.

Control

Incubate as for the test reaction but omit either the substrate or the NADP from the reaction medium.

Measurement

Both the granules and the diffuse stain of the neotetrazolium formazan can be measured accurately if the measurement is done at 585 nm (Butcher, 1972; Butcher and Altman (1973). These publications show how to convert the measured activity to units of hydrogen/unit tissue). The same wavelength can be used for measuring the formazan from nitroblue tetrazolium.

Solutions required for this method

Reaction medium

If neotetrazolium chloride is used as the final hydrogen acceptor, it is advisable to purify it by putting it into a Soxhlet apparatus and refluxing it with chloroform to remove impurities. When the refluxing indicates that most of the coloured impurity has been extracted, remove the solid material from the Soxhlet apparatus, spread the neotetrazolium chloride out onto a clean sheet and allow it to dry (to remove the chloroform). Then collect the powder and store in a closed jar.

The medium should contain the following:

Glucose 6-phosphate (disodium salt; Sigma)	5 mM (1.5 mg/ml)
NADP	3 mM (2.5 mg/ml)
Either: neotetrazolium chloride	5 mM (3 mg/ml)
Or: nitroblue tetrazolium	3 mM (5 mg/ml)

Phenazine methosulphate   0.7 mM (0.023 mg/ml)

Dissolved in 0.05 M glycylglycine buffer, pH 8.0, in 30% (w/v) of polyvinyl alcohol (or 40% Polypep 5115). [The phenazine methosulphate should be added just before the medium is to be used.]

To demonstrate type I hydrogen, leave out the phenazine methosulphate. For some tissues the activity is greatly enhanced by the addition of calcium to the reaction medium. Ibrahim *et al.* (1983) used 34 mmol/l calcium chloride for some malignant cells. Potassium cyanide (e.g. 10 mM) has also been required for some tissues (Chambers *et al.*, 1978).

6-PHOSPHOGLUCONATE DEHYDROGENASE

Method

Since this enzymatic activity generates carbon dioxide (as shown above) the histochemical reaction must be done either in polyvinyl alcohol (30% GO4/140) or in the carbon dioxide-free form of the collagen polypeptide (Polypeptide 8350; Molecular Design and Synthesis, Robens Institute, University of Surrey, Guildford, Surrey, UK).

1. Incubate in the reaction medium in rings in an atmosphere of nitrogen. About 20 min at 37 °C should be sufficient for most tissues although bone and cartilage may require 45 min. If the reaction is done in the absence of phenazine methosulphate, longer times will be required. As with reactions for glucose 6-phosphate dehydrogenase activity, the results should be recorded as the activity for a given time, for example for 10 min incubation.
2. Wash in distilled water.
3. Mount in an aqueous mountant such as Farrants' medium.

Result

A red-purple stain is obtained when nitroblue tetrazolium is used; blue grains and a red colour when neotetrazolium is the hydrogen acceptor.

Control

Incubate as for the test reaction but omit either the 6-phosphogluconate or the NADP from the reaction medium.

Measurement

Both the granules and the diffuse stain can be measured accurately if the measurement is done at 585 nm (as described for measuring G6PD activity, above; also Butcher, 1972; Butcher and Altman, 1973).

Solutions required for this method

Reaction medium

If neotetrazolium chloride is used as the final hydrogen acceptor, it should first be purified, as described for the G6PD reaction (above). The medium should contain the following:

6-Phosphogluconate (trisodium salt)	6 mM (2.5 mg/ml)
NADP	3 mM (2.5 mg/ml)
Either: neotetrazolium chloride	5 mM (3 mg/ml)
Or: nitroblue tetrazolium	3 mM (5 mg/ml)
Phenazine methosulphate	0.7 mM (0.023 mg/ml)

dissolved in 0.05 M glycylglycine buffer, pH 8.4 in 30% (w/v) of polyvinyl alcohol (or 50% of the carbon dioxide-free Polypep)

Because the phenazine methosulphate decomposes, it should be added just before the medium is to be used.

NADPH-DIAPHORASE

In the past, the existence of diaphorases which oxidize NADPH (and those that oxidize NADH) has been recognized. The diaphorase from pig heart is a powerful lipoic dehydrogenase (*Biochemists' Handbook*, p. 356). Other enzymes which may act as diaphorases are the NADH-cytochrome c reductase and the NADPH-cytochrome c reductase. However, more recently there has been considerable interest in the powerful NAD(P)H-oxidizing system, linked to a microsomal respiratory chain (Figure 36), which is involved in the detoxification

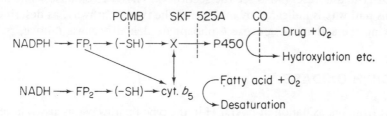

Figure 36. A simplified scheme showing the generalized microsomal respiratory system, as it is understood at present. Cytochrome *P*-450 is involved in drug detoxification and hydroxylation; it can be inhibited by carbon monoxide (CO). The inhibitor SKF 525A acts before this cytochrome in the respiratory chain, and *para*-chloromercuribenzoate (PCMB) inhibits sulphydryl components of the chain. There appear to be various points (indicated by arrows) at which reducing equivalents can be transferred between the NADPH and the NADH respiratory systems. FP_1 and FP_2 are two flavoproteins.

of drugs and many potentially damaging agents. This system passes hydrogen from both NADPH and NADH to react with atmospheric oxygen through a cytosolic (microsomal) chain where the final moiety that reacts with oxygen is either cytochrome *P*-450 or cytochrome *P*-448. This system is required for the detoxification of very many compounds such as barbiturates, codeine, aminopyrine and chlorpromazine; it is involved in the ω-oxidation of fatty acids; in the hydroxylation of many steroids and of some carcinogenic hydrocarbons; and in mixed function amine oxidase activity which uses NADPH in the oxygen-dependent detoxification of physiologically important amines such as tranquillizers, antihistamines, narcotics and many related compounds. Consequently the NAD(P)H-microsomal respiratory activity, with cytochrome *P*-450 or *P*-448 as the terminal component, has become of very considerable importance in toxicology and pharmacology. The NADH-microsomal pathway, which appears to link with the NADPH-dependent pathway, is noted particularly for its function in the desaturation of fatty acids.

A further important effect of this microsomal respiratory pathway is that it may hydroxylate otherwise inert compounds and convert them into necrotizing or carcinogenic substances.

The function of these microsomal respiratory systems has been widely reviewed (e.g. Chayen *et al.*, 1972a; Benford *et al.*, 1987).

Histochemically it must be remembered that the oxidation of glucose 6-phosphate or of 6-phosphogluconate will not produce a formazan from neotetrazolium unless there is some system that will oxidize the reduced coenzyme at the expense of the tetrazolium salt. In the absence of phenazine methosulphate (as used for demonstrating the total activity of the two relevant dehydrogenases), the hydrogen from the reduced coenzyme must pass to a relatively positive electrode potential before it can reduce neotetrazolium chloride. This is done by the microsomal respiratory pathway or by some other NADPH-diaphorase. (Also see the section on NADPH-oxidase.) The activity of this pathway is indicated by measuring the type I pathway, as described for studying the activity of glucose 6-phosphate dehydrogenase (above, p. 262).

NAD(P)H-OXIDASE

Apart from the oxidation of NADPH in the type I pathway (discussed in relation to G6PD activity, p. 257), this coenzyme can be oxidized directly by a specific oxidase. This enzyme is well known as bound to the plasma membrane of polymorphonuclear leucocytes; the membrane becomes invaginated during phagocytosis and the enzyme generates active oxygen species (Badway and Karnovsky, 1980). Patriarca *et al.* (1975a, b) gave evidence for the production of free radicals by this system. There is now some evidence that it may occur more widely. Consequently, in spite of the high cost of the substrate (NADPH), it may sometimes be worthwhile to consider whether or not it is operating in particular instances.

Method

Fresh cryostat sections must be used.

1. Incubate in the reaction medium in rings in an atmosphere of nitrogen. It may be necessary to react for 30–60 min.
2. Wash in distilled water.
3. Mount in a water-miscible medium such as Farrants' medium or Z5.

Result

A deposit of the coloured formazan indicates the activity of this enzyme.

Measurement

At 585 nm.

Noto

It will be significant if this activity is greater than that of the NADPH-diaphorase activity.

Solutions required for this method

Reaction medium

NAD(P)H	3 mM (2.5 mg/ml)
Neotetrazolium chloride	5 mM (3 mg/ml)

dissolved in 0.05 M glycylglycine buffer, pH 8.0 in 30% (w/v) of polyvinyl alcohol (or 40% Polypep 5115)

Note: neotetrazolium chloride is used in this reaction. Other tetrazoles, such as nitroblue tetrazolium, will react with NAD(P)H directly.

TRANSHYDROGENASES

Another way by which the hydrogen from NADPH can be transferred to a tetrazole is by transhydrogenation mechanisms. These have been extensively reviewed by Villee (1962). They are of considerable physiological interest in that they involve the direct intervention of steroid hormones. Other mechanisms by which hydrogen, liberated from substrate in the cytoplasm, can be transferred to the mitochondrial hydrogen transport system, have been discussed by Lehninger (1965). The work of Butcher and Chayen (1966b) indicated that histochemical methods may prove more valuable than more conventional biochemical procedures for studying such phenomena; this question has been discussed in some detail by Chayen and Bitensky (1968).

References

Adams, C. W. M. 1956. A stricter interpretation of the ferric ferricyanide reaction with particular reference to the demonstration of protein-bound sulphydryl and di-sulphide groups. *J. Histochem. Cytochem.*, **4**, 23–35.

Adams, C. W. M. 1961. A perchloric acid–naphthoquinone method for the histochemical localization of cholesterol. *Nature*, **192**, 331–2.

Adams, C. W. M., Smith, W. N. and Stoward, P. J. 1967. A fluorescence histochemical study of ketosteroids in gonadal and adrenal tissue of the rat. *J. Roy. Microsc. Soc.*, **87**, 47–52.

Ahlers, J. 1975. The mechanism of hydrolysis of β-glycerophosphate by kidney alkaline phosphatase. *Biochem. J.*, **149**, 535–46.

Aldridge, W. N. 1956. Organophosphorus compounds and esterases. *Ann. Rep. Chem. Soc.*, **53**, 294–305.

Aldridge, W. N. 1961. Esterases. In: *Biochemists' Handbook*. (Ed. C. Long). pp. 273–7. Spon, London.

Alfert, M. and Geschwind, I. I. 1953. A selective staining method for the basic proteins of cell nuclei. *Proc. Natl. Acad. Sci.*, **39**, 991–9.

Allison, A. C. 1968. Lysosomes. In: *The Biological Basis of Medicine*. (Eds E. E. Bittar and N. Bittar). Vol. 1. pp. 209–42. Academic Press, London.

Altman, F. P. 1969a. The quantitative elution of nitro-blue formazan from tissue sections. *Histochemie*, **17**, 319–26.

Altman, F. P. 1969b. The use of eight different tetrazolium salts for a quantitative study of pentose-shunt dehydrogenation. *Histochemie*, **19**, 363–74.

Altman, F. P. 1972a. *An Introduction to the Use of Tetrazolium Salts in Quantitative Enzyme Cytochemistry*. Koch-Light, Bucks.

Altman, F. P. 1972b. Quantitative dehydrogenase histochemistry with special reference to pentose-shunt dehydrogenases. *Progr. Histochem. Cytochem.* Fisher, Stuttgart.

Altman, F. P. and Barrnett, R. J. 1975. The ultra-structural localisation of enzyme activity in unfixed tissue sections. *Histochemistry*, **41**, 179–83.

Altman, F. P. and Chayen, J. 1965. Retention of nitrogenous material in unfixed sections during incubation for histochemical demonstration of enzymes. *Nature*, **207**, 1205–6.

Altman, F. P. and Chayen, J. 1966. The significance of a functioning hydrogen-transport system for the retention of 'soluble' dehydrogenases in unfixed sections. *J. Roy. Microsc. Soc.*, **85**, 175–80.

Altman, F. P., Bitensky, L., Butcher, R. G. and Chayen, J. 1970. Integrated cellular chemistry applied to malignant cells. In: *Cytology Automation*. (Ed. D. M. D. Evans). pp. 82–97. Livingstone, Edinburgh.

Altman, F. P., Moore, D. S. and Chayen, J. 1975. The direct measurement of cytochrome *P*-450 in unfixed tissue sections. *Histochemistry*, **41**, 227–32.

Andersen, H., Møllgard, K. and von Bülow, F. A. 1970. On the specificity of staining by Alcian blue in the study of human foetal adenohypophysis. *Histochemie*, **22**, 362–75.

Anderson, P. J. 1967. Purification and quantitation of glutaraldehyde and its effect on several enzyme activities in skeletal muscle. *J. Histochem. Cytochem.*, **15**, 652–61.

Appleton, T. C. 1964. Autoradiography of soluble labelled compounds. *J. Roy. Microsc. Soc.*, **83**, 277.

Arborgh, B., Ericsson, J. L. and Helminen, H. 1971. Inhibition of renal acid phosphatase and aryl sulfatase activity by glutaraldehyde fixation. *J. Histochem. Cytochem.*, **19**, 449–51.

Armstrong, J. A. 1956. Histochemical differentiation of nucleic acids by means of induced fluorescence. *Exptl. Cell Res.*, **11**, 640–3.

Asahina, E. 1956. The freezing process of plant cell. *Contr. Inst. Low Temp. Sci., Japan*, **10**, 83–126.

Bachelet, M., Bader, C., Merlot, A. M., Laborde, K., Snarska, J. and Ulmann, A. 1983. Cellular utilization of cytosolic NADPH in kidney and liver cells from rats fed a normal or a vitamin D-deficient diet. *Cell Biochem. Funct.*, **1**, 25–9.

Bachelet, M., Bourdeau, A., Lair, M., Bader, C., Ben Nasr, L., Thomas, M. and Ulmann, A. 1985. Effect of plasma levels of parathyroid hormone on NADPH pathways in kidney and liver. *Kidney Intl.*, **27**, 401–4.

Bachrach, U. 1984. Physiological aspects of ornithine decarboxylase. *Cell Biochem. Funct.*, **2**, 6–10.

Bachrach, U., Kaye, A. and Chayen, R. (Eds). 1983. *Advances in Polyamine Research*. Vol. 4. Raven Press, New York.

Badway, J. A. and Karnovsky, M. L. 1980. Active oxygen species and the functions of phagocytic leukocytes. *Ann. Rev. Biochem.*, **49**, 695–726.

Baker, J. R. 1947. The histochemical recognition of certain guanidine derivatives. *Q. J. Microsc. Sci.*, **88**, 115–21.

Baker, J. R. 1956. The histochemical recognition of certain phenols, especially tyrosine. *Q. J. Microsc. Sci.*, **97**, 161–4.

Baker, J. R. 1958. *Principles of Biological Microtechnique*. Methuen, London.

Baker, P. F. 1972. The sodium pump in animal tissues and its role in the control of metabolism and function. In: *Metabolic Transport*. (Ed. L. E. Hokin). 3rd edn. pp. 243–68. Academic Press, New York, London.

Baillie, A. H., Ferguson, M. M. and Hart, D. McK. 1966. *Developments in Steroid Histochemistry*. Academic Press, London.

Balduini, C., Brovelli, A., De Luca, G., Galligani, L. and Castellani, A. A. 1973. Uridine diphosphate glucose dehydrogenase from cornea and epiphysial-plate cartilage. *Biochem. J.*, **133**, 243–9.

Baldwin, E. 1959. *Dynamic Aspects of Biochemistry*. Methuen, London.

Ball, J. and Jackson, D. S. 1954. Histological, chromatographic and spectrophotometric studies of toluidine blue. *Stain Technol.*, **28**, 33–40.

Balogh, K. 1966. Histochemical demonstration of 3α-hydroxysteroid dehydrogenase activity. *J. Histochem. Cytochem.*, **14**, 77–83.

Balogh, K. and Cohen, R. B. 1961. Histochemical localization of uridine diphosphoglucose dehydrogenase in cartilage. *Nature*, **192**, 1199–200.

Barka, T. 1960. A simple azo-dye method for histochemical demonstration of acid phosphatase. *Nature*, **187**, 248–9.

Barka, T. and Anderson, P. J. 1963. *Histochemistry*. Harper and Row, New York.

Barrett, A. T. and Heath, M. F. 1977. Lysosomal enzymes. In: *Lysosomes, A Laboratory Handbook*. (Ed. J. T. Dingle). 2nd edn. pp. 119–45. North-Holland, Amsterdam.

Barrnett, R. J. and Seligman, A. M. 1951. Histochemical demonstration of esterases by production of indigo. *Science*, 114, 579.

Barrnett, R. J. and Seligman, A. M. 1952. Histochemical demonstration of protein-bound sulfhydryl groups. *Science*, 116, 323–7.

Barrnett, R. J. and Seligman, A. M. 1954. Histochemical demonstration of sulfhydryl and disulfide groups of protein. *J. Natl. Cancer Inst.*, 14, 769–803.

Barter, R., Danielli, J. F. and Davies, H. G. 1955. A quantitative cytochemical method for estimating alkaline phosphatase activity. *Proc. Roy. Soc. London, B*, 144, 412–26.

Beaufay, H. and De Duve, C. 1954. Le système hexose-phosphatasique. VI. Essais de démembrement des microsomes portent de glucose-6-phosphatase. *Bull. Soc. Chim. Biol. Paris*, 36, 1551–61.

Beevers, H. 1969. Glyoxysomes of castor bean endosperm and their relation to gluconeogenesis. *Ann. N.Y. Acad. Sci.*, 168, 313–24.

Behal, F. J., Little, G. H. and Klein, R. A. 1969. Arylamidases of human liver. *Biochim. Biophys. Acta*, 178, 118–27.

Bell, L. G. E. 1956. Freeze-drying. In: *Physical Techniques in Biological Research*. (Eds G. Oster and A. W. Pollister). Vol. III, *Cells and Tissues*. Academic Press, New York.

Bendall, D. S. and De Duve, C. 1960. Tissue fractionation studies. 14. The activation of latent dehydrogenases in mitochondria from rat liver. *Biochem. J.*, 74, 444–50.

Benford, D. J., Bridges, J. W. and Gibson, G. E. (Eds). 1987. *Drug Metabolism—from Molecules to Man*. Taylor and Francis, London, New York, Philadelphia.

Bensley, R. R. and Gersh, I. 1933. Studies on cell structure by the freezing-drying method. II. The nature of mitochondria in the hepatic cell of amblyostoma. *Anat. Rec.*, 57, 217–33.

Berenbaum, M. C. 1958. The histochemistry of bound lipids. *Q. J. Microsc. Sci.*, 99, 231–42.

Berenbom, M., Yokoyama, H. O. and Stowell, R. E. 1952. Chemical and enzymatic changes in liver following freeze-drying and acetone fixation. *Proc. Soc. Exptl. Biol. Med.*, 81, 125–8.

Berg, N. O. 1951. A histological study of masked lipids. *Acta. Pathol. Microbiol. Scand., Suppl.* 40.

Bergmann, M. 1942. A classification of proteolytic enzymes. *Adv. Enzymol.*, 2, 49–68.

Bergmann, M. and Fruton, J. S. 1941. The specificity of proteinases. *Adv. Enzymol.*, 1, 63–98.

Bernfeld, P., Bernfeld, H. C., Nisselbaum, J. S. and Fishman, W. H. 1954. Dissociation and activation of β-glucuronidase. *J. Am. Chem. Soc.*, 76, 4872–7.

Biochemists' Handbook. 1961. (Ed. C. Long). Spon, London.

Bitensky, L. 1962. The demonstration of lysosomes by the controlled temperature freezing–sectioning method. *Q. J. Microsc. Sci.*, 104, 193–6.

Bitensky, L. 1963a. The reversible activation of lysosomes in normal cells and the effects of pathological conditions. In: *Ciba Symp. on 'Lysosomes'*. (Eds A. V. S. de Reuck and M. P. Cameron). Churchill, London.

Bitensky, L. 1963b. Cytotoxic action of antibodies. *Br. Med. Bull.*, 19, 241–4.

Bitensky, L. 1963c. Modifications to the Gomori acid phosphatase technique for controlled-temperature frozen sections. *Q. J. Microsc. Sci.*, 104, 193–6.

Bitensky, L. 1967a. Histochemistry of liver disease. In: *The Liver*. (Ed. A. E. Read). Colston Res. Soc. Symp. Butterworths, London.

Bitensky, L. 1967b. Histochemistry in experimental immunology. In: *Handbook of Experimental Immunology*. (Ed. D. M. Weir). Blackwell, Oxford.

Bitensky, L. 1980. Microdensitometry. In: *Trends in Enzyme Histochemistry and Cytochemistry*. (Eds D. Evered and M. O'Connor). Ciba Foundation Symposium, 73, 181–202. Excerpta Medica, Amsterdam.

Bitensky, L. and Chayen, J. 1977. Histochemical methods for the study of lysosomes. In: *Lysosomes, A Laboratory Handbook*. (Ed. J. T. Dingle). 2nd edn. pp. 209–43. North-Holland, Amsterdam.

Bitensky, L., Ellis, R., Silcox, A. A. and Chayen, J. 1962. Histochemical studies on carbohydrate material in liver. *Ann. Histochim.*, **1**, 9–14.

Bitensky, L., Chayen, J., Cunningham, G. J. and Fine, J. 1963. Behaviour of lysosomes in haemorrhagic shock. *Nature*, **199**, 493–4.

Bitensky, L., Butcher, R. G. and Chayen, J. 1973. Quantitative cytochemistry in the study of lysosomal function. In: *Lysosomes in Biology and Pathology*. (Ed. J. T. Dingle). Vol. 3. pp. 465–510. North-Holland, Amsterdam.

Bitensky, L., Alaghband-Zadeh, J. and Chayen, J. 1974a. Studies on thyroid stimulating hormone and the long-acting thyroid stimulating hormone. *Clin. Endocrinol.*, **3**, 363–74.

Bitensky, L., Butcher, R. G., Johnstone, J. J. and Chayen, J. 1974b. Effect of glucocorticoids on lysosomes in synovial lining cells in human rheumatoid arthritis. *Ann. Rheum. Dis.*, **33**, 57–61.

Bitensky, L., Cashman, B., Johnstone, J. and Chayen, J. 1977. Effect of glucocorticoids on the hexose monophosphate pathway in human rheumatoid synovial lining cells *in vitro* and *in vivo*. *Ann. Rheum. Dis.*, **36**, 448–52.

Bitensky, L., Chayen, J. and Husain, O. A. N. 1984. Cytochemical detection of cancer: a review. *J. Roy. Soc. Med.*, **77**, 677–81.

Björklund, A. and Stenevi, U. 1970. Acid catalysis of the formaldehyde condensation reaction for sensitive histochemical demonstration of tryptamines and 3-methoxylated phenylethylamines. I. Model experiments. *J. Histochem. Cytochem.*, **18**, 794–882.

Björklund, A., Ehinger, B. and Falch, B. 1968. A method for differentiating dopamine from noradrenaline in tissue sections by microspectrofluorometry. *J. Histochem. Cytochem.*, **16**, 263–70.

Blaschko, H. 1952. Amine oxidase and amine metabolism. *Pharmacol. Rev.*, **4**, 415–58.

Blaschko, H. 1961. Amine oxidases. In: *Biochemists' Handbook*. (Ed. C. Long). pp. 373–5. Spon, London.

Bonham, D. G. 1964. A new test for the diagnosis of gynaecological cancer: 6-phosphogluconate dehydrogenase activity in vaginal fluid. *Triangle*, **6**, 157–62.

Bonham, D. G. and Gibbs, D. F. 1962. A new enzyme test for gynaecological cancer: 6-phosphogluconate dehydrogenase activity in vaginal fluid. *Br. Med. J.*, **ii**, 823–4.

Bonner, J. 1950. *Plant Biochemistry*. Academic Press, New York.

Borgers, M. and Thone, F. 1975. The inhibition of alkaline phosphatase by L-*p*-bromotetramisole. *Histochemistry*, **44**, 277–80.

Borgers, N. and Verheyen, A. 1983. Electron microscopic cytochemistry of alkaline phosphatase. *Cell Biochem. Funct.*, **1**, 77–80.

Boxer, G. E. 1965. Glycolytic enzymes in human tumours: difficulties, methods and results. *Eur. J. Cancer*, **1**, 161.

Brachet, J. 1954. The use of basic dyes and ribonuclease for the cytochemical detection of ribonucleic acid. *Q. J. Microsc. Sci.*, **94**, 1–10.

Bradford, M. M. 1976. A rapid and sensitive method for the quantitation of microgram quantities of protein utilizing the principle of protein-dye binding. *Anal. Biochem.*, **72**, 248–54.

Braimbridge, M. V., Darracott, S., Chayen, J., Bitensky, L. and Poulter, L. W. 1967. Possibility of a new infective aetiological agent in congestive cardiomyopathy. *Lancet*, **i**, 171–6.

Braimbridge, M. V., Canković-Darracott, S. and Hearse, D. J. 1982. Crystalloid cardioplegia — experience with the St. Thomas' solution. In: *A Textbook of Clinical Cardioplegia*. (Eds R. M. Engelman and S. Levitsky). pp. 177–98. Future Publishing, Mount Kisco, New York.

Brown, D. M. and Todd, A. R. 1955. Evidence on the nature of the chemical bonds in nucleic acids. In: *The Nucleic Acids*. (Eds E. Chargaff and J. N. Davidson). Vol. I. pp. 409–46. Academic Press, New York.

Budd, C. C. and Pelc, S. R. 1964. *Stain Technol.*, **39**, 295.

Bullock, G. R. and Petruzs, P. (Eds). 1982. *Techniques in Immunocytochemistry*. Academic Press, New York.

Burdon, K. L. 1947. *Textbook of Microbiology*. 3rd edn. Macmillan, New York.

Burstone, M. S. 1959. New histochemical techniques for the demonstration of tissue oxidase (cytochrome oxidase). *J. Histochem. Cytochem.*, **7**, 112–22.

Burstone, M. S. 1960. Histochemical demonstration of cytochrome oxidase with new amine reagents. *J. Histochem. Cytochem.*, **8**, 63–70.

Burstone, M. S. 1962. *Enzyme Histochemistry and its Application in the Study of Neoplasms*. Academic Press, New York.

Burton, K. 1961. Free energy data and oxidation-reduction potentials. In: *Biochemists' Handbook*. (Ed. C. Long). pp. 90–5. Spon, London.

Butcher, R. G. 1968. The estimation of the nucleic acids of tissue sections and its use as a unit of comparison for quantitative histochemistry. *Histochemie*, **13**, 263–75.

Butcher, R. G. 1971a. The chemical determination of section thickness. *Histochemie*, **28**, 131–6.

Butcher, R. G. 1971b. Tissue stabilization during histochemical reactions: the use of collagen peptides. *Histochemie*, **28**, 231–5.

Butcher, R. G. 1972. Precise cytochemical measurement of neotetrazolium formazan by scanning and integrating microdensitometry. *Histochemie*, **32**, 171–90.

Butcher, R. G. 1979. The oxygen insensitivity phenomenon as a diagnostic aid in carcinoma of the bronchus. In: *Quantitative Cytochemistry and its Applications*. (Eds J. R. Pattison, L. Bitensky and J. Chayen). Academic Press, London.

Butcher, R. G. and Altman, F. P. 1973. Studies on the reduction of tetrazolium salts. II. The measurement of the half-reduced and fully reduced formazans of neotetrazolium chloride in tissue sections. *Histochemie*, **37**, 351–63.

Butcher, R. G. and Chayen, J. 1966a. Quantitative studies on the alkaline phosphatase reaction. *J. Roy. Microsc. Soc.*, **85**, 111–17.

Butcher, R. G. and Chayen, J. 1966b. Dehydrogenase interactions in integrated tissue biochemistry. *Biochem. J.*, **100**, 47P.

Butcher, R. G., Diengdoh, J. V. and Chayen, J. 1964. A study of the histochemical demonstration of cytochrome oxidase. *Q. J. Microsc. Sci.*, **105**, 497–502.

Butcher, R. G., Chayen, J. and Labrum, A. H. 1965. Oxidation of L-ascorbic acid by cells of carcinoma of the human cervix. *Nature*, **207**, 992–3.

Butterworth, P. J. 1983. Biochemistry of mammalian alkaline phosphatases. *Cell Biochem. Funct.*, **1**, 66–70.

Cain, A. J. 1950. The histochemistry of lipoids in animals. *Biol. Rev.*, **25**, 73–112.

Callingham, B. A. 1986. Some aspects of monoamine oxidase pharmacology. *Cell Biochem. Funct.*, **4**, 99–108.

Cameron, C. B. and Husain, O. A. N. 1965. 6-Phosphogluconate dehydrogenase activity in vaginal fluid: limitations as a screening test for genital cancer. *Br. Med. J.*, **i**, 1529–30.

Canković-Darracott, S. 1982. Methods of assessing preservation techniques — invasive methods (enzymatic, cytochemical). In: *A Textbook of Clinical Cardioplegia*. (Eds R. M. Engelman and S. Levitsky). pp. 43–61. Futura Publishing, Mount Kisco, New York.

Canković-Darracott, S., Braimbridge, M. V., Williams, B. T., Bitensky, L. and Chayen, J. 1977. Myocardial preservation during aortic valve surgery: assessment of five techniques by cellular chemical and biophysical methods. *J. Thorac. Cardiovasc. Surg.*, **73**, 699–706.

Carafoli, E. (Ed.) 1982a. *Membrane Transport of Calcium*. Academic Press, London.
Carafoli, E. 1982b. The regulation of the cellular functions of Ca^{2+}. In: *Disorders of Mineral Metabolism*. Vol. II. (Eds F. Bronner and J. W. Coburn). pp. 2–42. Academic Press, London.
Carafoli, E. 1982c. The transport of calcium across the inner membrane of mitochondria. In: *Membrane Transport of Calcium*. (Ed. E. Carafoli). pp. 109–40. Academic Press, London.
Caro, L. G. 1962. High resolution autoradiography. *J. Cell Biol.*, **15**, 189.
Carver, M. J., Brown, F. C. and Thomas, L. E. 1953. An arginine histochemical method using Sakaguchi's new reagent. *Stain Technol.*, **28**, 89–91.
Casselman, W. G. B. 1959. *Histochemical Technique*. Methuen, London.
Chain, E. B. 1959. Recent studies on carbohydrate metabolism. *Br. Med. J.*, **ii**, 707–19.
Chambers, D. J., Dunham, J., Zanelli, J. M., Parsons, J. A., Bitensky, L. and Chayen, J. 1978. A sensitive bioassay of parathyroid hormone in plasma. *Clin. Endocrinol.*, **9**, 375–9.
Chambers, D. J., Braimbridge, M. V., Frost, G. T. B., Nahir, A. M. and Chayen, J. 1982. A quantitative cytochemical method for the measurement of β-hydroxyacyl CoA dehydrogenase activity in the rat heart muscle. *Histochemistry*, **75**, 67–76.
Chambers, M. G. 1991. Quantitative cytochemistry of natural murine osteoarthritis. *Ph.D. Thesis*, Brunel University.
Charlton, J. A. and Bayliss, P. H. 1989. The stimulation of ornithine decarboxylase activity by arginine vasopressin in the rat medullary thick ascending limb of Henle's loop. *J. Endocrinol.*, **120**, 195–9.
Charnock, J. S. and Opit, L. J. 1968. Membrane metabolism. *Br. Med. J.*, **ii**, 707–19.
Chayen, J. 1952. The methyl green–pyronin method. *Exptl. Cell Res.*, **3**, 652–5.
Chayen, J. 1953. Ascorbic acid and its intracellular localization, with special reference to plants. *Int. Rev. Cytol.*, **2**, 78–131.
Chayen, J. 1968a. Histochemistry of phospholipids and its significance in the interpretation of the structure of cells. In: *The Interpretation of Cell Structure*. (Eds K. F. Ross and S. McGee-Russell). Arnold, London.
Chayen, J. 1968b. Quantitative histochemistry: cell structure revealed through cellular function. In: *The Interpretation of Cell Structure*. (Eds K. F. Ross and S. McGee-Russell). Arnold, London.
Chayen, J. 1978a. The cytochemical approach to hormone assay. *Int. Rev. Cytol.*, **53**, 333–96.
Chayen, J. 1978b. Microdensitometry. In: *Biochemical Mechanisms of Liver Injury*. (ed. T. F. Slater). pp. 257–91. Academic Press, London, New York, San Francisco.
Chayen, J. 1980. *The Cytochemical Bioassay of Polypeptide Hormones*. Monographs on Endocrinology, Springer, Berlin.
Chayen, J. 1982. Concerning the possibility of redox drugs. *Agents Actions*, **12**, 530–5.
Chayen, J. 1983. Polarized light microscopy: principles and practice for the rheumatologist. *Ann. Rheum. Dis.*, **42** (Suppl. 1), 64–7.
Chayen, J. 1984. Quantitative cytochemistry: a precise form of cellular biochemistry. *Biochem. Soc. Trans.*, **12**, 887–98.
Chayen, J. and Bitensky, L. 1968. The multiphase chemistry of cell injury. In: *The Biological Basis of Medicine*. (Eds E. E. Bittar and N. Bittar). Vol. I. Academic Press, London.
Chayen, J. and Bitensky, L. 1971. Lysosomal enzymes and inflammation with particular reference to rheumatoid diseases. *Ann. Rheum. Dis.*, **30**, 522–36.
Chayen, J. and Bitensky, L. 1982. The effects of pharmacologically active agents on cytosolic reducing equivalents. *Rev. Pure Appl. Pharmacol. Sci.*, **3**, 271–317.

Chayen, J. and Bitensky, L. (Eds). 1983. *Cytochemical Bioassays: Techniques and Clinical Applications*. Marcel Dekker, New York.

Chayen, J. and Denby, E. F. 1968. *Biophysical Technique as Applied to Cell Biology*. Methuen, London.

Chayen, J., Gahan, P. B. and La Cour, L. F. 1959. The masked lipids of nuclei. *Q. J. Microsc. Sci.*, **100**, 325–37.

Chayen, J., Jones, G. R. N., Bitensky, L. and Cunningham, G. J. 1961. Histological variation as a source of biochemical error. *Biochem. J.*, **79**, 34P.

Chayen, J., Bitensky, L., Aves, E. K., Jones, G. R. N., Silcox, A. A. and Cunningham, G. J. 1962. Histochemical demonstration of 6-phosphogluconate dehydrogenase in proliferating and malignant cells. *Nature*, **195**, 714–15.

Chayen, J., Altmann, F. P., Bitensky, L., Braimbridge, M. V. and Kadas, T. 1966a. A study of the changes in hydrogen transport in an isolated rat heart preparation. *J. Roy. Microsc. Soc.*, **86**, 151–8.

Chayen, J., Bitensky, L. and Poulter, L. W. 1966b. Fluorescence histochemistry of steroids and carotenoids. *Proc. Roy. Microsc. Soc.*, **1**, 212.

Chayen, J., Darracott, S. and Kirkby, W. W. 1966c. A re-interpretation of the role of the mast cell. *Nature*, **209**, 887–8.

Chayen, J., Bitensky, L. and Wells, P. J. 1966d. Mitochondrial enzyme latency and its significance in histochemistry and biochemistry. *J. Roy. Microsc. Soc.*, **86**, 69–74.

Chayen, J., Bitensky, L., Butcher, R. G., Poulter, L. W. and Ubhi, G. S. 1970. Methods for the direct measurement of anti-inflammatory action on human tissue maintained *in vitro*. *Br. J. Dermatol.*, **82** (Suppl. 6), 62–81.

Chayen, J., Bitensky, L., Butcher, R. G. and Cashman, B. 1971. Evidence for altered lysosomal membranes in synovial lining cells from human rheumatoid joints. *Beitr. Path. Bd.*, **142**, 137–49.

Chayen, J., Altman, F. P. and Butcher, R. G. 1972a. The production and possible utilization of reducing equivalents outside the mitochondria and the effects of certain drugs. In: *Fundamentals of Cell Pharmacology*. (Ed. S. Dikstein). Thomas, Springfield, Illinois.

Chayen, J., Bitensky, L. and Ubhi, G. S. 1972b. The experimental modification of lysosomal dysfunction by anti-inflammatory drugs acting *in vitro*. *Beitr. Path. Bd.*, **147**, 6–20.

Chayen, J., Daly, J. R., Loveridge, N. and Bitensky, L. 1976. The cytochemical bioassay of hormones. *Recent Progr. Hormone Res.*, **32**, 33–79.

Chayen, J., Frost, G. T. B., Dodds, R. A., Bitensky, L., Pitchfork, J., Baylis, P. H. and Barrnett, R. J. 1981. The use of a hidden metal-capture reagent for the measurement of Na^+-K^+-ATPase activity: a new concept in cytochemistry. *Histochemistry*, **71**, 533–41.

Chayen, J., Howat, D. W. and Bitensky, L. 1986. Cellular biochemistry of glucose 6-phosphate and 6-phosphogluconate dehydrogenase activities. *Cell Biochem. Funct.*, **4**, 249–53.

Chayen, J., Pitsillides, A. A., Bitensky, L., Muir, I. H., Taylor, P. M. and Askonas, B. A. 1990. T-cell mediated cytolysis: evidence for target-cell suicide. *J. Exptl. Pathol.*, **71**, 197–208.

Chayen, R. 1984. Clinical chemistry of polyamines. *Cell Biochem. Funct.*, **2**, 15–20.

Chayen, R. and Roberts, E. R. 1955. Some observations on the metachromatic reaction. *Sci. J. Roy. Coll. Sci.*, **25**, 50–6.

Chèvremont, M. and Fréderic, J. 1943. Une nouvelle méthode histochimique de mise en evidence des substances à fraction sulfhydrile. *Arch. Biol.*, **54**, 589.

Chiquoine, A. D. 1953. The distribution of glucose-6-phosphatase in the liver and kidney of the mouse. *J. Histochem. Cytochem.*, **1**, 429–35.

Chiquoine, A. D. 1955. Further studies on the histochemistry of glucose-6-phosphatase. *J. Histochem. Cytochem.*, **3**, 471–8.

Cohen, S. and Way, S. 1966. Histochemical demonstration of pentose-shunt activity in smears from the uterine cervix. *Br. Med. J.*, **i**, 88–9.

Cohen, S., Bitensky, L. and Chayen, J. 1965. The study of monoamine oxidase activity by histochemical procedures. *Biochem. Pharmacol.*, **14**, 223–6.

Constantine, V. S. and Mowry, R. W. 1968. The selective staining of human dermal collagen. II. The use of Picrosirius Red F3BA with polarization microscopy. *J. Invest. Dermatol.*, **50**, 419–23.

Corrodi, H. and Jonsson, G. 1967. The formaldehyde fluorescence method for the histochemical demonstration of biogenic monoamines. *J. Histochem. Cytochem.*, **15**, 65–78.

Cotson, S. and Holt, S. J. 1958. Studies in enzyme cytochemistry. IV. Kinetics of aerial oxidation of indoxyl and some of its halogen derivatives. *Proc. Roy. Soc., B*, **148**, 506–19.

Coulton, L. A., Henderson, B., Bitensky, L. and Chayen, J. 1980. DNA synthesis in human rheumatoid and nonrheumatoid synovial lining. *Ann. Rheum. Dis.*, **39**, 241–7.

Coulton, L. A., Henderson, B. and Chayen, J. 1981. The assessment of DNA-synthetic activity. *Histochemistry*, **72**, 91–9.

Cuello, A. C. (Ed.). 1983. *Immunohistochemistry*. IBRO, Oxford.

Culling, C. F. A. 1963. *Handbook of Histopathological Techniques*. 2nd edn. Butterworths, London.

Dance, N., Price, R. G. and Robinson, D. 1970. Differential assay of human hexosaminidases A and B. *Biochim. Biophys. Acta*, **222**, 662–4.

Danielli, J. F. 1953. *Cytochemistry: A Critical Approach*. Wiley, New York.

Danpure, C. J. 1984. Lactate dehydrogenase and cell injury. *Cell Biochem. Funct.*, **2**, 144–8.

Daoust, R. 1965. Histochemical localization of enzyme activities by substrate film methods: ribonucleases, deoxyribonucleases, protease, amylase and hyaluronidase. *Int. Rev. Cytol.*, **8**, 191–221.

Darlington, C. D. and La Cour, L. F. 1947. *The Handling of Chromosomes*. 2nd edn. Allen and Unwin, London.

Darracott-Canković, S., Braimbridge, M. V. and Chayen, J. 1983. Biopsy assessment of preservation during open-heart surgery with cold cardioplegic arrest. *Adv. Myocardiol.*, **4**, 497–504.

Darracott-Canković, S., Bitensky, L. and Chayen, J. 1986. Histochemistry of monoamine oxidase activity. *Cell Biochem. Funct.*, **4**, 109–10.

Darracott-Canković, S., Wheeldon, D., Cory-Pearce, R., Wallwork, J. and English, T. A. H. 1987. Biopsy assessment of fifty hearts during transplantation. *J. Thorac. Cardiovasc. Surg.*, **93**, 95–102.

Darracott-Canković, S., Stovin, P. G. I., Wheeldon, D., Wallwork, J., Wells, F. and English, T. A. H. 1989. Effect of donor heart damage on survival after transplantation. *Eur. J. Cardio-thorac. Surg.*, **3**, 525–32.

Davidson, D. 1937. The Prussian blue paradox. *J. Chem. Education*, **14**, 233–41.

Davidson, J. N. and Waymouth, C. 1944. The histochemical demonstration of ribonucleic acid in mammalian liver. *Proc. Roy. Soc., Edinburgh*, **62**, 96–8.

Davis, W. L., Jones, R. G., Farmer, G. R., Cortinas, E., Matthews, J. L. and Goodman, D. B. P. 1989a. The glyoxylate cycle in rat epiphyseal cartilage: the effect of vitamin K_3 on the activity of the enzymes isocitrate lyase and malate synthase. *Bone*, **10**, 201–6.

Davis, W. L., Jones, R. G., Farmer, G. R., Matthews, J. L. and Goodman, D. B. P. 1989b. Glyoxylate cycle in the epiphyseal growth plate: isocitrate lyase and malate synthase identified in mammalian cartilage. *Anat. Rec.*, **223**, 357–62.

De Bruyn, P. P. H., Farr, R. S., Banks, H. and Morthland, F. W. 1953. *In vivo* and *in vitro* affinity of diaminoacridines for nucleoproteins. *Exptl. Cell Res.*, **4**, 174.

De Cosse, J. J. and Aiello, N. 1966. Feulgen hydrolysis: effect of acid and temperature. *J. Histochem. Cytochem.*, **14**, 601–4.

De Duve, C. 1959. Lysosomes, a new group of cytoplasmic particles. In: *Subcellular Particles*. (Ed. T. Hayashi). Ronald, New York.

De Duve, C. 1969. The lysosome in retrospect. In: *Lysosomes in Biology and Pathology*. Vol. I. (Eds J. T. Dingle and H. B. Fell). pp. 3–42. Elsevier, Amsterdam.

De Duve, C., Pressman, B. C., Gianetto, R., Wattiaux, R. and Appelmans, F. 1955. Tissue fractionation studies. VI. Intracelluar distribution patterns of enzymes in rat liver tissue. *Biochem. J.*, **60**, 604–17.

De Luca, G., Speziale, P., Rindi, S., Balduini, C. and Castellani, A. A. 1976. Effect of some nucleotides on the regulation of glycosaminoglycan biosynthesis. *Connective Tissue Res.*, **4**, 247–54.

De Pierre, J. W. and Karnovsky, M. L. 1973. Plasma membranes of mammalian cells: a review of methods for their characterisation and isolation. *J. Cell Biol.*, **56**, 275–303.

Deeley, E. M. 1955. An integrating microdensitometer for biological cells. *J. Sci. Instr.*, **32**, 263–7.

Defendi, V. and Pearson, B. 1955. Quantitative estimation of succinic dehydrogenase activity in a single microscope tissue section. *J. Histochem. Cytochem.*, **3**, 61–9.

Desai, I. D. and Tappel, A. L. 1963. Damage to proteins by peroxidized lipids. *J. Lipid Res.*, **4**, 204–7.

Deutsch, D. J. 1983. Glucose 6-phosphate dehydrogenase. In: *Methods of Enzymatic Analysis*. (Ed. H. U. Bergmeyer), Vol. 3, 3rd edn, pp. 190–7. Verlag Chemie, Basel

Dickens, F. 1956. The hexosemonophosphate oxidative pathway of yeast and animal tissues. *3rd Int. Congr. Biochem., Brussels*, 1955. (Ed. C. Liebecq). Academic Press, New York.

Dixon, M. and Webb, E. C. 1964. *Enzymes*. 2nd edn. Longmans, London.

Dodds, R. A. and Chayen, J. 1984. Histochemical and cytochemical demonstration of ornithine decarboxylase. *Cell Biochem. Funct.*, **2**, 10–11.

Dodds, R. A., Pitsillides, A. A. and Frost, G. T. B. 1990. A quantitative cytochemical method for ornithine decarboxylase activity. *J. Histochem. Cytochem.*, **38**, 123–7.

Dodgson, K. S., Spencer, B. and Thomas, J. 1955. Studies on sulphatases. 9. The aryl sulphatases of mammalian livers. *Bicochem. J.* **59**, 29–37.

Doniach, I. and Pelc, S. R. 1950. Autoradiograph technique. *Br. J. Radiol.*, **23**, 184–92.

Doyle, W. L. 1948. Effects of dehydrating agents on phosphatases in the lymphatic nodules of the rabbit appendix. *Proc. Soc. Exptl. Biol. Med.*, **69**, 43–4.

Drummond, G. J. and Yamamoto, M. 1971. Nucleoside cyclic phosphate diesterases. In: *The Enzymes*. Vol. 4. 3rd edn. (Ed. D. Boyer). pp. 337–54. Academic Press, New York.

Dumonde, D. C., Bitensky, L., Cunningham, G. J. and Chayen, J. 1965. The effects of antibodies on cells. 1. Biochemical and histochemical effects of antibodies and complement on ascites tumour cells. *Immunology*, **8**, 25–36.

Dunham, E. T. and Glynn, I. M. 1961. Adenosine triphosphatase activity and the active movements of alkali metal ions. *J. Physiol.*, **156**, 274–93.

Dunham, J., Dodds, R. A., Nahir, A. M., Frost, G. T. B., Catterall, A., Bitensky, L. and Chayen, J. 1983. Aerobic glycolysis of bone and cartilage: the possible involvement of fatty acid oxidation. *Cell Biochem. Funct.*, **1**, 168–72.

Dunham, J., Shackleton, D. R., Nahir, A. M., Billingham, M. E. J., Bitensky, L., Chayen, J. and Muir, I. H. 1985. Altered orientation of glycosaminoglycans and cellular changes in the tibial cartilage in the first two weeks of experimental canine osteoarthritis. *J. Orthop. Res.*, **3**, 258–68.

Dunham, J., Shackleton, D. R., Bitensky, L., Chayen, J., Billingham, M. E. J. and Muir, I. H. 1986. Enzymatic heterogeneity of canine articular cartilage. *Cell Biochem. Funct.*, **4**, 43–6.

Dunham, J., Shackleton, D. R., Billingham, M. E. J., Bitensky, L., Chayen, J. and Muir, I. H. 1988. A reappraisal of the structure of normal canine articular cartilage. *J. Anat.*, **157**, 89–99.

Dunham, J., Chambers, M. G., Jasani, M. K., Bitensky, L. and Chayen, J. 1990. Changes in the orientation of proteoglycans during the early development of natural murine oestoarthritis. *J. Orthop. Res.*, **8**, 101–4.

Dyer, J. R. 1956. The use of periodate oxidations in biochemical analysis. In: *Methods of Biochemical Analysis*. (Ed. D. Glick). Vol. III. Interscience, New York.

Ealey, P. A., Henderson, B. and Loveridge, N. 1984. A quantitative study of peroxidase activity in unfixed tissue sections of the guinea-pig thyroid gland. *Histochem. J.*, **16**, 111–22.

Ebel, J. P. 1952. Recherches sur les polyphosphates contenus dans diverses cellules vivantes. IV. Localisation cytologique et rôle physiologique des polyphosphates dans la cellule vivante. *Bull. Soc. Chim. Biol.*, **34**, 498–505.

Ellis, S. and Perry, M. 1966. Pituitary arylamidases and peptidases. *J. Biol. Chem.*, **241**, 3479.

Emmel, V. M. 1946. The intracellular distribution of alkaline phosphatase activity following various methods of histologic fixation. *Anat. Rec.*, **95**, 159.

Eränkö, O. 1967. The practical histochemical demonstration of catecholamines by formaldehyde-induced fluorescence. *J. Roy. Microsc. Soc.*, **87**, 259–76.

Eränkö, O. and Härkönen, M. 1965. Monoamine-containing small cells in the superior cervical ganglion of the rat and an organ composed of them. *Acta Physiol. Scand.*, **63**, 511.

Ernst, S. A. and Philpott, C. W. 1970. Preservation of Na-K-activated and Mg-activated adenosine triphosphatase activities of avian salt gland and teleost gill with formaldehyde as fixative. *J. Histochem. Cytochem.*, **18**, 251–63.

Esterbauer, H. 1972. Beitrag zum quantitativen histochemischen Nachweis von Sulfhydrylgruppen mit der DDD-Färbung. I. Untersuchung der Farbstoffe. *Acta Histochem.*, **42**, 351–5.

Esterbauer, H. 1973. Beitrag zum quantitativen histochemischen Nachweis von Sulfhydrylgruppen mit der DDD-Färbung. II. Bestimmung von SH-Gruppen in unlöslichen Proteinen. *Acta Histochem.*, **47**, 95–105.

Fahimi, H. D. and Drochmans, P. 1968. Purification of glutaraldehyde: its significance for preservation of acid phosphatase activity. *J. Histochem. Cytochem.*, **16**, 199–204.

Farber, E. and Bueding, E. 1956. Histochemical localization of specific oxidative enzymes. V. The dissociation of succinic dehydrogenase from carriers by lipase and the specific histochemical localization of the dehydrogenase with phenazine metho-sulfate and tetrazolium salts. *J. Histochem. Cytochem.*, **4**, 357–62.

Farber, E., Sternberg, W. H. and Dunlap, C. E. 1956a. Histochemical localization of specific oxidative enzymes. I. Tetrazolium stains for diphosphopyridine nucleotide diaphorase and triphosphopyridine nucleotide diaphorase. *J. Histochem. Cytochem.*, **4**, 254–66.

Farber, E., Sternberg, W. H. and Dunlap, C. E. 1956b. Histochemical localization of specific oxidative enzymes. III. Evaluation studies of tetrazolium staining methods for diphosphopyridine nucleotide diaphorase, triphosphopyridine nucleotide diaphorase and the succindehydrogenase system. *J. Histochem. Cytochem.*, **4**, 284–94.

Fernley, H. N. 1971. Mammalian alkaline phosphatases. *The Enzymes*. (Ed. P. D. Boyer). Vol. 4. pp. 417–47. Academic Press, New York, London.

Feulgen, R. and Rossenbeck, H. 1924. Mikroskopisch-chemischen Nachweis einer Nucleinsaure vom Typus der Thymonucleinsaure und die darauf beruhende elektive Farbung von Zellkernen in mikroskopischen Praparaten. *Hoppe-Seyler's Z. Physiol. Chem.*, **135**, 203–48.

Fiala, S. and Fiala, A. E. 1959. On the correlation between metabolic and structural changes during carcinogenesis in rat liver. *Br. J. Cancer*, **13**, 136–51.

Fialkow, P. J. and Fishman, W. H. 1961. Studies on a liver activator of β-glucuronidase. *J. Biol. Chem.*, **236**, 2169–71.

Fieser, L. F. and Fieser, M. 1953. *Organic Chemistry*. 2nd edn. Harrap, London.

Findley, J., Levvy, G. A. and Marsh, C. A. 1958. Inhibition of glycosidoses by aldonolactones of corresponding configuration: two inhibitors of β-*N*-acetylglucosaminidase. *Biochem. J.*, **69**, 467–72.

Fisher, E. R. and Lillie, R. D. 1954. The effect of methylation of basophilia. *J. Histochem. Cytochem.*, **2**, 81.

Fishman, W. H. 1951. The relationship of the enzyme beta-glucuronidase to cancer of the cervix uteri. *Bull. N. Engl. Med. Center*, **13**, 12–19.

Fishman, W. H. 1961. Renal β-glucuronidase response to steroids of the androgen series. In: *Mechanism of Action of Steroid Hormones*. Pergamon, Oxford.

Fishman, W. H. 1963. Recent studies on β-glucuronidase. *Farmaco*, **18**, 397–407.

Fishman, W. H. and Anlyan, A. J. 1947. β-Glucuronidase activity in human tissues: some correlations with processes of malignant growth and with the physiology of reproduction. *Cancer Res.*, **7**, 808–17.

Fishman, W. H. and Baker, J. R. 1956. Cellular localization of β-glucuronidase in rat tissues. *J. Histochem. Cytochem.*, **4**, 570–87.

Fishman, W. H. and Bigelow, R. 1950. A comparative study of the morphology and glucuronidase activity in 44 gastrointestinal neoplasms. *J. Natl. Cancer Inst.*, **10**, 1115–22.

Fishman, W. H. and Goldman, S. S. 1965. A postcoupling technique for β-glucuronidase employing the substrate, naphthol AS-BI-β-*O*-glucosiduronic acid. *J. Histochem. Cytochem.*, **13**, 441–7.

Fishman, W. H. and Lipkind, J. B. 1958. Comparative ability of some steroids and their esters to enhance the renal β-glucuronidase activity of mice. *J. Biol. Chem.*, **232**, 729–36.

Fishman, W. H., Goldman, S. S. and Green, S. 1964. Several biochemical criteria for evaluating β-glucuronidase localization. *J. Histochem. Cytochem.*, **12**, 239–51.

Fitzgerald, P. J. 1959. Autoradiography in cytology. In: *Analytical Cytology*. (Ed. R. C. Mellors). 2nd edn. McGraw-Hill, New York.

Flint, F. O. 1982. Light microscopy preparation techniques for starch and lipid containing snack foods. *Food Microstruct.*, **1**, 145–50.

Folch, J. and Lees, M. 1951. Proteolipides, a new type of tissue lipoprotein: their isolation from brain. *J. Biol. Chem.*, **191**, 807–17.

Folch, J., Arsove, S. and Meath, J. A. 1951. Isolation of brain strandin, a new type of large molecule tissue component. *J. Biol. Chem.*, **191**, 819–31.

Folch, J., Ascoli, I., Lees, M., Meath, J. A. and Lebaron, F. N. 1951. Preparation of lipide extracts from brain tissue. *J. Biol. Chem.*, **191**, 833–41.

Folin, O. and Marenzi, A. D. 1929. Tyrosine and typtophane determinations in onetenth gram of protein. *J. Biol. Chem.*, **83**, 89–102.

Fredricsson, B. 1958. Preservation of activities of alkaline phosphatase and naphthyl splitting esterase during freeze-drying, embedding in poly-ethylene glycol and post fixation. *Acta Histochem.*, **6**, 165–73.

Frohwein, Y. Z. and Gatt, S. 1967. Isolation of β-*N*-acetyl-hexosaminidase, β-*N*-acetyl-glucosaminidase and β-*N*-acetyl-galactosaminidase from calf brain. *Biochemistry (N.Y.)*, **6**, 2775–82.

Fukuda, M., Bohm, N. and Fujita, S. 1978. Cytophotometry and its biological application. *Progr. Histochem. Cytochem.*, **11**, 1–119.

Gad, A. and Sylvén, B. 1969. On the nature of the high iron diamine method for sulfomucins. *J. Histochem. Cytochem.*, **17**, 156–60.

Gahan, P. B. 1967a. Histochemistry of lysosomes. *Int. Rev. Cytol.* **21**, 1–63.

Gahan, P. B. 1967b. Lysosomes. In: *Plant Cell Organelles*. (Ed. J. B. Pridham). Academic Press, New York.

Gahan, P. B. (Ed.). 1972. *Autoradiography for Biologists*. Academic Press, New York.

Gahan, P. B. and Carmignac, D. F. 1989. Determination of vascular tissue in roots of dicotyledonous plants. In: *Structural and Functional Aspects of Transport in Roots*. (Ed. B. C. Loughman). pp. 25–8. Kluwer Academic Publishers, Boston.

Garrett, J. R. 1966a. The innervation of salivary glands. 1. Cholinesterase-positive nerves in normal glands of the cat. *J. Roy. Microsc. Soc.*, **85**, 135–48.

Garrett, J. R. 1966b. The innervation of salivary glands. 3. The effects of certain experimental procedures on cholinesterase positive nerves in glands of the cat. *J. Roy. Microsc. Soc.*, **86**, 1–14.

Gau, G. and Chard, T. 1976. Location of the protein hormones of the placenta by the immunoperoxidase technique. *Br. J. Obstet. Gynaecol.*, **83**, 876–8.

Giacobini, E. 1969. Value and limitations of quantitative chemical studies in individual cells. *J. Histochem. Cytochem.*, **17**, 139–55.

Gibbs, H. D. 1927. Phenol tests. II. Nitrous acid tests. The Millon and similar tests. Spectrophotometric investigations. *J. Biol. Chem.*, **71**, 445–59.

Gibbs, D. A., Hauschildt, S. and Watts, R. W. E. 1977. Glyoxylate oxidation in rat liver and kidney. *J. Biochem.*, **82**, 221–30.

Glenner, G. G., Burtner, H. J. and Brown, G. W. 1957. The histochemical demonstration of monoamine oxidase activity by tetrazolium salts. *J. Histochem. Cytochem.*, **5**, 591–600.

Glick, D. 1962. *Quantitative Chemical Techniques of Histo- and Cytochemistry*. Vol. I. Wiley, New York.

Glick, D. 1963. *Quantitative Chemical Techniques of Histo- and Cytochemistry*. Vol. II. Wiley, New York.

Glock, G. E. and McLean, P. 1953. Further studies on the properties and assay of glucose 6-phosphate dehydrogenase and 6-phosphogluconate dehydrogenase of rat liver. *Biochem. J.*, **55**, 400–8.

Glock, G. E. and McLean, P. 1954. Levels of enzymes of the direct oxidative pathway of carbohydrate metabolism in mammalian tissues and tumours. *Biochem. J.*, **56**, 171–5.

Glock, G. E. and McLean, P. 1958. Pathways of glucose utilization in mammary tissue. *Proc. Roy. Soc., B*, **149**, 354–62.

Glover, V. and Sandler, M. 1986. Clinical chemistry of monoamine oxidase. *Cell Biochem. Funct.*, **4**, 89–97.

Godlewski, H. G. 1960. Application of ethylene-diamine-tetracetic acid (EDTA) in the histochemical method for demonstration of phosphorylase and the branching enzyme. *Bull. l'Acad. Pol. Sci. (Sér. Biol.)*, **8**, 441–4.

Godlewski, H. G. 1962. The distribution of the histochemically demonstrable phosphorylases in various cancer tissues. *Acta Un. Int. Contra Cancer*, **8**, 63–5.

Godlewksi, H. G. 1963. Histochemical studies of phosphorylases in precancerous lesions of the uterine cervix and mammary gland. *Fol. Histochem. Cytochem.*, **1**, 231–5.

Godlewski, H. G. 1964. Histochemistry of the glycogen synthetase and phosphorylases in normal and pathologic tissues. *Acta Histochem., Suppl.*, **4**, 30–51.

Goldfischer, S. 1965. The cytochemical demonstration of lysosomal aryl sulfatase activity by light and electron microscopy. *J. Histochem. Cytochem.*, **13**, 520–3.

Gomori, G. 1952. *Microscopic Histochemistry*. University Press, Chicago.

Gomori, G. 1955. Histochemistry of human esterases. *J. Histochem. Cytochem.*, **3**, 479–84.

Gomori, G. 1956. Histochemical methods for protein-bound sulphydryl and disulphide groups. *Q. J. Microsc. Sci.*, **97**, 1–9.

Graham, R. C. and Karnovsky, M. J. 1965. The histochemical demonstration of monoamine oxidase activity by coupled peroxidative oxidation. *J. Histochem. Cytochem.*, **13**, 604–5.

Graham, R. C. and Karnovsky, M. J. 1966. The early stages of absorption of injected horseradish peroxidase in the proximal tubules of mouse kidney: ultrastructural cytochemistry by a new technique. *J. Histochem. Cytochem.*, **14**, 291–302.

Habeeb, A. F. F. A. 1973. A sensitive method for localisation of disulphide containing peptides in column effluents. *Anal. Biochem.*, **56**, 60–5.

Hale, C. W. 1946. Histochemical demonstration of acid polysaccharides in animal tissues. *Nature*, **157**, 802.

Hamberger, B. and Norberg, K. A. 1964. Histochemical demonstration of monoamines in fresh-frozen sections. *J. Histochem. Cytochem.*, **12**, 48.

Hansson, H. P. J. 1967. Histochemical demonstration of carbonic anhydrase activity. *Histochemie*, **11**, 112–28.

Hardonk, M. J. 1968. 5'-Nucleotidase. I. Distribution of 5'-nucleotidase in tissues of rat and mouse. *Histochemie*, **12**, 1–17.

Hardonk, M. J. and Koudstaal, J. 1968. 5'-Nucleotidase. II. The significance of 5'-nucleotidase in the metabolism of the nucleotides studied by histochemical and biochemical methods. *Histochemie*, **12**, 18–28.

Hatefi, Y. and Stiggall, D. L. 1976. Metal-containing flavoprotein dehydrogenases. In: *The Enzymes.* (Ed. P. D. Boyer). Vol. XIII. pp. 328–9. Spon, London

Hayashi, M. 1965. Histochemical demonstration of *N*-acetyl-β-glucosaminidase employing naphthol AS-BI-acetyl-β-glucosaminidase as a substrate. *J. Histochem. Cytochem.*, **13**, 355–60.

Hayashi, M., Nakajima, Y. and Fishman, W. H. 1964. The cytologic demonstration of β-glucuronidase employing naphthol AS-BI glucuronide and hexazonium pararosanilin; a preliminary report. *J. Histochem. Cytochem.*, **12**, 293–7.

Hayhoe, F. G. J. 1983. Alkaline phosphatase in haematology. *Cell Biochem. Funct.*, **1**, 74–6.

Henderson, B. 1976. Quantitative cytochemical measurement of glyceraldehyde 3-phosphate dehydrogenase activity. *Histochemistry*, **48**, 191–204.

Henderson, B. and Glynn, L. E. 1981. Metabolic alterations in the synoviocytes in chronically inflamed knee joints in immune arthritis in the rabbit: comparison with rheumatoid arthritis. *Br. J. Exptl. Pathol.*, **62**, 27–33.

Henderson, B., Loveridge, N. and Robertson, W. R. 1978. A quantitative study of the effects of different grades of polyvinyl alcohol on the activities of certain enzymes in unfixed tissue sections. *Histochem. J.*, **10**, 453–63.

Henderson, B., Johnstone, J. J. and Chayen, J. 1980. 5'-Nucleotidase activity in the human synovial lining in rheumatoid arthritis. *Ann. Rheum. Dis.*, **39**, 248–52.

Henderson, B., Glynn, L. E., Bitensky, L. and Chayen, J. 1981. Evidence for cell division in synoviocytes in acutely inflamed rabbit joints. *Ann. Rheum. Dis.*, **40**, 177–81.

Heppel, L. A. and Hilmoe, R. J. 1951. Purification and properties of 5-nucleotidase. *J. Biol. Chem.*, **188**, 665–78.

Hewitt, L. F. 1950. *Oxidation–Reduction Potentials in Bacteriology and Biochemistry.* 6th edn. Livingstone, Edinburgh.

Hewitt, L. F. 1961. Standard oxidation–reduction potentials of inorganic systems and dyes. In: *Biochemists' Handbook.* (Ed. C. Long). pp. 85–9. Spon, London.

Heyden, G. 1979. Enzymatic changes associated with malignancy with special references to aberrant G 6-PD activity. In: *Quantitative Cytochemistry and its Applications*. (Eds J. R. Pattison, L. Bitensky and J. Chayen). Academic Press, London.

Hofer, H. W. and Bauer, H. P. 1987. 6-Phosphogluconolactonase. *Cell Biochem. Funct.*, **5**, 97–9.

Hokin, L. E. and Dahl, J. L. 1972. The sodium–potassium adenosine triphosphate. In: *Metabolic Transport*. (Ed. L. E. Hokin). 3rd edn. pp. 269–315. Academic Press, New York, London.

Hoile, R. W. 1983. Technique and clinical relevance of a cytochemical bioassay for gastrin-like activity. In: *Cytochemical Bioassays*. (Eds J. Chayen and L. Bitensky). pp. 189–224. Marcel Dekker, New York.

Holt, S. J. 1958. Indigogenic staining methods for esterases. In: *General Cytochemical Methods*. (Ed. J. F. Danielli). Academic Press, New York.

Holt, S. J. 1959. Factors governing the validity of staining methods for enzymes, and their bearing upon the Gomori acid phosphatase technique. *Exptl. Cell. Res., Suppl.*, **7**, 1–27.

Holt, S. J. 1963. Discussion. In: *Ciba Symp. on 'Lysosomes'*. (Eds A. V. S. de Reuck and M. P. Cameron). pp. 375–8. Churchill, London.

Holt, S. J. and O'Sullivan, D. G. 1958. Studies in enzyme cytochemistry. 1. Principles of cytochemical staining methods. *Proc. Roy. Soc., B*, **148**, 465–80.

Holt, S. J. and Sadler, P. W. 1958. Studies in enzyme cytochemistry. III. Relationships between solubility, molecular association and structure in indigoid dyes. *Proc. Roy. Soc., B*, **148**, 495–505.

Holt, S. J. and Withers, R. F. J. 1958. Studies in enzyme cytochemistry. V. An appraisal of indigogenic reactions for esterase localization. *Proc. Roy. Soc., B*, **148**, 530–2.

Holt, S. J., Hobbiger, E. E. and Pawan, G. L. S. 1960. Preservation of integrity of rat tissues for cytochemical staining purposes. *J. Biophys. Biochem. Cytol.*, **7**, 383–6.

Holter, H. and Li, S. O. 1950. Phosphamidase activity of some proteolytic enzymes and rennin. *Acta Chem. Scand.*, **4**, 1321–2.

Holter, H. and Li, S. O. 1951. Determination and properties of phosphamidase. *Compt. Rend. Trav. Lab. Carlsberg (Sér. Chim.)*, **27**, 393–407.

Horecker, B. L. and Mehler, A. H. 1955. Carbohydrate metabolism. *Ann. Rev. Biochem.*, **24**, 207–74.

Hosoya, T., Matsukawa, S. and Kurata, Y. 1972. Cytochemical localisation of peroxidase in follicular cells of pig thyroid gland. *Endocrinol. Jap.*, **19**, 359–69.

Hotchkiss, R. D. 1948. A microchemical reaction resulting in the staining of polysaccharide structures in fixed tissue preparations. *Arch. Biochem.*, **16**, 131–41.

Howe, A. 1959. The distribution of arginine in the pituitary gland of the rat with particular reference to its presence in neurosecretory material. *J. Physiol.*, **149**, 519–25.

Husain, O. A. N. and Cameron, C. B. 1966. Automated methods for cancer screening. *Proc. Roy. Soc. Med.*, **59**, 982–6.

Ibrahim, K. S., Husain, O. A. N., Bitensky, L. and Chayen, J. 1983. A modified tetrazolium reaction for identifying malignant cells from gastric and colonic cancer. *J. Clin. Pathol.*, **36**, 133–6.

Jacobson, W. and Webb, M. 1952. The two types of nucleoproteins during mitosis. *Exptl. Cell Res.*, **3**, 163–83.

Jacoby, F. and Martin, B. F. 1949. The histochemical test for alkaline phosphatase. *Nature*, **163**, 875–6.

Jardetsky, C. D. and Glick, D. 1956. Studies in histochemistry. XXXVIII. Determination of succinic dehydrogenase in microgram amounts of tissue and its distribution in rat adrenal. *J. Biol. Chem.*, **216**, 283–92.

Jensen, W. A. 1955. The histochemical localization of peroxidase in roots and its induction by indolacetic acid. *Plant Physiol.*, **38**, 426–32.

Joftes, D. L. and Warren, S. 1955. *J. Biol. Phot. Assoc.*, **23**, 145.

Johansen, D. A. 1940. *Plant Microtechnique*. McGraw-Hill, New York.

Jones, G. R. N. 1963. Succinic dehydrogenase levels in livers of rats during early feeding of 4-dimethylaminobenzene: a reinterpretation of the biochemical data. *Br. J. Cancer*, **17**, 153–61.

Jones, G. R. N., Bitensky, L., Chayen, J. and Cunningham, G. J. 1961. Tissue dilution artefact: a re-interpretation of variations in levels of succinic dehydrogenase during chemical carcinogenesis. *Nature*, **191**, 1203.

Jones, G. R. N., Maple, A. J., Aves, E. K., Chayen, J. and Cunningham, G. J. 1963. Quantitative histochemistry of succinate dehydrogenase in tissue sections. *Nature*, **197**, 568–70.

Jordanov, J. 1963. On the transition of desoxyribonucleic acid to apurine acid and the loss of the latter from tissues during Feulgen reaction hydrolysis. *Acta Histochem.*, **15**, 135–52.

Juhlin, L. 1967. Determination of histamine in small biopsies and histological sections. *Acta Physiol. Scand.*, **71**, 30.

Junqueira, L. C. U., Cossermelli, W. and Brentani, R. 1978. Differential staining of collagens type I, II and III by Sirius Red and polarization microscopy. *Arch. Histol. Jap.*, **41**, 267–74.

Junqueira, L. C. U., Bignolas, G. and Brentani, R. R. 1979a. A simple and sensitive method for the quantitative estimation of collagen. *Anal. Biochem.*, **94**, 96–9.

Junqueira, L. C. U., Bignolas, G. and Brentani, R. R. 1979b. Picrosirius staining plus polarization microscopy, a specific method for collagen detection in tissue sections. *Histochem. J.*, **11**, 447–55.

Junqueira, L. C. U., Montes, G. S. and Sanchez, E. M. 1982. The influence of tissue section thickness on the study of collagen by the picrosirius–polarization method. *Histochemistry*, **74**, 153–6.

Kasdon, S. C., Fishman, W. H. and Homburger, F. 1950. Beta-glucuronidase studies in women. 2. Cancer of the cervix uteri. *J. Am. Med. Assoc.*, **144**, 892–6.

Kasdon, S. C., Homburger, F., Yorshis, E. and Fishman, W. H. 1953. Beta-glucuronidase studies in women. VI. Premenopausal vaginal fluid values in relation to invasive cervical cancer. *Surg. Gynecol. Obstet.*, **97**, 579–83.

Kashiwa, H. K. and Atkinson, W. B. 1963. The applicability of a new Schiff base, glyoxal bis (2-hydroxy-anil), for the cytochemical localization of ionic calcium. *J. Histochem. Cytochem.*, **11**, 258–64.

Kashiwa, H. K. and Marshall House, C., Jr. 1964. The glyoxal bis (2-hydroxyanil) method modified for localizing insoluble calcium salts. *Stain Technol.*, **39**, 359–67.

Katz, A. I., Doucet, A. and Morel, F. 1979. Na-K-ATPase activity along the rabbit, rat and mouse nephron. *Am. J. Physiol.*, **237**, F114–20.

Kaye, A. M. 1984. Purification and properties of ornithine decarboxylase. *Cell Biochem. Funct.*, **2**, 2–6.

Kent, G. N., Dodds, R. A., Bitensky, L., Chayen, J., Klenerman, L. and Watts, R. W. E. 1983. Changes in crystal size and orientation of acidic glycosaminoglycans at the fracture site in fractured neck of femur. *J. Bone Joint Surg.*, **65B**, 189–94.

King, J. 1965. *Practical Clinical Enzymology*. Van Nostrand, London.

Kirkby, W. W. 1965. The use of histochemical examination of maintenance-cultured tissue in assessing potential cytotoxic substances. *Biochem. J.*, **94**, 24–5P.

Kirschke, H., Langner, J., Wiedesanders, B., Ansorge, S., Bohley, P. and Hanson, H. 1977. Cathepsin H: an endopeptidase from rat liver lysosomes. *Acta Biol. Med. Germ.*, **36**, 185–99.

Klotz, I. M. 1950. The nature of some ion–protein complexes. *Cold Spring Harbor Symp. Quant. Biol.*, **14**, 97–112.

Koelle, G. B. and Friedenwald, J. S. 1949. A histochemical method for localizing cholinesterase activity. *Proc. Soc. Exptl. Biol. Med.*, **70**, 617–22.

Koelle, G. B. and Gromadzki, C. G. 1966. Comparison of the gold–thiocholine and gold–thiolacetic acid methods for the histochemical localization of acetylcholinesterase and cholinesterase. *J. Histochem. Cytochem.*, **14**, 443–54.

Koenig, C. S. and Vial, J. D. C. 1970. A histochemical study of adenosine triphosphatase in the toad (*Bufo spinulosus*) gastric mucosa. *J. Histochem. Cytochem.*, **18**, 340–53.

Kornberg, A. 1957. Pathways of enzymatic synthesis of nucleotides and polynucleotides. In: *The Chemical Basis of Heredity*. (Eds W. D. McElvoy and B. Glass). Johns Hopkins, Baltimore.

Kornberg, H. L. 1961. The breakdown and synthesis of fatty acids. In: *Biochemists' Handbook*. (Ed. C. Long), pp. 558–63. Spon, London.

Krebs, H. A. and Eggleston, L. V. 1974. The regulation of the pentose phosphate cycle in rat liver. *Adv. Enzyme Regul.*, **12**, 421–34.

Krebs, H. A. and Kornberg, H. L. 1957. *Energy Transformations in Living Matter*. Springer, Berlin.

Kurnick, N. B. 1950a. Methyl green–pyronin. 1. Basis of selective staining of nucleic acids. *J. Gen. Physiol.*, **33**, 243–64.

Kurnick, N. B. 1950b. The quantitative estimation of desoxyribonucleic acid based on methyl green staining. *Exptl. Cell Res.*, **14**, 469–85.

Kurnick, N. B. and Mirsky, A. E. 1950. Methyl green–pyronin. II. Stoichiometry of reaction with nucleic acids. *J. Gen. Physiol.*, **33**, 265–74.

La Cour, L. F., Chayen, J. and Gahan, P. B. 1958. Evidence for lipid material in chromosomes. *Exptl. Cell Res.*, **14**, 469–85.

Lawson, J. G. and Watkins, D. K. 1965. Vaginal fluid enzymes in relation to cervical cancer. *J. Obstet. Gynaecol. Br. Commonw.*, **72**, 1–8.

Lehninger, A. L. 1965. *The Mitochondrion: Molecular Basis of Structure and Function*. Benjamin, New York.

Lessler, M. A. 1953. The nature and specificity of the Feulgen nucleal reaction. *Int. Rev. Cytol.*, **2**, 231–47.

Leuchtenberger, C. 1954. Critical evaluation of Feulgen microspectrophotometry for estimating the amount of DNA in cell nuclei. *Science*, **120**, 1022–3.

Leuchtenberger, C. 1958. Quantitative determination of DNA in cells by Feulgen microspectrophotometry. In: *General Cytochemical Methods*. (Ed. J. F. Danielli). pp. 220–78. Academic Press, New York.

Levi, H. and Rogers, A. W. 1963. On the quantitative evaluation of autoradiographs. *Nat. Fys. Medd. Dan. Vid. Selsk.*, **33**, 5–51.

Levine, N. D. 1939. The determination of apparent iso electric points of cell structures by staining at controlled reactions. *Stain Technol.*, **15**, 91.

Lillie, R. D. and Mowry, R. W. 1949. Histochemical studies on absorption of iron by tissue reactions. *Bull. Int. A M Mus.*, **30**, 91–4.

Lima-De-Faria, A. 1961. Progress in tritium autoradiography. *Progr. Biophys. Biophys. Chem.*, **12**, 281–317.

Lojda, Z., Vecerek, B. and Pelichova, H. 1964. Some remarks concerning the histochemical detection of acid phosphatase by azo-coupling reactions. *Histochemie*, **3**, 428–54.

Lojda, Z., Van Der Ploeg, M. and Van Duijn, P. 1967. Phosphates of the naphthol AS series in the quantitative determination of alkaline and acid phosphatase activities 'in situ' studied by polyacrylamide membrane model systems and by cytophotometry. *Histochemie*, **11**, 13–32.

Lovelock, J. E. 1957. The denaturation of lipid–protein complexes as a cause of damage by freezing. *Proc. Roy. Soc., B*, **147**, 427–33.

Loveridge, N. 1978. A quantitative cytochemical method for measuring carbonic anhydrase activity. *Histochem. J.*, **10**, 361–72.

Loveridge, N., Alaghband-Zadeh, J., Daly, J. R. and Chayen, J. 1975. The nature of the redox change measured in the cytochemical bioassay of corticotrophin. *J. Endocrinol.*, **67**, 28P.

Loveridge, N., Bitensky, L., Chayen, J., Hausamen, T. U., Fisher, J. M., Taylor, K. B., Gardner, J. D., Bottazzo, G. F. and Doniach, D. 1980a. Inhibition of parietal cell function by human gamma-globulin containing gastric parietal-cell antibodies. *Clin. Exptl. Immunol.*, **41**, 264–70.

Loveridge, N., Hoile, R. W., Johnson, A. G., Gardner, J. D. and Chayen, J. 1980b. The cytochemical section-bioassay of gastrin-like activity. *J. Immunoassay*, **1**, 195–209.

Lowry, O. H. 1964. Microanalysis for histochemical purposes. *Proc. 2nd Int. Congr. Histochem. Cytochem.* Springer, Berlin.

Luyet, B. J. 1951. Survival of cells and organisms after ultrarapid freezing. In: *Freezing and Drying*. (Ed. R. J. C. Harris). pp. 77–98. Institute of Biologists, London. Also in discussion, *ibid*, p. 201.

Luyet, B. J. 1960. On various phase transitions occurring in aqueous solutions at low temperature. *Ann. N.Y. Acad. Sci.*, **85**, 549–69.

Luzzatto, L. 1987. Glucose 6-phosphate dehydrogenase: genetic and haematological aspects. *Cell Biochem. Funct.*, **5**, 101–7.

Lynch, R., Bitensky, L. and Chayen, J. 1966. On the possibility of super-cooling in tissues. *J. Roy. Microsc. Soc.*, **85**, 213–22.

Maclagan, N. F. 1964. Diseases of the liver and biliary tract. In: *Biochemical Disorders in Human Disease*. (Eds R. H. S. Thompson and E. J. King). pp. 122–5. Churchill, London.

Mahadevan, S. and Tappel, A. L. 1967. Arylamidases of rat liver and kidney. *J. Biol. Chem.*, **242**, 2369.

Malmgren, H. and Sylvén, B. 1955. On the chemistry of the thiocholine method of Koelle. *J. Histochem. Cytochem.*, **3**, 441–5.

Mancini, R. E. 1948. Histochemical study of glycogen in tissues. *Anat. Rec.*, **101**, 149–59.

Maren, T. H. 1967. Carbonic anhydrase: chemistry, physiology and inhibition. *Physiol. Rev.*, **47**, 595–781.

Markert, C. L. 1984. Biochemistry and function of lactate dehydrogenase. *Cell Biochem. Funct.*, **2**, 131–4.

Marks, N., Datta, R.K. and Lajtha, A. 1968. Partial resolution of brain arylamidases and aminopeptidases. *J. Biol. Chem.*, **243**, 2882–9.

Massart, L., Peeters, G., De Ley, J., Vercauteren, R. and Von Honcke, A. 1947. The mechanism of the biochemical activity of acridines. *Experientia*, **3**, 288.

Materazzi, G. and Ferretti, E. 1970. Histochemical and histophysical investigations on the acetylation blockade of carboxylic groups of polysaccharides. *J. Histochem. Cytochem.*, **18**, 504–9.

Matschinsky, F. M., Passonneau, J. V. and Lowry, O. H. 1968. Quantitative histochemical analysis of glycolytic intermediates and co-factors with an oil well technique. *J. Histochem. Cytochem.*, **16**, 29–39.

McCabe, M. and Chayen, J. 1965. The demonstration of latent particulate aminopeptidase activity. *J. Roy. Microsc. Soc.*, **84**, 361–71.

McDonald, J. K., Reilly, J. T., Zeitman, B. B. and Ellis, S. 1968. Dipeptidyl arylamidase II of the pituitary. *J. Biol. Chem.*, **243**, 2028–37.

McGarry, A. and Gahan, P. B. 1985. A quantitative cytochemical study of UDP-D-glucose : NAD-oxidoreductase (EC 1.1.1.22) activity during stelar differentiation in *Pisum sativum* L. cv Meteor. *Histochemistry*, **83**, 552–4.

McGee-Russell, S. M. 1955. A new reagent for the histochemical and chemical detection of calcium. *Nature*, **175**, 301–2.

McGee-Russell, S. M. 1958. Histochemical methods for calcium. *J. Histochem. Cytochem.*, **6**, 22–42.

McLean, P. and Brown, J. 1966. Activities of some enzymes concerned with citrate and glucose metabolism in transplanted rat hepatomas. *Biochem. J.*, **98**, 874–82.

McLeish, J. 1959. Comparative microphotometric studies of DNA and arginine in plant nuclei. *Chromosoma*, **10**, 686–710.

McLeish, J. and Sherratt, H. S. A. 1958. The use of the Sakaguchi reaction for the cytochemical determination of combined arginine. *Exptl. Cell Res.*, **14**, 625–8.

McLeish, J. L., Bell, L. G. E., La Cour, L. F. and Chayen, J. 1957. The quantitative cytochemical estimation of arginine. *Exptl. Cell Res.*, **12**, 120–5.

McMillan, P. J. 1967. Differential demonstration of muscle and heart type lactic dehydrogenase of rat muscle and kidney. *J. Histochem. Cytochem.*, **15**, 21–31.

Meijer, A. E. F. H. 1980. Semipermeable membrane techniques in quantitative enzyme histochemistry. In: *Trends in Enzyme Histochemistry and Cytochemistry*. Ciba Foundation Symposium 73. (Eds D. Evered and M. O'Connor). Amsterdam, Oxford, New York.

Meyer, J. and Weinman, J. P. 1955. A modification of Gomori's method for demonstration of phosphamidase in tissue sections. *J. Histochem. Cytochem.*, **3**, 134–40.

Millett, J. A., Chin, U., Bitensky, L., Chayen, J. and Husain, O. A. N. 1980. Lysosomal naphthylamidase activity as a possible aid in cytological screening. *J. Clin. Pathol.*, **33**, 684–7.

Millett, J. A., Husain, O. A. N., Bitensky, L. and Chayen, J. 1982. Feulgen-hydrolysis profiles in cells exfoliated from the cervix uteri: a potential aid in the diagnosis of malignancy. *J. Clin. Pathol.*, **35**, 345–9.

Mircheff, A. K., Walling, M. W., van Os, C. H. and Wright, E. M. 1977. Distribution of alkaline phosphatase and Ca-ATPase in intestinal epithelial cell plasma membranes: differential response to 1,25 $(OH_2)D_3$. In: *Vitamin D: Biochemical, Chemical and Aspects Related to Calcium Metabolism*. (Eds A. W. Norman, K. Schaefer, J. W. Coburn, H. F. DeLuca, D. Fraser, H. G. Grigoleit and D. von Herrath). pp. 281–3. de Gruyter, Berlin.

Modis, L. 1974. Topo-optical investigations of mucopolysaccharides (acid glycosaminoglycans). In: *Handbuch der Histochemie*. Vol. II. Polysaccharides Part 4. (Eds W. Graumann and K. Neumann). Fischer, Stuttgart.

Moline, S. W. and Glenner, G. G. 1964. Ultrarapid tissue freezing in liquid nitrogen. *J. Histochem. Cytochem.*, **12**, 777–83.

Monet, J. D., Thomas, M., Dautigny, N., Brami, M. and Bader, C. A. 1987. Effects of 17β-estradiol and R5020 on glucose 6-phosphate dehydrogenase activity in MCF-7 human breast cancer cells: a cytochemical assay. *Cancer Res.*, **47**, 5116–19.

Moses, H. L., Rosenthal, A. S., Beaver, D. L. and Schuffman, S. S. 1966. Lead ion and phosphatase histochemistry. II. Effect of adenosine triphosphate hydrolysis by lead ion on the histochemical localization of adenosine triphosphatase activity. *J. Histochem. Cytochem.*, **14**, 702–10.

Moss, D. W. 1983. Clinical biochemistry of alkaline phosphatase. *Cell. Biochem. Funct.*, **1**, 70–4.

Muther, T. F. 1972. A critical evaluation of the histochemical methods for carbonic anhydrase. *J. Histochem. Cytochem.*, **20**, 319–30.

Nachlas, M. M. and Seligman, A. M. 1949. The histochemical demonstration of esterase. *J. Natl. Cancer Inst.*, **9**, 415–25.

Nachlas, M. M., Prinn, W. and Seligman, A. M. 1956. Quantitative estimation of lyo- and desmoenzymes in tissue sections, with and without fixation. *J. Biophys. Biochem. Cytol.*, **2**, 487–502.

Nachlas, M. M., Crawford, D. T. and Seligman, A. M. 1957a. The histochemical demonstration of leucine aminopeptidase. *J. Histochem. Cytochem.*, **5**, 264–78.

Nachlas, M. M., Tsou, K., De Souza, E., Cheng, C. and Seligman, A. M. 1957b. Cytochemical demonstration of succinic dehydrogenase by the use of a new *p*-nitrophenyl substituted ditetrazole. *J. Histochem. Cytochem.*, **5**, 420–36.

Nachlas, M. M., Walker, D. G. and Seligman, A. M. 1958a. A histochemical method for the demonstration of diphosphopyridine nucleotide diaphorase. *J. Biophys. Biochem. Cytol.*, **4**, 29.

Nachlas, M. M., Walker, D. G. and Seligman, A. M. 1958b. The histochemical localization of triphosphopyridine nucleotide diaphorase. *J. Biophys. Biochem. Cytol.*, **4**, 467.

Nakane, P. and Pierce, G. B. 1967. Enzyme-labelled antibodies for light and electron microscopic localisation of tissue antigens. *J. Cell Biol.*, **33**, 307–18.

Narumi, S. and Kanno, M. 1973. Effects of gastric acid stimulants and inhibitors on the activities of HCO_3 stimulated Mg^{2+} dependent ATPase and carbonic anhydrase in rat gastric mucosa. *Biochim. Biophys. Acta*, **311**, 80–9.

Neely, J. R. and Morgan, H. E. 1974. Relationship between carbohydrate and lipid metabolism and the energy balance of heart muscle. *Ann. Rev. Physiol.*, **36**, 413–59.

Neely, J. R., Rovetto, M. J. and Oram, J. F. 1972. Myocardial utilisation of carbohydrate and lipid. *Progr. Cardiovasc. Dis.*, **15**, 289–329.

Neufeld, E. F., Lim, T. W. and Shapiro, L. J. 1975. Inherited disorders of lysosomal metabolism. *Ann. Rev. Biochem.*, **44**, 357–76.

Niemi, M. and Sylvén, B. 1969. The naphthylamidase reaction as a diagnostic tool for the demonstration of cellular injury and autophagy. *Histochemie*, **18**, 40–6.

Niles, N. R., Bitensky, L., Chayen, J., Cunningham, G. J. and Braimbridge, M. V. 1964a. The value of histochemistry in the analysis of myocardial dysfunction. *Lancet*, i, 963–5.

Niles, N. R., Chayen, J., Cunningham, G. J. and Bitensky, L. 1964b. The histochemical demonstration of adenosine triphosphatase activity in myocardium. *J. Histochem. Cytochem.*, **12**, 740–3.

Niles, N. R., Bitensky, L., Braimbridge, M. V. and Chayen, J. 1966. Histochemical changes related to oxidation and phosphorylation in human heart muscle. *J. Roy. Microsc. Soc.*, **86**, 159–66.

Nöhammer, G., Schauenstein, E., Bajardi, F. and Unger-Ulmann, A. 1977. Microphotometrical quantification of protein thiols in morphologically intact cells of the cervical epithelium. *Acta Cytol.*, **21**, 341–4.

Novikoff, A. B. 1952. Histochemical demonstration of nuclear enzymes. *Exptl. Cell Res. Suppl.*, **2**, 123–39.

Novikoff, A. B. 1963. Discussion. In: *Ciba Symp. on 'Lysosomes'*. (Eds A. V. S. de Reuck and M. P. Cameron). pp. 375–8. Churchill, London.

Odell, L. D. and Burt, J. C. 1949. Beta-glucuronidase activity in human female genital cancer. *Cancer Res.*, **9**, 362–5.

Olsen, I., Dean, M. F., Harris, G. and Muir, H. 1981. Direct transfer of a lysosomal enzyme from lymphoid cells to deficient fibroblasts. *Nature*, **291**, 244–7.

Ordman, A. B. and Kirkwood, S. 1977. Mechanism of action of uridine diphosphoglucose dehydrogenase. *J. Biol. Chem.*, **252**, 1320–6.

Ostrowski, K. and Barnard, E. A. 1961. Application of isotopically-labelled specific inhibitors as a method in enzyme cytochemistry. *Exptl. Cell Res.*, **25**, 465–8.

Ostrowski, K., Komender, J., Koscianek, H. and Kwarecki, K. 1962a. Quantitative investigation of the P and N loss in the rat liver when using various media in the 'freeze-substitution' technique. *Experientia*, **18**, 142–6.

Ostrowski, K., Komender, J., Koscianek, H. and Kwarecki, K. 1962b. Quantitative studies on the influence of the temperature applied in freeze-substitution on P, N, and dry mass losses in fixed tissue. *Experientia*, **18**, 227–30.

Ostrowski, K., Komender, J., Koscianek, H. and Kwarecki, K. 1962c. Elution of some substances from the tissues fixed by the 'freeze-substitution' method. *Acta Biochim. Pol.*, **9**, 125–30.

Ostrowski, K., Barnard, E. A., Stocka, Z. and Darzynkiewicz, Z. 1963. Autoradiographic methods in enzyme cytochemistry. 1. Localization of acetylcholinesterase activity using a ^{3}H-labeled irreversible inhibitor. *Exptl. Cell Res.*, **31**, 89–99.

Padykula, H. A. and Herman, E. 1955a. Factors affecting the activity of adenosinetriphosphatase and other phosphatases as measured by histochemical techniques. *J. Histochem. Cytochem.*, **3**, 161–9.

Padykula, H. A. and Herman, E. 1955b. The specificity of the histochemical methods for adenosinetriphosphatase. *J. Histochem. Cytochem.*, **3**, 170–83.

Palmoski, M. J. and Brandt, K. D. 1983. *In vivo* effect of aspirin on canine osteoarthritic cartilage. *Arthritis Rheum.*, **26**, 994–1001.

Patriarca, P., Dri, P., Kakinuma, K., Tedesco, F. and Rossi, F. 1975a. Studies on the mechanisms of metabolic stimulation in polymorphonuclear leucocytes during phagocytosis. I. Evidence for superoxide anion involvement in the oxidation of NADPH$_2$. *Biochim. Biophys. Acta*, **385**, 380–6.

Patriarca, P., Cramer, R., Tedesco, F. and Kakinuma, K. 1975b. Studies on the mechanism of metabolic stimulation in polymorphonuclear leucocytes during phagocytosis. II. Presence of NADPH$_2$ oxidizing activity in a myeloperoxidase-deficient subject. *Biochim. Biophys. Acta*, **385**, 387–93.

Patterson, E. K., Hsiao, S. H. and Keppel, A. 1963. Studies on dipeptidases and aminopeptidases. I. Distinction between leucine aminopeptidase and enzymes that hydrolyze L-leucyl-β-naphthylamide. *J. Biol. Chem.*, **38**, 3611.

Patterson, E. K., Hsiao, S. H., Keppel, A. and Sorof, S. 1965. Studies on dipeptidases and aminopeptidases. II. Zonal electrophoretic separation of rat liver peptidases. *J. Biol. Chem.*, **240**, 710.

Pattison, J. R., Bitensky, L. and Chayen, J. (Eds). 1979. *Quantitative Cytochemistry and its Applications*. Academic Press, London.

Pearse, A. G. E. 1960. *Histochemistry: Theoretical and Applied*. 2nd edn. Churchill, London.

Pelc, S. R. 1956. The stripping-film technique of autoradiography. *Int. J. Appl. Radiation Isotopes*, **1**, 172–5.

Pelc, S. R. 1958. Autoradiography as a cytochemical method with special reference to C^{14} and S^{35}. In: *General Cytochemical Methods*. (Ed. J. F. Danielli). Academic Press, New York.

Pelc, S. R., Coombes, J. D. and Budd, C. C. 1961. On the adaptation of autoradiographic techniques for use with the electron microscope. *Exptl. Cell Res.*, **24**, 192–207.

Pérez-González De La Manna, M., Proverbio, F. and Whittembury, G. 1980. ATPases and salt transport in the kidney tubule. *Curr. Topics Membr. Transport*, **13**, 315–35.

Peters, T. J. 1976. Analytical subcellular fractionation of jejunal biopsy specimens: methodology and characterization of the organelles in normal tissue. *Clin. Sci. Mol. Med.*, **51**, 557–74.

Pilz, R. B., Randall, C. W. and Seegmiller, J. E. 1982. Regulation of human lymphoblast membrane 5′-nucleotidase by zinc. *J. Biol. Chem.*, **257**, 13544–9.

Pitsillides, A. A., Taylor, P. M., Bitensky, L., Chayen, J., Muir, I. H. and Askonas, B. A. 1988. Rapid changes in target cell lysosomes induced by cytotoxic T-cells: indication of target suicide? *Eur. J. Immunol.*, **18**, 1203–8.

Pocker, Y. and Watamori, N. 1973. Stopped-flow studies of high pH activity and acetazolamide inhibition of bovine carbonic anhydrase. *Biochemistry*, **12**, 2475–82.

Pollister, A. W. and Ornstein, L. 1959. The photometric chemical analysis of cells. In: *Analytical Cytology*. 2nd edn. (Ed. R. C. Mellors). McGraw-Hill, London.

Popjak, G. 1961. Biosynthesis of sterols. In: *Biochemists' Handbook*. (Ed. C. Long). pp. 566–81. Spon, London.

Potter, V. R., Price, J. M., Miller, E. C. and Miller, J. A. 1950. Studies on the intracellular composition of livers from rats fed various aminoazo dyes. III. Effects on succinoxidase and oxaloacetic acid oxidase. *Cancer Res.*, **10**, 28–35.

Puchter, H., Meloan, S. N. and Terry, M. S. 1969. On the history and mechanism of alizarin and alizarin red S stains for calcium. *J. Histochem. Cytochem.*, **17**, 110–24.

Quarles, R. H. and Dawson, R. M. C. 1969. A shift in the optimum pH of phospholipase D produced by activating long chain anions. *Biochem. J.*, **112**, 795–9.

Quattropani, S. K. and Weisz, J. 1973. Conversion of progesterone to oestrone and oestradiol *in vitro* by the ovary of the infantile rat in relation to the development of its interstitial tissue. *Endocrinology*, **93**, 1269–76.

Racker, E. 1961. Glyceraldehyde 3-phosphate dehydrogenase. In: *Biochemists' Handbook*. (Ed. C. Long). pp. 328–9. Spon, London.

Richards, B. M. 1960. Redistribution of nuclear proteins during mitosis. In: *The Cell Nucleus*. (Ed. J. S. Mitchell). Butterworths, London.

Ris, H. and Mirsky, A. E. 1949. The state of the chromosomes in the interphase nucleus. *J. Gen. Physiol.*, **32**, 489–502.

Robertson, W. R. 1979. A quantitative cytochemical method for the demonstration of $\Delta 5,3\beta$-hydroxysteroid dehydrogenase activity in unfixed tissue sections of rat ovary. *Histochemistry*, **59**, 271–85.

Robertson, W. R. 1980. A quantitative study of N-acetyl-β-glucosaminidase activity in unfixed tissue sections of the guinea-pig thyroid gland. *Histochem. J.*, **12**, 87–96.

Robinson, D. and Stirling, J. L. 1968. N-Acetyl-β-glucosaminidases in human spleen. *Biochem. J.*, **107**, 321–7.

Rogers, A. W. 1967. *Techniques of Autoradiography*. Elsevier, Amsterdam.

Roodyn, D. B. 1965. The classification and partial tabulation of enzyme studies on subcellular fractions isolated by differential centrifuging. *Int. Rev. Cytol.*, **18**, 99–190.

Rosemeyer, M. A. 1987. The biochemistry of glucose-6-phosphate dehydrogenase, 6-phosphogluconate dehydrogenase and glutathione reductase. *Cell Biochem. Funct.*, **5**, 79–95.

Rosen, S. 1972. Observations on the specificity of newer histochemical methods for the demonstration of carbonic anhydrase activity. *J. Histochem. Cytochem.*, **20**, 951–4.

Rosene, D. L. and Mesulam, M. 1978. Fixation variables in horseradish peroxidase neurohistochemistry. 1. The effects of fixation time and perfusion procedures upon enzyme activity. *J. Histochem. Cytochem.*, **19**, 571–5.

Roy, A. B. 1960. The synthesis and hydrolysis of sulfate esters. *Adv. Enzymol.*, **22**, 205–35.

Roy, A. B. 1962. The histochemical detection of arysulphatases. *J. Histochem. Cytochem.*, **10**, 106–7.

Ryder, T. A., Mackenzie, M. L., Pryse-Davies, J., Glover, V., Lewinsohn, R. and Sandler, M. 1979. A coupled peroxidatic oxidation technique for the histochemical localization of monoamine oxidase A and B and benzylamine oxidase. *Histochemistry*, **62**, 93–100.

Sacktor, B. and Dick, A. R. 1960. Alpha-glycerophosphate and lactic dehydrogenases of hematopoietic cells from leukemic mice. *Cancer Res.*, **20**, 1408.

Sakaguchi, S. 1925. Uber eine neue Farbreaktion von Protein und Arginin. *J. Biochem. (Tokyo)*, **5**, 25.

Salmon, D. M., Azria, M. and Zanelli, J. M. 1983. Quantitative cytochemical responses to exogenously administered calcitonins in rat kidney and bone cells. *Mol. Cell. Endocrinol.*, **33**, 293–304.

Salpeter, M. M. and Bachmann, L. 1964. Autoradiography with the electron microscope: a procedure for improving resolution, sensitivity and contrast. *J. Cell Biol.*, **22**, 469–79.

Sandritter, W. and Fischer, R. 1962. Der DNS-Gehalt des normalen Plattenepithels, des Carcinoma *in situ* und des Invasiven Carcinoms der Portio. *Proc. 1st Int. Congr. Exfol. Cytol.* pp. 189–95. Lippincott, Philadelphia.

Sandritter, W. and Kleinhans, D. 1964. Uber das Trockengewicht, den DNS- und Histonproteingehalt von menschlichen Tumoren. *Z. Krebsforsch.*, **66**, 333–48.

Sandritter, W. and Krygier, A. 1959. Cytophotometrische Bestimmungen von proteingebundenen Thiolen in der Mitose und Interphase von HeLa-Zellen. *Z. Krebsforsch.*, **62**, 596–610.

Sanger, F. 1945. The free amino group of insulin. *Biochem. J.*, **39**, 507–15.

Savard, K. 1961. Steroid interconversions. In: *Biochemists' Handbook*. (Ed. C. Long). pp. 581–6. Spon, London.

Schäfer, H. 1979. *Cellular Calcium and Cell Function*. Progress in Pathology Series, No. 109. Fischer, Stuttgart.

Schaffner, W. and Weissmann, C. 1973. A rapid, sensitive and specific method for the determination of protein in dilute solution. *Anal. Biochem.*, **56**, 502–14.

Schatzmann, H. J. 1966. ATP-dependent Ca^{++}—extrusion from human red cells. *Experientia*, **22**, 364–5.

Schatzmann, H. J. 1982. The plasma membrane calcium pump of erythrocytes and other animal cells. In: *Membrane Transport of Calcium*. (Ed. E. Carafoli). pp. 109–40. Academic Press, London.

Schauenstein, E., Dosoye, G. and Nöhammer, G. 1980. Quantitative Proteinbestimmung in Einzelzellen mit Amidschwarz. *Histochemistry*, **68**, 75–90.

Schauenstein, E., Bajardi, F., Benedetto, C., Nohammer, G. and Slater, T. F. 1983. Histophotometrical investigations on the contents of protein and protein thiols of the epithelium and stroma of the human cervix. I. Cases with no apparent neoplastic alterations of the epithelium. *Histochemistry*, **77**, 465–72.

Schümmelfeder, N. 1958. Histochemical significance of the polychromatic fluorescence induced in tissues stained with acridine orange. *J. Histochem. Cytochem.*, **6**, 392–3.

Scott, J. E. 1970. Histochemistry of Alcian blue. I. Metachromasia of Alcian blue, Astrablau and other cationic phthalocyanin dyes. *Histochemie*, **21**, 277–85.

Scott, J. E. and Dorling, J. 1965. Differential staining of acid glycosaminoglycans (mucopolysaccharides) by Alcian blue in salt solutions. *Histochemie*, **5**, 221–33.

Scott, J. E. and Mowry, R. W. 1970. Alcian blue—a consumer's guide. *J. Histochem. Cytochem.*, **18**, 842.

Seligman, A. M. 1963. Developmental histochemistry: the introduction of new methods for light and electron microscopy by the design and preparation of appropriate substrates and reagents. In: *Histochemistry and Cytochemistry*. (Ed. R. Wegmann). Pergamon, Oxford.

Seligman, A. M., Chauncey, H. H. and Nachlas, M. M. 1951. Effect of formalin fixation on the activity of five enzymes of rat liver. *Stain Technol.*, **26**, 19–23.

Shannon, A. D. 1975. A post-coupling method for the demonstration of *N*-acetyl-β-glucosaminidase activity. *J. Histochem. Cytochem.*, **23**, 424–30.

Shedden, R., Dunham, J., Bitensky, L., Catterall, A. and Chayen, J. 1976. Changes in alkaline phosphatase activity in periosteal cells in healing fractures. *Calcif. Tissue Res.*, **22**, 19–25.

Shelley, W. B., Öhman, S. and Parnes, H. M. 1968. Mast cell stain for histamin in freeze-dried embedded tissue. *J. Histochem. Cytochem.*, **16**, 433–9.

Sherlock, S. 1958. *Diseases of the Liver and Biliary System*. 2nd edn. Oxford University Press, Oxford.

Shirazi, S. P., Beechey, R. B. and Butterworth, P. J. 1981. Potent inhibition of membrane-bound intestinal alkaline phosphatase by a new series of phosphate analogues. *Biochem. J.*, **194**, 797–802.

Shnitka, T. K. and Seligman, A. M. 1960. Evidence for the role of esteratic inhibition in producing certain beautiful localization artefacts. *J. Histochem. Cytochem.*, **8**, 344.

Shnitka, T. K. and Seligman, A. M. 1961. Role of esteratic inhibition on localization of esterase and the simultaneous cytochemical demonstration of inhibitor sensitive and resistant enzyme species. *J. Histochem. Cytochem.*, **9**, 504–27.

Shnitka, T. K. and Seligman, A. M. 1971. Ultrastructural localization of enzymes. *Ann. Rev. Biochem.*, **40**, 375–96.

Sibatani, A. and Naora, H. 1952. Enhancement of colour intensity in the histochemical Feulgen reaction: method and quantitative estimation. *Experientia*, **8**, 263–4.

Siekevitz, P. 1962. The relationship of cell structure to metabolic activity. In: *The Molecular Control of Cellular Activity*. (Ed. J. M. Allen). McGraw-Hill, New York.

Silcox, A. A., Poulter, L. W., Bitensky, L. and Chayen, J. 1965. An examination of some factors affecting histological preservation in frozen sections of unfixed tissue. *J. Roy. Microsc. Soc.*, **84**, 559–64.

Simpson, W. L. 1941. An experimental analysis of the Altmann technic of freeze-drying. *Anat. Rec.*, **80**, 173–89.

Singer, T. P. and Kearney, E. B. 1954. Solubilization, assay and purification of succinic dehydrogenase. *Biochim. Biophys. Acta*, **15**, 151–3.

Skillen, A. W. 1984. Clinical biochemistry of lactate dehydrogenase. *Cell Biochem. Funct.*, **2**, 140–4.

Skou, J. C. 1965. Enzymatic basis for active transport of Na^+ and K^+ across cell membranes. *Physiol. Rev.*, **45**, 596.

Slater, E. C. 1961. Succinic dehydrogenase. In: *Biochemists' Handbook*. (Ed. C. Long). pp. 369–71. Spon, London.

Slater, T. F. 1966. Necrogenic action of carbon tetrachloride in the rat: a speculative mechanism based on activation. *Nature*, **209**, 36–40.

Smith, E. L. 1951. Proteolytic enzymes. In: *The Enzymes*. (Eds J. B. Sumner and K. Myrbäck). Vol. I. Academic Press, New York.

Smith, E. L. 1960. Peptide bond cleavage. In: *The Enzymes*. (Eds P. D. Boyer, H. Lardy and K. Myrbäck). Vol. 4. Academic Press, New York.

Smith, E. L. and Spackman, D. H. 1955. Leucine aminopeptidase. V. Activation, specificity and mechanism of action. *J. Biol. Chem.*, **212**, 271–99.

Smith, M. T. and Wills, E. D. 1981. The effects of dietary lipid and phenobarbitone on the production and utilization of NADPH in the liver. *Biochem. J.*, **200**, 691–9.

Smith, M. T., Loveridge, N., Wills, E. D. and Chayen, J. 1979. The distribution of glutathione in the rat liver lobule. *Biochem. J.*, **182**, 103–8.

Smith, R. E. and Van Frank, R. M. 1975. The use of amino acid derivatives of 4-methoxy-β-naphthylamine for the assay and subcellular localization of tissue proteinases. In: *Lysosomes in Biology and Pathology*. (Eds J. T. Dingle and R. T. Dean). Vol. 4. pp. 173–249. North-Holland, Amsterdam.

Spicer, S. S. 1965. Diamine methods for differentiating mucosubstances histochemically. *J. Histochem. Cytochem.*, **13**, 211–34.

Stafford, R. O. and Atkinson, W. B. 1948. Effect of acetone and alcohol fixation and paraffin embedding on activity of acid and alkaline phosphatases in rat tissues. *Science*, **107**, 279–80.

Sternberg, W. H., Farber, E. and Dunlap, C. E. 1956. Histochemical localization of specific oxidative enzymes. II. Localization of diphosphopyridine nucleotide diaphorase and triphosphopyridine nucleotide diaphorase and the succindehydrogenase system in the kidney. *J. Histochem. Cytochem.*, **4**, 266–83.

Sternberger, L. A. 1979. *Immunocytochemistry*. 2nd edn. Wiley, New York.

Strugger, S. 1938. Die Vitalfarbund des Protoplasmas mit Rhodamin B und 6G. *Protoplasma*, **30**, 85–100.

Stuart, J. and Simpson, J. S. 1970. Dehydrogenase enzyme cytochemistry of unfixed leucocytes. *J. Clin. Pathol.*, **23**, 517–21.

Stuart, J., Simpson, J. S. and Mann, J. R. 1970. Intracellular hydrogen transport systems in acute leukaemia. *Br. J. Haematol.*, **19**, 739–48.

Stumpf, W. E. and Roth, L. J. 1966. High resolution autoradiography with dry mounted freeze-dried frozen sections. *J. Histochem. Cytochem.*, **14**, 274–87.

Sugimura, K. and Mizutani, A. 1979. The inhibitory effect of xanthine derivatives on alkaline phosphatase in rat brain. *Histochemie*, **61**, 131–7.

Sweat, F., Puchtler, H. and Rosenthal, S. I. 1964. Sirius red F3BA as a stain for connective tissue. *Arch. Pathol.*, **78**, 69–72.

Swift, H. H. 1953. Quantitative aspects of nuclear nucleoproteins. *Int. Rev. Cytol.*, **2**, 1–78.

Sylvén, B. 1954. Metachromatic dye–substrate interactions. *Q. J. Microsc. Sci.*, **95**, 327–58.

Sylvén, B. and Malmgren, H. 1957. The histological distribution of proteinase and peptidase activity in solid mouse tumour transplants: a histochemical study on the enzymic characteristics of different cell types. *Acta Radiol., Suppl.*, **154**, 1–124.

Sylvén, B. and Lippi, U. 1965. The suggested lysosomal localization of aminoacylnaphthylamide splitting enzymes. *Exptl. Cell Res.*, **40**, 145–7.

Symposium of the Institute of Biology. 1952. Freezing and Drying.

Takeuchi, T. and Kuriaki, H. 1953. Histochemical detection of phosphorylase in animal tissues. *J. Histochem. Cytochem.*, **3**, 153–60.

Tammann, G. 1898. *Ann. Phys.*, **2**, 1424–33; in Luyet, B. J. 1951. Survival of cells, tissues and organisms after ultra-rapid freezing. In: *Freezing and Drying*. (Ed. R. J. C. Harris). Institute of Biologists, London.

Tappel, A. L. 1969. Lysosomal enzymes and other components. In: *Lysosomes in Biology and Pathology*. (Eds J. T. Dingle and H. B. Fell). Vol. II. pp. 207–44. North-Holland, Amsterdam.

Tappel, A. L., Sawant, P. L. and Shibko, S. 1963. Lysosomes: distribution in animals, hydrolytic capacity and other properties. In: *Ciba Symp. on 'Lysosomes'*. (Eds A. V. S. de Reuck and M. P. Cameron). Churchill, London.

Taylor, J. H. 1956. Autoradiography at the cellular level. In: *Physical Techniques in Biological Research*. (Eds G. Oster and A. W. Pollister). Academic Press, New York.

Thornburg, W. and Mengers, P. E. 1957. An analysis of frozen section techniques. I. Sectioning fresh frozen tissue. *J. Histochem. Cytochem.*, **5**, 47–52.

Tipton, K. F. 1986. Enzymology of monoamine oxidase. *Cell. Biochem. Funct.*, **3**, 79–87.

Ullberg, S. 1962. Autoradiographic localization in the tissues of drugs and metabolites. *Biochem. Pharmacol.*, **9**, 29–38.

Underhay, E., Holt, S. J., Beaufay, H. and De Duve, C. 1956. Intracellular localization of esterase in rat liver. *J. Biophys. Biochem. Cytol.*, **2**, 635–7.

Van der Sluis, P. J. and Boer, G. J. 1986. The relevance of various tests for the study of specificity in immunocytochemical staining: a review. *Cell Biochem. Funct.*, **4**, 1–17.

Viala, R. and Gianetto, R. 1955. The binding of sulphatase by rat-liver particles as compared to that of acid sulphatase. *Can. J. Biochem. Physiol.*, **33**, 839–44.

Villee, C. A. 1962. The role of steroid hormones in the control of metabolic activity. In: *The Molecular Control of Cellular Activity*. (Ed. J. M. Allen). McGraw-Hill, New York.

Volk, B. W. and Popper, H. 1944. Microscopic demonstration of fat in urine and stool by means of fluorescence microscopy. *Am. J. Clin. Pathol.*, **14**, 234–8.

Wachstein, M. and Meisel, E. 1956. Histochemical demonstration of substrate specific phosphatases at a physiological pH. *Am. J. Clin. Pathol.*, **4**, 424–5.

Wachstein, M. and Meisel, E. 1964. Demonstration of peroxidase activity in tissue sections. *J. Histochem. Cytochem.*, **12** 538–44.

Walker, P. M. B. 1958. Ultraviolet microspectrophotometry. In: *General Cytochemical Methods*. (Ed. J. F. Danielli). Vol. 1. pp. 164–219. Academic Press, New York.

Walker, P. M. B. and Richards, B. M. 1959. Microscopical techniques for single cells. In: *The Cell*. (Eds J. Brachet and A. E. Mirsky). Vol. 1. pp. 91–138. Academic Press, New York, London.

Warren, W. A. 1970. Catalysis of both oxidation and reduction of glyoxylate by pig heart lactate dehydrogenase isozyme I. *J. Biol. Chem.*, **245**, 1675–81.

Watts, R. W. E. 1973. Oxaluria. *J. Roy. Coll. Physicians London*, **7**, 161–74.

Watts, R. W. E. 1987. Purine enzymes and immune function. *Clin. Biochem. Rev.*, 239–65.

Watts, R. W. E., Scott, J. T., Chalmers, R. A., Bitensky, L. and Chayen, J. 1971. Microscopic studies on skeletal muscle in gout patients treated with allopurinol. *Q. J. Med.*, **40**, 1–14.

West, E. S. and Todd, W. R. 1961. *Textbook of Biochemistry*. Macmillan, New York.

Whitt, G. S. 1984. Genetic, developmental and evolutionary aspects of the lactate dehydrogenase isozyme system. *Cell Biochem. Funct.*, **2**, 134–9.

Wied, G. L. and Bahr, G. F. (Eds). 1970. *Introduction to Quantitative Cytochemistry II*. Academic Press, New York.

Wigglesworth, V. B. 1952. The role of iron in histological staining. *Q. J. Microsc. Sci.*, **93**, 105–18.

Willighagen, R. G. J. and Planteijdt, H. T. 1959. Aminopeptidase activity in cancer cells. *Nature*, **183**, 4653–4.

Willighagen, R. G. J., Van Der Heul, R. O. and van Rijssel, Th. G. 1963. Enzyme histochemistry of human lung tumours. *J. Pathol. Bacteriol.*, **85**, 279–80.

Wilson, I. B., Hatch, M. A. and Ginsburg, S. 1960. Carbamylation of acetylcholinesterase. *J. Biol. Chem.*, **235**, 2312–15.

Wolman, M. 1955. Problems of fixation in cytology, histology and histochemistry. *Int. Rev. Cytol.*, **4**, 79–102.

Wood, T. 1973. NADPH-D-glyceraldehyde 3-phosphate oxidoreductase activity in muscle and other tissues of the rat. *Biochem. Biophys. Res. Commun.*, **54**, 176–81.

Wood, T. 1985. *The Pentose Phosphate Pathway*. Academic Press, New York.

Wood, T. 1986a. Distribution of the pentose phosphate pathway in living organisms. *Cell Biochem. Funct.*, **4**, 235–40.

Wood, T. 1986b. Physiological functions of the pentose phosphate pathway. *Cell Biochem. Funct.*, **4**, 241–7.

Woohsmann, H. and Hardrodt, W. 1965. Der Nachweis einer phosphatempfindlichen Sulfatase mit Naphthol-AS-sulfaten. *Histochemie.*, **4**, 336–44.

Yoshida, A. 1966. Glucose 6-phosphate dehydrogenase of human erythrocytes. I. Purification and characterization of normal (B^+) enzymes. *J. Biol. Chem.*, **241**, 4966–76.

Zaman, G. and Chayen, J. 1981. An aqueous mounting medium. *J. Clin. Pathol.*, **34**, 567–8.

Zekri, M., Harb, J., Bernard, S. and Meflah, K. 1988. Purification of bovine liver cytosolic 5′-nucleotidase: kinetic and structural studies as compared with the membrane isoenzyme. *Eur. J. Biochem.*, **172**, 93–9.

Zemel, E. and Nahir, A. M. 1989. Uridine diphosphoglucose dehydrogenase activity in normal and osteoarthritic human chondrocytes. *J. Rheumatol.*, **16**, 825–7.

Zoller, L. C. and Weisz, J. 1980. A demonstration of regional differences in lysosomal membrane permeability in the membrana granulosa of the Graafian follicles in cryostat sections of the rat ovary: a quantitative cytochemical study. *Endocrinology*, **80**, 871–7.

Appendix 1

Effect of Fixation on Enzymes

Type	Time	Temperature (°C)	Other treatment	Final activity (% of unfixed)	Reference
Alkaline phosphatase					
Formalin					
Buffered	15 min	4	—	76	Nachlas *et al.*, 1956
Buffered	30 min	4	—	71	Nachlas *et al.*, 1956
Buffered	60 min	4	—	80	Nachlas *et al.*, 1956
Buffered	2 h	4	—	53	Nachlas *et al.*, 1956
Not specified	2–4 h	4	—	75	Pearse, 1960
10% pH 7	2 h	4	—	73	Seligman *et al.*, 1951
10% pH 7	24 h	4	—	26	Seligman *et al.*, 1951
Buffered	30 min	25	—	52	Nachlas *et al.*, 1956
10% pH 7	2 h	25	—	35	Seligman *et al.*, 1951
Buffered	30 min	37	—	14	Nachlas *et al.*, 1956
10% pH 7	2 h	37	—	13	Seligman *et al.*, 1951
Formol–saline					
10%	2 h	—	—	9–23	Emmel, 1946
10%	48 h	—	—	0–24	Emmel, 1946
Acetone					
Absolute	15 min	4	—	85	Nachlas *et al.*, 1956
Absolute	30 min	4	—	95	Nachlas *et al.*, 1956
Absolute	60 min	4	—	107	Nachlas *et al.*, 1956
Absolute	2 h	4	—	101	Nachlas *et al.*, 1956
Absolute	24 h	4	—	70	Pearse, 1960
Absolute	28 h	Cold	—	71	Stafford and Atkinson, 1948
Absolute	50 h	Cold	—	65	Stafford and Atkinson, 1948
Absolute	72 h	Cold	—	64	Stafford and Atkinson, 1948
Absolute	30 min	25	—	80	Nachlas *et al.*, 1956
Absolute	30 min	37	—	79	Nachlas *et al.*, 1956
80%	24 h	Cold	—	75	Doyle, 1948
Absolute	28–72 h	Cold	Paraffin 2 h, 56 °C	29	Stafford and Atkinson, 1948
Absolute or 80%	24 h	—	Paraffin 1 h, 60 °C	28	Doyle, 1948

(*continued*)

Type	Time	Temperature (°C)	Other treatment	Final activity (% of unfixed)	Reference
Ethanol					
Absolute	15 min	4	—	53	Nachlas *et al.*, 1956
Absolute	30 min	4	—	62	Nachlas *et al.*, 1956
Absolute	60 min	4	—	57	Nachlas *et al.*, 1956
Absolute	2 h	4	—	66	Nachlas *et al.*, 1956
Absolute	30 min	25	—	38	Nachlas *et al.*, 1956
Absolute	30 min	37	—	27	Nachlas *et al.*, 1956
Absolute	24 h	—	—	75	Doyle, 1948
80%	24 h	—	—	75	Doyle, 1948
80%	28 h	Cold	—	74	Stafford and Atkinson, 1948
80%	50 h	Cold	—	88	Stafford and Atkinson, 1948
80%	72 h	Cold	—	91	Stafford and Atkinson, 1948
Absolute or 80%	24 h	—	Paraffin 1 h, 60 °C	40	Doyle, 1948
80%	28–72 h	Cold	Paraffin 2 h, 56 °C	20	Stafford and Atkinson, 1948
Freeze-drying					
—	—	—	—	85	Doyle, 1948
—	—	—	—	87	Berenbom *et al.*, 1952
—	—	—	Paraffin ½ h, 60°C	29	Berenbom *et al.*, 1952
—	—	—	Polyethylene glycol; then alcohol fixation	90	Fredricsson, 1958
5-Nucleotidase					
Formalin					
Not specified	24 h	Cold	—	40–60	Barka and Anderson, 1963
Acetone					
Absolute	Not specified	Cold	Paraffin 20–40 min, 56 °C	47	Novikoff, 1952
Acid phosphatase					
Formalin					
Buffered	15 min	4	—	67	Nachlas *et al.*, 1956
Buffered	30 min	4	—	58	Nachlas *et al.*, 1956
Buffered	60 min	4	—	67	Nachlas *et al.*, 1956
Buffered	2 h	4	—	42	Nachlas *et al.*, 1956
10% pH 7	2 h	4	—	79	Seligman *et al.*, 1951
Formol–calcium	24 h	2	—	8	Holt, 1959
10% pH 7	24 h	4	—	55	Seligman *et al.*, 1951
10% pH 7	48 h	4	—	21	Seligman *et al.*, 1951
Buffered	30 min	25	—	47	Nachlas *et al.*, 1956
10% pH 7	2 h	25	—	40	Seligman *et al.*, 1951

Type	Time	Temperature (°C)	Other treatment	Final activity (% of unfixed)	Reference
Formalin (contd)					
Buffered	30 min	37	—	24	Nachlas *et al.*, 1956
10% pH 7	2 h	37	—	27	Seligman *et al.*, 1951
Formol–calcium	24 h	2	Washed 12 h in running water	38–41	Holt, 1959
Formol–calcium	24 h	2	Gum-sucrose, 24 h, 2 °C	40–43	Holt, 1959
Formol–calcium	24 h	2	Gum-sucrose, 7 days, 2 °C	55–60	Holt, 1959
Formol–calcium	24 h	2	Gum-sucrose, 30 days, 2 °C	55–57	Holt, 1959
Formol–calcium	24 h	2	Gum-sucrose, 2 year, 2 °C	42–43	Holt, 1959
Formol–calcium	24 h	2	Alcohol 50%, 70%, 90%, then absolute each 30 min; toluene; paraffin, two changes at 45 °C; then dewaxed in benzene	1.7–1.8	Holt, 1959
Formol–saline					
10%	14 h	—	—	30	Emmel, 1946
10%	41 h	—	—	20	Emmel, 1946
Acetone					
Absolute	15 min	4	—	80	Nachlas *et al.*, 1956
Absolute	30 min	4	—	77	Nachlas *et al.*, 1956
Absolute	60 min	4	—	81	Nachlas *et al.*, 1956
Absolute	2 h	4	—	82	Nachlas *et al.*, 1956
Absolute	7 h	Cold	—	28	Stafford and Atkinson, 1948
Absolute	24 h	2	—	74–76	Holt, 1959
Absolute	28 h	Cold	—	20	Stafford and Atkinson, 1948
Absolute	50 h	Cold	—	22	Stafford and Atkinson, 1948
Absolute	72 h	Cold	—	40	Stafford and Atkinson, 1948
Absolute	72 h	Cold	—	25	Doyle, 1948
Absolute	30 min	25	—	67	Nachlas *et al.*, 1956
Absolute	30 min	37	—	70	Nachlas *et al.*, 1956
Absolute	7–72 h	Cold	Paraffin, 2 h, 56 °C	4	Stafford and Atkinson, 1948
Absolute	24 h	2	Paraffin two changes, 30 min, 45 °C, then dewaxed in benzene	20–28	Holt, 1959

(*continued*)

Type	Time	Temperature (°C)	Other treatment	Final activity (% of unfixed)	Reference
Ethanol					
Absolute	15 min	4	—	50	Nachlas *et al.*, 1956
Absolute	30 min	4	—	41	Nachlas *et al.*, 1956
Absolute	60 min	4	—	50	Nachlas *et al.*, 1956
Absolute	2 h	4	—	38	Nachlas *et al.*, 1956
Absolute or 80%	7 h	Cold	—	29	Stafford and Atkinson, 1948
Absolute or 80%	24 h	Cold	—	2	Doyle, 1948
Absolute or 80%	28 h	Cold	—	18	Stafford and Atkinson, 1948
Absolute or 80%	50 h	Cold	—	20	Stafford and Atkinson, 1948
Absolute or 80%	72 h	Cold	—	26	Stafford and Atkinson, 1948
Absolute	30 min	25	—	32	Nachlas *et al.*, 1956
Absolute	30 min	37	—	22	Nachlas *et al.*, 1956
Absolute or 80%	7–72 h	Cold	Paraffin 2 h, 56 °C	7	Stafford and Atkinson, 1948
Freeze-drying					
—	—	—	—	85	Doyle, 1948
—	—	—	Paraffin 60 °C, ½ h	13	Berenbom *et al.*, 1952
—	—	—	Paraffin 45 °C, 3 min, then dewaxed	98	Holt, 1959
β-Glucuronidase					
Formalin					
Buffered	15 min	4	—	38	Nachlas *et al.*, 1956
Buffered	30 min	4	—	23	Nachlas *et al.*, 1956
Buffered	60 min	4	—	16	Nachlas *et al.*, 1956
Buffered	2 h	4	—	4	Nachlas *et al.*, 1956
Buffered	30 min	25	—	6	Nachlas *et al.*, 1956
Buffered	30 min	37	—	2	Nachlas *et al.*, 1956
Acetone					
Absolute	15 min	4	—	76	Nachlas *et al.*, 1956
Absolute	30 min	4	—	73	Nachlas *et al.*, 1956
Absolute	60 min	4	—	60	Nachlas *et al.*, 1956
Absolute	2 h	4	—	71	Nachlas *et al.*, 1956
Absolute	30 min	25	—	80	Nachlas *et al.*, 1956
Absolute	30 min	37	—	74	Nachlas *et al.*, 1956
Absolute	24 h	4	Paraffin embedding	0	Pearse, 1960
Ethanol					
Absolute	15 min	4	—	33	Nachlas *et al.*, 1956
Absolute	30 min	4	—	26	Nachlas *et al.*, 1956
Absolute	60 min	4	—	30	Nachlas *et al.*, 1956
Absolute	2 h	4	—	14	Nachlas *et al.*, 1956
Absolute	30 min	25	—	9	Nachlas *et al.*, 1956
Absolute	30 min	37	—	1	Nachlas *et al.*, 1956

Type	Time	Temperature (°C)	Other treatment	Final activity (% of unfixed)	Reference
Esterase					
Formalin					
Buffered	15 min	4	—	45	Nachlas *et al.*, 1956
Buffered	30 min	4	—	54	Nachlas *et al.*, 1956
Buffered	60 min	4	—	43	Nachlas *et al.*, 1956
Buffered	2 h	4	—	15	Nachlas *et al.*, 1956
Formol–calcium	24 h	2	—	20–24	Holt *et al.*, 1960
Buffered	30 min	25	—	35	Nachlas *et al.*, 1956
Buffered	30 min	37	—	22	Nachlas *et al.*, 1956
Formol–calcium	24 h	2	Washed 12 h in running water	39–42	Holt *et al.*, 1960
Formol–calcium	24 h	2	Gum-sucrose, 24 h, 2 °C	48–52	Holt *et al.*, 1960
Formol–calcium	24 h	2	Gum-sucrose, 7 days, 2 °C	55–62	Holt *et al.*, 1960
Formol–calcium	24 h	2	Gum-sucrose, 30 days, 2 °C	50–56	Holt *et al.*, 1960
Formol–calcium	24 h	2	Gum-sucrose, 2 years, 2 °C	33–34	Holt *et al.*, 1960
Formol–calcium	24 h	2	Alcohol 50%, 70%, 90% then absolute each 30 min; two changes of paraffin at 45 °C, then dewaxed in benzene	4.6–7	Holt *et al.*, 1960
Acetone					
Absolute	15 min	4	—	61	Nachlas *et al.*, 1956
Absolute	30 min	4	—	68	Nachlas *et al.*, 1956
Absolute	60 min	4	—	62	Nachlas *et al.*, 1956
Absolute	2 h	4	—	59	Nachlas *et al.*, 1956
Absolute	24 h	2	—	62–70	Holt *et al.*, 1960
Absolute	24 h	2	Toluene; two changes of paraffin at 45 °C then dewaxed in benzene	26–30	Holt *et al.*, 1960
Ethanol					
Absolute	24 h	Cold	—	0	Nachlas and Seligman, 1949
Freeze-drying					
—	—	—	Paraffin 5 min, 45 °C, benzene, four changes, 30 min each at room temp	99	Holt *et al.*, 1960

(*continued*)

Type	Time	Temperature (°C)	Other treatment	Final activity (% of unfixed)	Reference
Freeze-drying (contd)					
—	—	—	—	91	Berenbom *et al.*, 1952
—	—	—	Paraffin 60 °C, ½ h	10	Berenbom *et al.*, 1952
Leucine aminopeptidase					
Formalin					
Buffered	15 min	4	—	90	Nachlas *et al.*, 1956
Buffered	30 min	4	—	82	Nachlas *et al.*, 1956
Buffered	60 min	4	—	85	Nachlas *et al.*, 1956
Buffered	2 h	4	—	54	Nachlas *et al.*, 1956
Buffered	30 min	25	—	63	Nachlas *et al.*, 1956
Buffered	30 min	37	—	14	Nachlas *et al.*, 1956
Acetone					
Absolute	15 min	4	—	90	Nachlas *et al.*, 1956
Absolute	30 min	4	—	76	Nachlas *et al.*, 1956
Absolute	60 min	4	—	84	Nachlas *et al.*, 1956
Absolute	2 h	4	—	79	Nachlas *et al.*, 1956
Absolute	30 min	25	—	80	Nachlas *et al.*, 1956
Absolute	30 min	37	—	77	Nachlas *et al.*, 1956
Ethanol					
Absolute	15 min	4	—	0	Nachlas *et al.*, 1956

Other enzymes

Enzyme	Fixation	Final activity	Reference
DPNH-tetrazolium reductase	Cold formol–calcium overnight	5–25	Novikoff, 1963
TPNH-tetrazolium reductase	Cold formol–calcium overnight	4–10	Novikoff, 1963
DPNH-cytochrome *c* reductase	Cold formol–calcium overnight	1–6	Novikoff, 1963
Cytochrome oxidase	Freeze-drying only	56	Berenbom *et al.*, 1952
Cytochrome oxidase	Freeze-drying, then embedded in paraffin ½ h, 60 °C	0	Berenbom *et al.*, 1952
Cytochrome oxidase	Formalin, 2–4 h, 4 °C	0	Pearse, 1960
Cytochrome oxidase	Acetone 24 h, 4 °C	traces	Pearse, 1960
Cytochrome oxidase	Cold acetone + embedding	0	Pearse, 1960
Succinic dehydrogenase	Formalin 2–4 h, 4 °C	0	Pearse, 1960
Succinic dehydrogenase	Acetone 24 h, 4 °C	60	Pearse, 1960
Succinic dehydrogenase	Cold acetone + embedding	0	Pearse, 1960
Succinoxidase	Freeze-drying only	37	Berenbom *et al.*, 1952
Succinoxidase	Freeze-drying, then embedded in paraffin ½ h, 60 °C	0	Berenbom *et al.*, 1952

Effects of glutaraldehyde fixation

Enzyme	Location	Final activity (% of unfixed) in: Glutaraldehyde, Distilled	Glutaraldehyde, Charcoal purified	Glutaraldehyde, Stock	Formaldehyde
Lactate dehydrogenase	Supernatant	10	9	8	20
Lactate dehydrogenase	Sediment	12	5	4	12
α-Hydroxybutyrate dehydrogenase	Supernatant	30	28	20	40
Isocitrate dehydrogenase	Supernatant	18	15	12	20
Aspartate aminotransferase	Supernatant	32	20	12	53
Alanine aminotransferase	Supernatant	48	40	28	40
Creatine phosphokinase	Supernatant	82	55	32	90
Creatine phosphokinase	Sediment	12	5	4	12
Cholinesterase	Supernatant	72	50	24	70
Cholinesterase	Sediment	75	45	20	65
Acid phosphatase	Supernatant	40	25	18	42
Acid phosphatase	Sediment	40	35	20	50

Tissue: Skeletal muscle mince (Anderson, 1967).
Fixative: 4% glutaraldehyde or 4% formaldehyde in 0.135 M Sörensen buffer, pH 7.3.
Conditions: 5 g minced muscle in 100 ml test solution for 1 h at 4 °C then washed three times in distilled water.

Type of glutaraldehyde	Final activity (% of unfixed) Free	Total
Distilled	43	33
Distilled (heated)	44	33
Union carbide	38	28
Fluka	33	24
Fisher (biological grade)	16	13

Tissue: Rat liver homogenate in 0.25 M sucrose (Fahimi and Drochmans, 1968).
Fixative: 5% glutaraldehyde in water.
Conditions: 4 ml homogenate and 1 ml fixative for 30 min at 4 °C.
Enzyme: Acid phosphatase.

Enzyme	Time of fixation	Time of washing	Final activity (% of unfixed)
Acid phosphatase	2 min perfusion	Perfusion rinse	30
Acid phosphatase	5 min perfusion	Perfusion rinse	12
Acid phosphatase	5 min perfusion 1 h immersion	—	10
Acid phosphatase	5 min perfusion 24 h immersion	24 h	10
Acid phosphatase	5 min perfusion 24 h immersion	7 days	12
Aryl sulphatase	5 min perfusion	Perfusion rinse	30
Aryl sulphatase	5 min perfusion 24 h immersion	—	30
Aryl sulphatase	5 min perfusion 24 h immersion	24 h	53
Aryl sulphatase	5 min perfusion 24 h immersion	7 days	60

Tissue: Rat kidney (Arborgh *et al.*, 1971).
Fixative: 1.5% glutaraldehyde (distilled) in Sörensen phosphate buffer, pH 7.4.
Conditions: On anaesthetized animals, kidney perfused for 30 s with isotonic saline, and then with fixative for 5 min at 20 °C. Then small slices (1 mm thick) immersed at 0–4 °C washed with 0.1 M Tris–maleate buffer, pH 7.4.

Type of adenosine triphosphatase	Tissue	Fixative	Time	Temperature (°C)	Final activity (% of unfixed)	Reference
Na⁺-K⁺ activated	Kidney membrane fractions	Glutaraldehyde	Brief	0	0	Moses et al., 1966
Na⁺-K⁺ activated	Kidney membrane fractions	Formaldehyde	Brief	0	0	Moses et al., 1966
Na⁺-K⁺ activated	Duck salt gland or teleost gill	Buffered glutaraldehyde 0.5%, pH 7.2	10 min	4	>10	Ernst and Philpott, 1970
Na⁺-K⁺ activated	Duck salt gland or teleost gill	Buffered formaldehyde 2%, pH 7.2	30 min	4	100*	Ernst and Philpott, 1970
Na⁺-K⁺ activated	Duck salt gland or teleost gill	Buffered formaldehyde 2%, pH 7.2	90 min	4	65	Ernst and Philpott, 1970
Mg⁺ activated	Duck salt gland or teleost gill	Buffered glutaraldehyde 0.5%, pH 7.2	60 min	4	15	Ernst and Philpott, 1970
Mg⁺ activated	Duck salt gland or teleost gill	Buffered formaldehyde 2%, pH 7.2	60 min	4	25–30*	Ernst and Philpott, 1970
Mg⁺ activated	Toad gastric mucosa	Buffered glutaraldehyde 6.25%, pH 7.4	2 h	—	35	Koenig and Vial, 1970

*It is difficult to reconcile these results

Appendix 2

Buffers

0.1 M phosphate buffer, pH 5.6–8.0

pH	A (ml)	B (ml)
5.6	5.5	94.5
5.8	8.0	92.0
6.0	12.0	88.0
6.2	18.0	82.0
6.4	28.0	72.0
6.6	38.0	62.0
6.8	50.0	50.0
7.0	62.0	38.0
7.2	72.0	28.0
7.4	82.0	18.0
7.6	88.0	12.0
7.8	92.0	8.0
8.0	94.5	5.5

A: 0.1 M disodium hydrogen orthophosphate, anyhdrous
Na₂HPO₄ (molecular weight 141.96).
B: 0.1 M potassium dihydrogen orthophosphate, KH₂PO₄
(molecular weight 136.09).

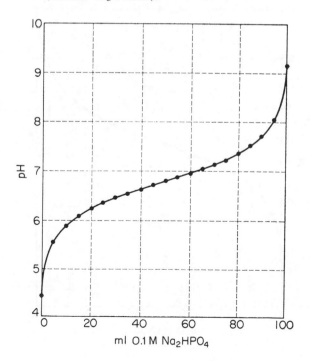

0.05 M glycylglycine buffer, pH 7.0–9.6

pH	A (ml)	B (ml)
7.0	50.0	2.5
7.2	50.0	4.0
7.4	50.0	6.0
7.6	50.0	9.0
7.8	50.0	12.5
8.0	50.0	17.0
8.2	50.0	23.0
8.4	50.0	28.0
8.6	50.0	33.0
8.8	50.0	37.5
9.0	50.0	40.5
9.2	50.0	43.0
9.4	50.0	45.0
9.6	50.0	46.0

A: 0.1 M glycylglycine (molecular weight 132.12).
B: 0.1 M sodium hydroxide, NaOH (molecular weight 40.00).
Made up to 100 ml with distilled water.

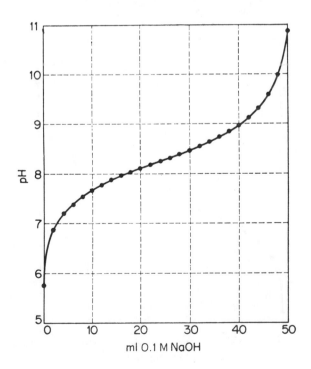

0.05 M Tris buffer, pH 7.0–9.2

pH	A (ml)	B (ml)
7.0	25.0	45.5
7.2	25.0	44.0
7.4	25.0	42.0
7.6	25.0	38.5
7.8	25.0	34.5
8.0	25.0	29.0
8.2	25.0	23.0
8.4	25.0	17.5
8.6	25.0	13.0
8.8	25.0	9.0
9.0	25.0	6.5
9.2	25.0	4.0

A: 0.2 M Tris-(hydroxymethyl)-aminomethane (molecular weight
   121.14).
B: 0.1 N hydrochloric acid.
Made up to 100 ml with distilled water.

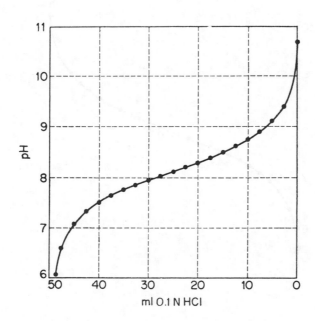

0.05 M barbitone buffer, pH 6.8–9.0

pH	A (ml)	B (ml)
6.8	50.0	45.5
7.0	50.0	43.5
7.2	50.0	41.0
7.4	50.0	37.0
7.6	50.0	32.5
7.8	50.0	27.5
8.0	50.0	22.5
8.2	50.0	17.5
8.4	50.0	13.0
8.6	50.0	9.5
8.8	50.0	7.0
9.0	50.0	5.0

A: 0.1 M barbitone-sodium (sodium 5,5-diethylbarbiturate)
   (molecular weight 206.18).
B: 0.1 N hydrochloric acid.
Made up to 100 ml with distilled water.

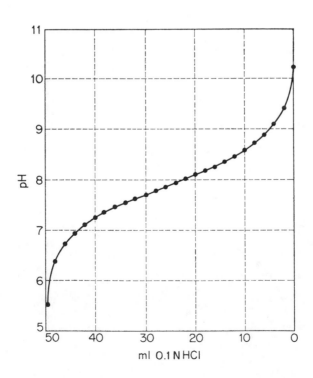

McIlvaine buffer, pH 2.2–7.8

pH	A (ml)	B (ml)
2.2	98.0	2.0
2.4	93.5	6.5
2.6	89.0	11.0
2.8	84.0	16.0
3.0	79.0	21.0
3.2	75.0	25.0
3.4	71.0	29.0
3.6	67.5	32.5
3.8	64.0	36.0
4.0	60.5	39.5
4.2	58.0	42.0
4.4	55.5	44.5
4.6	52.5	47.5
4.8	50.0	50.0
5.0	48.0	52.0
5.2	46.0	54.0
5.4	44.0	56.0
5.6	42.0	58.0
5.8	40.0	60.0
6.0	37.0	63.0
6.2	34.0	66.0
6.4	31.0	69.0
6.6	27.5	72.5
6.8	23.0	77.0
7.0	18.0	82.0
7.2	13.5	86.5
7.4	10.0	90.0
7.6	7.0	93.0
7.8	5.0	95.0

A: 0.1 M citric acid (molecular weight 210.14)
B: 0.2 M disodium hydrogen orthophosphate, anhydrous,
   Na_2HPO_4 (molecular weight 141.96).

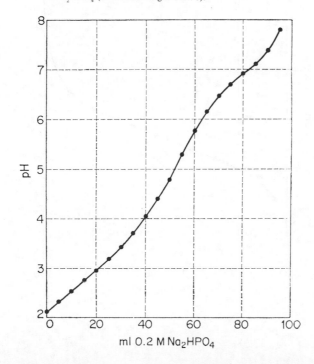

0.2 M acetate buffer, pH 3.6–5.6

pH	A (ml)	B (ml)
3.6	8.0	92.0
3.8	13.0	87.0
4.0	20.0	80.0
4.2	30.0	70.0
4.4	40.0	60.0
4.6	50.0	50.0
4.8	60.0	40.0
5.0	70.0	30.0
5.2	80.0	20.0
5.4	86.0	14.0
5.6	90.0	10.0

A: 0.2 M sodium acetate, anhydrous (molecular weight 82.03).
B: 0.2 M acetic acid (molecular weight 60.00).

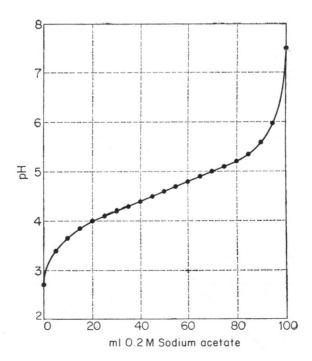

Index

Note: detailed methods are given in **bold** type.

Acetazolamide, inhibition of carbonic anhydrase, 205, 207
Acetic anhydride, for acetylation, 105
Acetic–ethanol fixatives, 36–37
Acetonitrile, for benzoylation, 104
Acetylation, for PAS reaction, **105**
Acetylcholinesterase, 177–178, 179
  method, **183–185**
Acetyl β-glucosaminidase
  biochemical background, 198–199
  control, 200
  histochemical background, 199
  measurement, 201
  method, **199–201**
Acid haematein, method for phospholipids, **118–121**
Acidic polysaccharides, stain for, 108
Acidic moieties, 94, 97
Acidity, test for, 65
Acid phosphatase, 151–154
  azo-dye method, **154–156**
  biochemical background, 151
  fragility test, **154**
  histochemical background, 151–152
  in lysosomes, 152, 154
  in macrophages, 152
  in osteoclasts, 152
  inhibition by fluoride, 151, **153**
  latency of, 154
  lead phosphate method, **152–154**
  post-coupling method, **156–157**
  prostatic, 151
Acridine orange, for DNA and RNA, 98
  false positive reactions, 99

response of cartilage, 99
response with keratin, 99
staining of mast cells, 99
Acyl coenzyme A, 250
Acyl dehydrogenase, 250
Adenosine diphosphate, and mitochondrial respiration, 226
Adenosine monophosphate, activator of phosphorylase a, 201–202
Adenosine triphosphatase
  activation by Ca^{2+} and dinitrophenol, 165
  biochemical background, 164
  histochemical background, 164–165
  importance in open-heart surgery, 165
  inhibition by conventional methods, 164
  method, **165–166**
Adenosine triphosphatase (calcium-activated)
  biochemical background, 169–170
  histochemical background, 170
  in calcium transport, 169–170
  in erythrocytes, 170
  in various tissues, 170
  inhibition by vanadate, 169–170
  method, **170–171**
  special colloid stabilizer for, 171
Adenosine triphosphatase (magnesium activated) (also Na^+-K^+-ATPase)
  biochemical background, 166–167
  hidden-lead trapping agent for, 167
  histochemical background, 167
  inhibition by ouabain, 167
  instability of, 167
  LACA method for, **167–169**

307

Adenosine triphosphatase (*cont.*)
  mitochondrial, 166–167
  skeletal muscle, 166–167
Adenosine triphosphate, formation of, 226
Adenosine 5'-phosphate, 158
Adenylate cyclase, 158
Adhesion of sections to slide, 8
Adrenaline, and DOPA-oxidase, 217
  for monoamine oxidase activity, 220
  structure of, 220
Adrenochrome, 217
Alcian blue
  definition, 106
  for oriented proteoglycans, 33
  induced birefringence of, 33
  method, **105–107**
Aldehydes, and the PAS reaction, 103
Aldolase activity, 242, 243
Alizarin red, for calcium, 123
Alkaline phosphatase, 145 *et seq.*
  biochemical background, 145–146
  electron microscopy of, 146
  Gomori–Takamatsu method for, **147–149**
  histochemical background, 146–147
  in clinical chemistry, 146
  in haematology, 146
  in liver biopsies, 146
  in membranes, 146
  in various tissues, 147
  inhibition by zinc, 145
  inhibitors of, 147
  intestinal, 145
  intestinal and placental, 147
  localization of, 148
  naphthol phosphate method, **149–150**
  nature of, 146
  rationale of histochemical methods, 147
Amido black, quantitative method for
    proteins, 59–61
Aminopeptidases
  biochemical background, 188
  effect of amidase, 188
  effect of leucine amide, 188
  of kidney, 188
  also see Naphthylamidases
Amphetamine, and monoamine oxidase, 219
  structure of, 221
  to inhibit histochemical reaction, 223
Amylases, to confirm PAS reaction, 111
Amyloid, method for, 43–44
Amytal, and hydrogen-transport chain, 226

Analysis of tissue components
  general, 46–47
  of acidic moieties, 49–51
  of carbohydrates
  of proteins, 47–49
Antimycin A, and hydrogen-transport
    chain, 226
Anti-inflammatory drugs, 189, 190
Anti-roll plate, adjustment, 6
Apatite, in bone, 30
Apurinic acid, 79
Arginine
  Fast green method for, **63–64**
  formula and pK, 65
  Sakaguchi test for, **66–68**
Arsenite, and hydrogen–transport chain,
    226
Aryl sulphatases
  biochemical background, 207
  histochemical background, 208
Arylamidases, *see* Naphthylamidases
Ascorbate, oxidation by DOPA-oxidase,
    218
  methods for, **125–126**
Aspirin, effect on G6PD, 258
Autofluorescence
  carotenoids, 122
  problems with, 126
Autofluorescent substances, 126–127
Automatic cutting of section, 7–8
Autoradiographs, staining methods, 40
Autoradiography
  developer for, **134**
  dipping film method, **133–134**
  of fixed compounds, 129
  localization by, 129
  methods for, 128–134
  resolving power of, 130
  of soluble compounds, 129
  stripping film method, **131–133**
Azure B, for metachromasy of RNA, 97

Basophilia
  check on, 93
  toluidine blue test of, **95**
Beer–Lambert Law, 22, 24–25, 70
Benzedrine
  structure of, 221
Benzoylation, for PAS reaction, **104**
Benzylamine oxidase, 220
  in histochemistry, 223

Benzypyrene, for lipids, **116–118**
Best's carmine, method for glycogen, **109–111**
Bioassay, of gastrin, 206
Biosynthetic 'hydrogen', 232
Birefringence
 definition, 30
 induced birefringence of proteoglycans, 107
 induced form of, 32–33
 of steroids, 121
 types of, 30–31
Borohydride, to disclose − SH, 73–74
Bound enzymes, 143–144
Bromination, for acid haematein test, 121
Bromotetramisole to inhibit alkaline phosphatase, 147
Buffers, McIlvaine's (pH 8), 64

Calcium activated ATPase, see Adenosine triphosphatase (calcium-activated)
Calcium ATPase, see under Adenosine triphosphatase (calcium-activated)
Calcium, detection of, **123–124**
Calcium ions, for stabilizing cells, 16
Calcium oxalate, 123
Calcium pretreatment of sections, 166
Calculation
 of hydrogen liberated, 21
 of optical retardation, 32
Callose, 100
Capture reactions, 142
Carbohydrate
 method for, **99–105**
 proof of, 103
 also see Polysaccharides
Carbohydrates, screening tests for, 51
Carbonic anhydrase
 biochemical background, 205–206
 cobalt in, 205
 functions of, 205
 and gastrin, 206
 histochemical background, 206
 in chloroplasts, 205
 inhibition by acetazolamide, 205
 method, **206–207**
 zinc in, 205
Carboxylesterase, distinction from other esterases, 180
Carotenoids, methods for, 121–122

Cartilage, and the glyoxylate cycle, 244
Catecholamines, induced fluorescence methods, 127–128
Cathepsin-C, 187
Cathepsin-H, 188
Cell damage
 enzymatic marker of, 244
 glutamate dehydrogenase in, 248–249
Cell membranes, marker of, 158
Celloidin, solution, 67
Cells
 cytochemical studies on, 15, 144
 method for, 15–16
Cellulose
 by polarized light, 32
 staining, 100
Cell-wall constituents, 100
Cerebrosides
 and PAS reaction, 101
 screening test for, 51
Ceroid, autofluorescence, 127
Cervical cancer, 194, 259
Chemical fixation, general, see under Fixation
Chilling
 cooling curves, 3–4
 of delicate tissues, 36
 method, 4, 5
 pretreatment for, 3, 36
 problems in, 2–3
Chloroplasts, and carbonic anhydrase activity, 205
Cholesterol, in the test for steroids, 121
Cholinesterase, 177–178, 179
 inhibition by ISO-OMPA, 184
 method, **183–185**
 rationale of method, 183
Chymotrypsin, 187
Chymotrypsin, and phosphamidase, 172
 inhibition of, 176–177
Classification of enzymes, 145
Clorgyline
 and MAO activity, 220
 use in histochemistry, 223
Collagen
 induced birefringence of, 61
 picrosirius red method for, 61–63
 van Gieson stain for, 40–41
Collagen polypeptides
 CO_2-free, 262
 for pentose-shunt dehydrogenases, 260

Collagen polypeptides (*cont.*)
  to stabilize sections, 234, 259
  also see Colloid stabilizers
Colloid stabilizers
  degraded collagen, 14, 137, 144, 234
  Ficoll, 14
  need for, 13
  Polypep 8350, 210–211
  polyvinyl alcohol, 14, 137, 144, 234
  selection of, 14
  test of efficacy, 13–14
  types of, 14
Congo red, for staining cellulose, 100
Coplin jar, reactions in, 9–10
Critical electrolyte method for proteoglycans, 105
Crystal violet, structure, 87
Crystalline fats, studied by polarized light, 32
Crystals
  birefringence of, 30
  characteristics of, 31–32
Cuticular substances, 100
Cyanide, inhibition of cytochrome, 226
Cysteine, formula and pK, 65
Cytochemical bioassay of gastrin, 206
Cytochemistry, and biochemical analysis, 195
Cytochrome *b*, in mitochondrial transport of hydrogen, 226
Cytochrome *c*
  in hydrogen transport, 226
  oxidase, 212–213
  structure of, 212
Cytochrome oxidase
  biochemical background, 211–213
  definition of, 211–212
  histochemical background, 213–214
  and hydrogen transport chain, 226
  in cardiac muscle, 214
  in kidney, 214
  in rat liver, 214
  in tumours, 214
  inhibition by cyanide, 215
  inhibitors of, 213
  lipid in, 213
  method, **214–215**
  mitochondrial, 215
  structure of cytochrome *c*, 212
Cytochrome *P*-448, 263–264
Cytochrome *P*-450, 257, 263–264

Cytochromes
  in hydrogen transport chain, 226
  nature of, 212
Cytosolic dehydrogenases, problems with, 233–234

DDD method for proteins, 69 *et seq.*
Deamination, method, 64
Decarboxylases, *see* Ornithine decarboxylase
Dehydrogenases
  cytosolic location, 233–234
  definition, 225
  electrode potentials of, 225 *et seq.*
  formula of neotetrazolium, 230
  general histochemistry of, 230–232
  glucose 6-phosphate, 255
  glutamate, 247–249
  glyceraldehyde 3-phosphate, 241–243
  α-glycerophosphate, 239–241
  hydroxyacyl CoA, 249–252
  lactate, 243–247
  liberation of energy, 227–230
  and mitochondrial hydrogen transport chain, 225 *et seq.*
  mitochondrial location, 233
  6-phosphogluconate, 255–263
  redox of tetrazoles for, 231
  steroid, 254–255
  succinate, 234–239
  tetrazolium salts for, 230–231
  uridine diphosphoglucose, 252–254
  utilization of hydrogen, from, 225 *et seq.*
Deoxyribofuranose, 79
Deoxyribonuclease
  for removing DNA, 89–91, **90**
  tests to confirm specificity, **90**
Deoxyribonucleic acid
  acridine orange method, **93–99**
  deoxyribonuclease for, **89–91**
  Feulgen reactions for, 77–85, **83–85**
  hydrolysis profile graphs, 78
  in malignant cells, 83
  Kurnick's method for, **86–89**
  methyl green–pyronin method, **91–93**
  rate of synthesis of, 83
  synthesis profiles, 85–86
  tests for, 49, 50
Deprenyl
  and MAO activity, 220
  in histochemistry, 223

Detoxification, and *P*-450 system, 257–264
Diabetes, and glucose 6-phosphatase, 162
Diamines, oxidation of, 213
Diaphorases, 137
Diaphorases
    forms of, 263
    in non-stabilized section, 234, 259
    oxidation of NADPH, 263
    also see NADPH-diaphorase
Diastase, to confirm PAS test, 111
Dichlorophenolindophenol
    as hydrogen acceptor, 236
    E_0 of, 229
Diffusion gradients, 140–142
Difluoromethylornithine, 208, 210
Dihydroxyacetone phosphate, in glycolysis, 241
Dihydroxyacetone phosphate shuttle, 239
    in malignancy, 240
Dimedone
    for blocking aldehydes and ketones, 81
    for PAS reaction, 104
Dimethyl formamide, for elution from sections, 19
Dinitrofluorobenzene method for proteins, **54–56**
Disulphides
    determination of, 68–74
    Prussian blue method, 71, 75
DNA-synthesis index, **85–86**
Dopamine, induced fluorescence of, 127–128
DOPA-oxidase
    biochemical background, 215–217
    diagnostic value, 217
    effect of copper inhibitors, 216
    histochemical background, 217–218
    in cancer, 217–218
    in infective agent, 218
    in mast cells, 218
    in plant wounding, 216
    and melanins, 216–217
    method, **218–219**
    oxidation of adrenaline, 217
    oxidation of ascorbate, 218
    rationale of histochemical method, 218

Ebel's test for polyphosphates, **98**, 152
Electrode potential
    in liberation of energy, 227–230
    meaning of, 228
    of cytochrome *c*, 212

of hydrogen, 226
of mitochondrial systems, 225 *et seq.*
of oxygen, 226
of respiratory chain, 226
of tetrazolium salts, 226
redox indicators for, 229, 230
Elution method, 19–21
Embden–Meyerhof pathway, 235, 241
Endo- and exopeptidases, 187
Energy
    and formation of ATP, 228
    from phosphoamide bond, 172
    production of, 227–230
Enzymatic activity
    characteristics of reactions, 138–140
    diffusion gradients in, 140–142
    loss in untreated sections, 11–12
    use of colloid stabilizers, 137
Enzyme classification, 145
Enzymology
    diffusion gradients in, 140–142
    elements of, 135–144
    general, 230–234
    and substantivity, 140
    also see under Histochemistry of enzymes
Epinephrine
    stimulation of phosphorylase kinase, 203
    structure of, 221
Eserine, on esterases, 177–178
Esterases
    biochemical background, 174–180
    chain-length of substrate, 176
    histochemical background, 179–180
    indoxyl acetate method for, **181–183**
    inhibition of, 176–178
    inhibitors of, 180
    in lysosomes, 179
    in plant tissues, 180
    in tumours, 180
    naphthol acetate method for, **180–181**
    nomenclature, 178
    orientation of substrate, 176
    physical chemistry of, 174–176
    prognostic value of, 180
    properties, 179
    relation to phosphatases, 175
    stain for nervous tissue, 180
    also see Cholinesterase

FAD and succinate dehydrogenase, 234
Fast green, method for basic proteins, **63**

Fats, oil red method, **114–115**
Fatty acid oxidation, 235
  in cardiac muscle, 250
Fatty acids, and Nile blue method, 115
Fatty acids
  energy from, 250
  synthesis of, 257
  utilization of, 250
Ferric iron, method for, **122**
Ferricyanide methods for −SH and glutathione, 71–76
Ferrous iron, method for, **122**
Feulgen reaction, 77–91
  for DNA synthesis, 78
  for metabolic DNA, 78
  methods, 82–83, **83–85**
  plasmals, 81
  problems with stoichiometry, 78
  rationale, 78–80
  removal of purines, 79
  stoichiometry of, 78, 82
Fibrinolysin, as endopeptidase, 187
Fixation, and diffusion of activities, 142
Fixation
  dangers of, 2, 11–15
  inhibition by, 11–12
  inhibition of acetyl glucosaminidase, 199
  loss of soluble matter, 1, 12
  masking physiological changes, 11–13
Fixatives
  acetic–ethanol, 36–37
  formol–calcium, 37
  formol–saline, 37
  Heidenhain's Susa, 37
  Lewitsky's fluid, 67
  picric–formalin, 36
  picro-acetic, 36–37
Flavoprotein
  as acceptor of hydrogen, 225
  in mitochondrial hydrogen-transport, 226, 227
Fluorescence, induced in catecholamines, 127–128
Fluorescence method for DNA and RNA, **98–99**
  for lipids, 118
  for phospholipids, 116
Folic acid, autofluorescence, 127
Form birefringence, 30
Formazans
  absorption maxima and extinction, 20

elution from section, 19
influence of substantivity, 233
localization of, 232–234
of neotetrazolium chloride, 230
quantification, 19–21, 261
redox for production of, 231
Fractured neck of femur, *see under* Osteoporosis
Fragility test for lysosomes, 154
Free radicals, and NAD(P)H oxidase, 264
Freeze-sectioning
  basis of, 2–3
  recommended method, 2–8
Frozen sections, storage of, 44
Fuchsin, 101

Gangliosides, in disease, 198–199
  screening test for, 51
Gastrin, 206
Glucagon
  and phosphorylase kinase, 203
  in respiration, 235
Glucocorticoids, influencing glucose 6-phosphatase, 162
Glucose 6-phosphatase
  biochemical background, 162
  effect of glucocorticoids *in vivo*, 162
  histochemical background, 162–163
  histochemical method, **163**
  in hepatic glycogen-storage disease, 162
  influence of insulin, 162
Glucose 6-phosphate dehydrogenase
  biochemical background, 256–260
  gene on X-chromosome, 258
  histochemical background, 259–260
  in erythrocytes, 258
  in growth plate, 260
  in malignant cells, 260
  in malignant growths, 257
  in protein synthesis, 257
  in rheumatoid arthritis, 260
  inhibition by fatty acid esters, 258
  loss from non-stabilized sections, 137, 234
  method, **260–261**
  occurrence, 257
  oxygen insensitivity in malignant cells, 260
  reactivation by spermine, 258
  solubilization of, 12–13

β-Glucuronidase
  biochemical background, 193–194
  control, 198
  dissociation of, 194
  dual function of, 193
  effect of testosterone, 194
  histochemical background, 194–195
  in neoplasms, 194
  influence of hormones, 194
  inhibition by fixatives, 194
  measurement, 196
  methodological problems, 194–195
  modified method, **196–198**
  occurrence, 193–194
  post-coupling method, **195–196**
  specific inhibitor, 194
Glutamate, formula and pK, 65
Glutamate dehydrogenase
  biochemical background, 247–248
  deamination by, 248
  effect of disruption, 248
  histochemical background, 248–249
  indicator of cell damage, 248–249
  method, **249**
  NAD or NADP dependence, 248
  and transaminase reactions, 248
Glutathione, method for, **125–126**
  inhibition of DOPA-oxidase, 216
Glutathione, method for, **75–76**
Glyceraldehyde 3-phosphate dehydrogenase
  and aldolase activity, 242
  biochemical background, 241–242
  histochemical background, 242–243
  inhibition by iodoacetate, 241, 242
  link with pentose-phosphate pathway,
    241–242
  method, **242–243**
Glycerine jelly, 198
Glycerine-iodine, 205
Glycerol, alkaline, 68
α-Glycerophosphate dehydrogenase
  biochemical background, 239–240
  effect of thyroidectomy, 240
  histochemical background, 240–241
  in cytoplasm, 239
  in malignancy, 240
  in mitochondria, 239–240
  in shuttle, 239–240
  methods, **240–241**
  two forms of, 239
Glycogen, and PAS method, 101

Best's carmine for, 52, **109**
  in respiration, 235
  synthesis of, 201
Glycogen phosphorylase, 201
Glycolysis and pyruvate, 243
Glycolytic pathway, 235
Glycosaminoglycans
  Alcian blue method for, **105–107**
  staining procedures, 46–47
  and UDPG-dehydrogenase activity, 252
  also see Proteoglycans
Glyoxal, method for calcium, **123–124**
Glyoxylate cycle, 236, 244
  enzymes of, 244
Glyoxysomes, 244
G-nadi reaction, 213
Gordon and Sweet's method for reticulum,
  41
Gout, urate crystals in, 30
Gum acacia, for removing fixative, 195

Haematoxylin–eosin staining methods
  for autoradiographs, 40
  formol–calcium method, 38
  modified method, 39
  normal method, 38–39
Heart
  birefringence for assessing human
    transplants, 32
  fatty acid oxidation in, 250
Heidenhain's fixative, 37
Hemicellulose, 252
Heparan sulphate, staining for, 105
Hexane, for chilling, 4
Hexosaminidase, 198
  also see Acetyl β-glucosaminidase
Hexose monophosphate pathway, 256
  also see under Pentose phosphate pathway
Histamine
  formula, 220
  reaction with phthalaldehyde, 128
Histidine, formula and pK, 65
Histochemistry
  advantages of, 135–136
  comparison with biochemistry, 136
  definition, xiii
  effect of endogenous substrates, 143
  for assay of hormones, 136
  in open-heart surgery, 135
  nature of a 'good' reaction, 138

Histochemistry (*cont.*)
  of bound enzymes, 143–144
  validity of reactions, 143
Histochemistry of enzymes
  characteristics of reactions for activity,
    138–140
  equilibrium constant, 139
  general, 230–234
  of soluble enzymes, 136–138
  solubility product, 139–140
  also see under specific enzymes
Histone, staining by Fast green, 63
Hydrogen
  acceptors, 226
  potential and kinetic energy, 226 *et seq.*
  transport and electrode potentials, 225
    et seq.
  transport in mitochondria, 226
  transport mechanism, 226
  type I and type II, 231–232
  utilization of, 225
Hydrogen acceptors, electrode potentials
  of, 226
Hydroxyacyl CoA dehydrogenase
  biochemical background, 249–250
  histochemical background, 250–251
  method, **251–252**
Hydroxyapatite, in bone, 30
Hydroxyisobutyrate, 209, 211
Hydroxylation, 257
Hydroxylation, of drugs, 264
  and *P*-450 system, 263–264
Hydroxytryptamine
  for monoamine oxidase activity, 220
  formula of, 220
  induced fluorescence of, 127

Immunoassays, problems with, xiv
Immunohistochemistry
  for detecting proteins, 76–77
  problems with, xiv, 77
Incubation
  box, 11
  chamber, 10
  in open rings, 10–11
  methods for, 9–13
  use of PVA and calcium, 11–13
Induced birefringence, 32–33, 107
Induced fluorescence, 127–128
  intensification, 128

Inflammation, and lysosomal activity, 189
Inhomogeneity error, 23–25
Inositol-lipids, non-response to PAS, 101
Intrinsic birefringence, 30
Iodoacetate, for blocking −SH, 72
Iodoacetate, inhibition by, 241, 242
Iron
  masked, 123
  methods for, **122**
Iron haematoxylin, 43
Isocitrate dehydrogenase, 256
Isocitrate lyase
  in epiphyseal cartilage, 245
  in growth plate, 245
  of glyoxysomes, 244
Isoenzymes of lactate dehydrogenase,
    243–244
Isolated cells
  methods for, 15–16
  studies on, 15
Isotopes, radioactive, ranges of, 129
ISO-OMPA, for cholinesterase, 184

Janus green, 229

Keratan sulphate, staining for, 105–107
Ketones, and the PAS reaction, 103
Krebs' cycle, 235
Kurnick's methyl green method, **86–89**

Lactate dehydrogenase
  biochemical background, 243–244
  diagnostic significance of, 244
  general method, **241–245**
  and glyoxylate pathway, 244
  histochemical background, 244–245
  in glycolysis, 243
  inhibition by oxamate, 244
  in plants, 244
  isoenzymes of, 243–244, 245
  malonate as inhibitor, 244
  recommended method, **246–247**
  specificity, 244
  subunits of, 243–244
  use of cyanide, 245
  use of phenazine methosulphate, 245–247
Latency
  measurement of, 191
  of lysosomal activity, 190–191

Lauth's violet, 229
Lead ammonium citrate/acetate
  for ATPase method, 167–169
  in 5'-nucleotidase method, 159–162
  reagent for 'hidden' lead, 159
Lead hydroxyisobutyrate, 209, 211
  preparation of, 210
Leucocytes, reaction of, 213
Levamisole, to inhibit alkaline phosphatase,
  147
Levelling, dyes, 66
Lewitsky's fluid, 67
Liebermann–Burchardt test for steroids, 121
Lignin, 100
Lipases
  as esterases, 179
  method, **186**
  rationale of method, 185
Lipids
  benzpyrene method, **116–118**
  fluorescent methods, 118
  identification by benzpyrene, 52
  identification with Sudan black, 52,
    112–114
  Nile blue method, **115–116**
  Sudan black methods, **112–114**
Lipofuscins, autofluorescence, 127
Lipoproteins, definition, 111
Localization, problems with, 142
Long-acting thyroid-stimulating γ-globulin,
  190
Lugol's iodine, 43
  for cytochrome oxidase method, 214–215
  test for polysaccharides, **110–111**
  to remove mercury, 73
Lymphocytes, and nucleotidase activity,
  159
Lysine, formula and pK, 65
Lysosomal enzymes,
  acid phosphatase method, **152–154**
  effect of membrane, 144, 154
  freely available; bound, 144, 191
  naphthylamidase method, **192–193**
  naphthylamidase test, 188–193
  test for, 143–144
  test for latency, 190–191
Lysosomal membranes, labilization of, 190
Lysosomal storage disease, 198–199
Lysosomes
  acid phosphatase histochemistry, 151–152
  in plants, 152

influence of disease, 151
latency, 154
manifest activity, 151, 154
membrane, 143
of macrophages, 152
permeability of, 144, 154
reactions for, 143–144
size, 143, 151, 192
and thyroid stimulating hormone, 144

Macrophages, and acid phosphatase, 152
Malate synthetase
  in epiphyseal cartilage, 245
  and glyoxysomes, 244
  in growth plate, 245
Maleimide, for blocking – SH, 72
Malic enzyme, 255
Malignancy
  decreased phosphorylase activity in, 203
  detection of uterine cancers, 257, 259
  β-glucuronidase in, 194
  and naphthylamidase activity, 189
  and ornithine decarboxylase in, 208
  pentose shunt in, 257, 259
  and phosphamidase activity, 172
  succinate dehydrogenase in, 237
Mannans, 100
Marsilid, structure of, 220
Masked lipids, 112, 119
Mast cells
  and DOPA-oxidase, 218
  pyronin staining of, 92
  staining with acridine orange, 99
  staining with Alcian blue, 105
Melanin, formation, 216–217
  in cancer, 217
Menadione
  as hydrogen carrier, 241
  for fatty acid dehydrogenase, 251
Mercuric chloride, for blocking – SH, 72–73
Metallophilia, 152
Methanol–chloroform, for lipids, 119,
  120–121
Methyl green
  purification, 89
  quantitative method for DNA, **86–89**
  stoichiometry with DNA, 87
  structure, 87
Methyl green–pyronin, method for nucleic
  acids, **91–93**

Methyl violet method for amyloid, 43–44
Methyl viologen, 229
Methylene blue, 229
Microdensitometer, 25–27
  calibration, 27–28
  also see under Microdensitometry
Microdensitometry
  absolute dehydrogenase activity by, 28–29
  assay by, 27–29
  background, 21–22
  calculation in absolute terms, 28–29
  calibration for, 27–28
  conversion of relative absorption to extinction, 28
  instrument, 25–29
  microdensitometer, 25–27
  potential errors in, 23
  scanning and integrating, 22–25
  size of scanning spot, 25
Microsomal respiratory pathway, 257, 263–264
Microspectrophotometry
  see Microdensitometry
Microtomy
  adhesives, 8
  anti-roll plate, 6
  automatic cutting, 7–8
  method for, 5–8
  speed of cutting, 8
  thickness, 8
Mitochondria, effect of disruption, 248
  monoamine oxidase in, 219
Mitochondrial hydrogen transport, 225 *et seq.*, 233
M-nadi reaction, 213
Monoamine oxidase
  biochemical background, 219–221
  changes during menstruation, 222
  classes of, 220
  clinical use of inhibitors of, 221
  distribution in tissues, 221
  effect of clorgyline, 220
  effect of deprenyl, 220
  histochemical background, 221–222
  in mitochondria, 222
  in Parkinson's disease, 219
  inhibition studies, 223
  iproniazid, as inhibitor, 220–221, 223–224
  peroxidase method, **222**
  stereochemical specificity of, 219

substrates for (formulae), 220
  tetrazolium method, **223–224**
Monoamines, induced fluorescence of, 127–128
Mounting medium, 44
Mucins
  screening test for, 51
  stain for, 108
  staining with Alcian blue, 105
Mucopolysaccharides,
  Alcian blue method, 105
  colloidal iron method, 108
  screening tests for, 51
  staining by the periodic acid–Schiff method, 103
Mucosubstances
  colloidal iron method, 108
  diamine method, 107
Muscle
  use of carnosine, 36
  fatty acid oxidation in, 250
  lactate dehydrogenase isoenzyme of, 245
  use of polarized light, 42
  pretreatment with PVA and calcium, 36
Myeloid leucocytes, reaction of, 213
Myocardium, fatty acid oxidation in, 250
  preservation of human myocardium, 32
  succinate dehydrogenase in, 237
Myosin, orientation of, in heart transplants, 165
Myosin-type ATPase, *see under* Calcium-activated ATPase

NAD
  bound, in mitochondria, 226
  and dehydrogenases, 225
  electrode potential of, 231
  in mitochondrial respiratory chain, 226
  and production of energy, 227
NADH
  and dihydroxyacetone phosphate shuttle, 239
  and formation of ATP, 226
NADH-diaphorase
  measurement of, **264**
  see NADPH-diaphorase
Nadi reaction, 213
NADP
  and dehydrogenases, 225, 255–264
  electrode potential of, 231
  and production of energy, 227

NADPH
  critical test for histochemistry, 259
  for biosynthetic activity, 232, 257
  for hydroxylation, 232
  oxidation by diaphorases, 263
  type I and II pathways, 257
  uses of, 257
NADPH-diaphorase, 263–264
  and drug hydroxylation, 263–264
  measurement of, **264**
  NAD(P)H-cytochrome *c* reductase, 263
NAD(P)H-oxidase
  and free radicals, 265
  method, **265**
  occurrence, 264
Naphthylamidases
  activity of cathepsin H, 188
  as diagnostic test for malignancy, 189
  biochemical background, 188
  freely available activity of, 191
  histochemical background, 189
  in autophagic vacuoles, 189
  in inflammation, 189
  in malignant growths, 189
  in rheumatoid arthritis, 189
  lysosomal location, 188, 189
  manifest activity of, 191
  measurement of, 192
  method, **192–193**
  pH for optimal activity, 188
  rationale of histochemistry, 189–190
  response to leucine amide, 188, 192
Neoplasms, *see* Malignancy
Neotetrazolium chloride
  competition with oxygen, 233
  for type I and type II hydrogen, 231–232
  no reaction with reduced coenzyme, 231
  purification, 261
  redox potential of, 231
  structure of, 230
Nerves, and esterase activity, 180
Neutral red, 229
Nile blue method, for lipids, **115–116**
Nile red, 115
Nitroblue tetrazolium, redox potential of, 231
Noradrenaline, induced fluorescence of, 127
Nuclear fast red, for calcium, 123
Nucleic acids
  isoelectric point, 91
  response to pH, 97
  staining for, 47, **91–93**
5′-Nucleotidase
  activation by manganese, 159
  biochemical background, 158
  calcium method, **160–162**
  histochemical background, 159–160
  in B-lymphocytes, 159
  in degradation of purines, 158–159
  in ischaemia, 158
  influence of zinc, 158
  inhibition by nickel and zinc, 159
  inhibitors of, 158, 159
  sequestered lead method, **161–162**

Oil red O, for colouring fats, **114–115**
Optical path difference, in polarized light microscopy, 30
Organophosphorus compounds and esterase, 176–178
Orientation of proteoglycans, 107
  of collagen with Sirius red, 61
Ornithine decarboxylase
  biochemical background, 208
  half-life, 208
  histochemical background, 209
  inhibitors, 208
  lead hydroxyisobutyrate as trapping agent, 209, 211
  method, **210–211**
  specificity, 210
Osteoarthritis
  orientation of proteoglycans in, 33
Osteoclasts, and acid phosphatase, 152
Osteoporosis, orientation of proteoglycans in, 33
Ouabain, on Na^+-K^+-ATPase, 167–169
Oxidases
  definition of, 211
  general methods for, 211
  also see under specific enzymes
Oxidation–reduction
  indicators of, 229
  potentials of tetrazoles, 231
  production of energy, 228

Pararosaniline
  in acid phosphate method, 155–156
  in Feulgen reaction, **80–81**

Parkinson's disease, 219
Paraffin wax, embedding frozen section, 44
Pectic substances, 100
Pentose phosphate pathway, 241, 256
  also see under glucose 6-phosphate
    dehydrogenase
Pepsin, and phosphamidase, 172
Peptidases, *see* Proteases
Periodate, for periodic acid–Schiff
    reaction, 103
Periodic acid–Schiff
  acetylation for, **105**
  benzoylation for, **104**
  controls, 103–104
  method, **99–104**
  rationale, 100–101
Perls' method for ferric iron, **122**
Peroxidases, 211
Peroxidases
  diaminobenzidine method, 224
  method, **224–225**
  method for plants, 224
Peroxisomes, 244
Phenazine methosulphate
  for glucose 6-phosphate dehydrogenase,
    261–262
  for glyceraldehyde dehydrogenase activity,
    243
  for glycerophosphate dehydrogenase
    activity, 240–241
  for lactate dehydrogenase, 245–247
  for 6-phosphogluconate dehydrogenase,
    263
  for succinate dehydrogenase, 234, 236
  general use of, 231, 239
  method for succinate dehydrogenase
    activity, 238–239
Phenolase, *see* DOPA-oxidase
Phenylmercuric chloride, to block −SH, 72
Phloroglucinol, for staining lignin, 100
Phosphamidase
  biochemical background, 171–172
  energy of phosphoamide bond, 172
  histochemical background, 172–173
  method, **173–174**
  and proteolytic enzymes, 172
  relation to malignancy, 172
Phosphatases
  acid phosphatase, 151 *et seq.*
  adenosine triphosphatases, 164–171
  alkaline phosphatase, 145 *et seq.*

glucose 6-phosphatase, 162–164
  and ornithine decarboxylase, 208
Phosphates, test for, 51
Phosphine 3R, for lipids, 118
6-Phosphogluconate dehydrogenase
  biochemical background, 256–259
  detection of uterine cancer, 257, 259
  histochemical background, 259–260
  inhibition by palmitoyl-CoA, 257
  in malignant growths, 257, 259
  loss from section, 234
  loss in untreated section, 12–13
  method, **262–263**
  need for CO_2-free medium, 262
  occurrence, 257
  reactivation by spermine, 257
  reversal by carbon dioxide, 257
Phosphogluconolactonase, 256
  biochemical background, 259
Phospholipids
  acid haematein method, **118–121**
  benzpyrene method, **116–118**
  bromination of, 121
  identification by the acid haematein
    method, 53
  Nile blue method, **115–116**
  unmasking of, 119
Phosphorylase kinase, 203–204
Phosphorylase phosphatase, 203
Phosphorylases
  activation by AMP, 201
  biochemical background, 201–203
  decrease in tumours, 203
  effect of insulin, 204
  histochemical background, 203–204
  interconversion of phosphorylases *a*
    and *b*, 203
  methods for phosphorylases, **204–205**
  phosphorylase *a*, 201–202
  phosphorylase *b*, 202
Phosphotungstic acid–haematoxylin,
    for myoblasts and cross striations, 42
Photometry, errors in, 22
Phthalaldehyde, for demonstrating
    histamine, 128
Picric acid fixatives, 36–37
  for polysaccharides, 102–103
Picric acid, use of, 103
Pigments, 124
pK values, 64–65
Planimetry, for elution methods, 19

Plant wounding, and DOPA-oxidase, 216
Plasmals, 81, 104
Plasmin
  as endopeptidase, 187
  inhibition of, 176–177
Polarized light microscopy
  apparatus for, 31
  applications of, 32
  characteristic features in, 31–32
  dichroic stain for cellulose, 100
  for calcium salts, 123
  for collagen, 32
  for defining crystals in food products, 32
  for defining crystals in tissue, 32
  for human heart transplants, 165
  for monitoring open-heart surgery, 32,
    165
  for muscle, 42
  for proteoglycans, 107
  general concepts, 29–30
  and induced birefringence, 32–33
  induced birefringence, 107
  types of birefringence, 30–31
Polyamine pathway, 208
Polypeptides, for stabilizing sections and
    cells, *see* Collagen stabilizers
Polyphenol oxidase, *see* DOPA-oxidase
Polyphosphates
  Ebel's test, **98**
  screening tests for, 51, 97
Polysaccharides
  acidic, stain for, 108
  fixation for, 101
  PAS method, **101–104**
  screening test for, 51
Polyuronides, 100
Polyvinyl alcohol
  addition of calcium, 3
  for glucuronidase activity, 197
  for pentose-shunt dehydrogenases, 260
  for pretreatment before chilling, 3, 5
  for 'soluble' enzymes, 259
  for stabilizing sections, 144
  preparation of stock solution, 18
  use with intact cells, 144
Poly-L-lysine, for section adhesion, 8
Porphyrins, autofluorescence, 127
Potassium iodide
  for periodate–Schiff reaction, 103
  for staining cellulose, 100
Precipitation of reaction products, 142

Propylene glycol, for Sudan black method,
    113
Prostigmine, on esterase, 177–178
Proteases
  biochemical background, 186–187
  histochemical background, 187
Proteins
  amido black method, **59–61**
  dinitrofluorobenzene method, **54–56**
  general stains for, 49
  method for basic proteins, **63–64**
  methods for disulphides, **68–75**
  methods for thiols, **68–74**
  tetra-azotized method, **56–58**
Proteoglycans
  induced birefringence method, 107
  staining with Alcian blue, **105–107**
  studied by polarized light microscopy,
    33, **107**
Proteolipid, definition, 111
Prussian blue method for thiols, 71–73
Pteridines, autofluorescence, 127
Putrescine, 208
Pyridoxal phosphate
  and alkaline phosphatase, 146, 147
  and phosphorylases, 202
Pyruvic acid, 243

Quantification
  elution techniques, 18–21
  of microdensitometry, 21–29
  of −SH groups, 70
  problems of, 17
  spectrophotometric studies, 17
Quantitative cytochemistry, relation with
    biochemistry, 195
Quantitative polarized light microscopy,
    for state of myocardium, 165

Radioactive isotopes, detection of, 129
Redox indicators, 229
Redox systems, 228–229
  also see under Oxidation–reduction
Reducing substances, methods for, 124–126
Refractive index
  in polarized light microscopy, 29–30
  of crystalline deposits, 32
Resazurin, 229

Resorcin blue, for callose, 100
Retention of soluble enzymes, 138
Reticulin, stain for, 41
Rheumatoid arthritis, 260
Rheumatoid arthritis, lysosomes in, 154, 189
Rhodamine B for lipids, 118
Riboflavin, autofluorescence, 127
Ribofuranose, 79
Ribonuclease, use of, **93**
RNA
    acridine orange method, **93–99**
    diagnostic stain for, 50
    hydrochloric acid test, 94
    methyl green–pyronin method, **91**
    use of ribonuclease, **93**
Ruthenium red, for pectic substances, 100

Sandhoff disease, 198–199
Scanning and integrating microdensitometry, *see* Microdensitometry
Scarlet H, for cuticular substances, 100
Schiff's reagent, 79–81
    reaction with polysaccharides, 101–104
    reactions with DNA, 80–81
Sectioning
    automatic, 20–21
    knives for, 7
    method, 7–8
    picking up sections, 6
    thickness of sections, 20
Sialomucin
    and Alcian blue, 105
    method for, 107
Sirius red
    method for collagen, 61–63
    stain for collagen, 32
    with polarized light, 32
Solubility product, 139–140
Soluble enzymes, loss in untreated section, 12–13, 137
Spermidine, 208
Spermine, 208
    reactions with G6PD activity, 258
Stabilizing sections
    in Polypep 5115, 137, 144
    in PVA, 16, 144
    use of calcium, 16
Starch, by polarized light microscopy, 32

Steroid dehydrogenases
    general background, 254–255
    methods, **255**
Steroids
    biosynthesis and NADPH, 257
    identification by autofluorescence, 52
    methods for identifying, 52, 121
Storing frozen sections, 44
Substantivity, 140
    of formazans, 233
Succinate dehydrogenase
    activation by ubiquinone, 234
    as marker of Krebs' cycle, 237
    biochemical background, 234–237
    FAD with, 234
    histochemical background, 237
    and hydrogen transport, 227, 236
    in myocardial dysfunction, 237
    inhibition by malonate, 234–237
    method, **237–238**
    phenazine methosulphate method, **238–239**
Sudan black method for lipids, **112–113**
    burnt method, 114
Sudan III, for cuticular substances, 100
Sulphates, test for, 97
Sulphomucins, method for, 107
Sulphydryl groups
    blockade for, **72**
    DDD method, **69–71**
    ferricyanide method, **71–73**
    unmasking, 68
Susa, fixative, 37

Tay–Sachs disease, 198–199
Tetrazolium salts
    influence of oxygen, 233
    localization of formazan of, 232
    reaction with reduced cofactor, 231
    redox potential of, 226, 230–231
    site of reaction, 233
    structure, 230
Tetra-azotized dianisidine, for proteins, 56–58
Thiols
    DDD method, 69–71
    ferricyanide method, 71–76
    methods for, 68–74
    Prussian blue method, 71–74
    unmasking procedure, 68
Thrombin, inhibition of, 176–177

Thionine, 229
Thyroid-stimulating antibodies, 144
Thyroid-stimulating γ-globulin, 190
Thyroid-stimulating hormone, 144
  assay of, 190
Tissue, chilled in hexane at −70 °C, 36
Tissue dilution artefact, 237
Tissue stabilizers
  need for, 11–13
  use of, 13–16, 144
Toluidine blue
  for rapid staining, 37–38
  metachromasy of, 96
  staining acid glycosaminoglycans, 105
  test for basophilia, **95**
  test of metachromasy, **97**
Transaminase, and glutamate dehydrogen-
    ase, 248
Transfer experiments, as test of protection
    of sections, 234
Transfer experiments, to test retention, 138
Transhydrogenases, 265–266
Transplanted human hearts, 165
Trapping agents, in enzyme histochemistry,
    141–142
Triphenyltetrazolium, redox potential of,
    231
Trypsin
  inhibition of, 175–177
  possible phosphamidase activity, 172,
    187
Tryptamine
  for monoamine oxidase activity, 220
  formula of, 220

Tumours, and phosphamidase activity, 172
  also see Malignancy
Turnbull's blue method, 71, **122**
Tweens, 185
Tyramine
  formula of, 220
  for monoamine oxidase activity, 220
Tyrosinase, *see* DOPA-oxidase
Tyrosine
  demonstration of, 58–59
  formula and pK, 65

Urates, crystals in gout, 30
Uridine diphosphoglucose dehydrogenase,
    UDPG dehydrogenase
  biochemical background, 252
  histochemical background, 253
  in cornea, 252
  in epiphyseal plate, 252
  in plants, 252
  inhibition by UDP-xylose, 252
  method, **253–254**
  problems of stabilizing section, 253
Uterine cancer, 257

Van Gieson, stain for collagen, 40–41
Vitamin A, fluorescence, 122

Wounding in plants, 216

Z5 mounting medium, 44